Infosys Science Foundation Series
Infosys Science Foundation Series in Applied Sciences
and Engineering

More information about this series at ►http://www.springer.com/series/13554

Dipankar Dasgupta
Arunava Roy
Abhijit Nag

Advances in User Authentication

 Springer

Dipankar Dasgupta
Department of Computer Science
The University of Memphis
Memphis, TN, USA

Abhijit Nag
Department of Computer Science
The University of Memphis
Memphis, TN, USA

Arunava Roy
Department of Computer Science
The University of Memphis
Memphis, TN, USA

Infosys Science Foundation Series
ISSN 2363-6149   ISSN 2363-6157 (electronic)
Infosys Science Foundation Series in Applied Sciences and Engineering
ISSN 2363-4995   ISSN 2363-5002 (electronic)
ISBN 978-3-319-58806-3   ISBN 978-3-319-58808-7   (eBook)
DOI 10.1007/978-3-319-58808-7

Library of Congress Control Number: 2017944211

Printed on acid-free paper

This Springer imprint is published by Springer Nature
The registered company is Springer International Publishing AG
The registered company address is: Gewerbestrasse 11, 6330 Cham, Switzerland

This book is dedicated to our parents who brought us to this wonderful world!

Preface

This book provides the state-of-art account of authentication technologies which covers not only basic authentication methodologies but also emerging technologies which are yet to be deployed and adopted by industry. This book is the outcome of last 5 years extensive literature review on authentication, research, and project implementation of specific authentication technologies. Specifically, the concept of negative authentication (described in ▶Chap. 3) and adaptive multi-factor authentication (▶Chap. 7) are outcomes of the first author's innovative research ideas which were submitted for US patent. The salient topics of each chapter are highlighted below:

▶ Chapter 1 introduces the basic authentication mechanism for identifying and authorizing legitimate users to computing systems and denying the use by others. It narrates different authentication categories, which are now widely used and highlights how a combination of these categories can provide strong authentication compared to a single category of authentication factor.

▶ Chapter 2 discusses various biometric authentication modalities such as fingerprint, face recognition, retina, etc., and provided details on various aspects of each biometrics. Different performance metrics of biometric modalities are also covered which determines their usage under various operating conditions.

▶ Chapter 3 describes a complementary approach of passwords, namely, the Negative Authentication Systems (NAS), the concept of which was introduced by the first author in 2008. This chapter highlights different Negative Authentication Algorithms and represent schemes followed by design and implementation details of a prototype system.

▶ Chapter 4 focuses on the other complementary approaches including Honeywords, Cracking-Resistant Password Vaults using Natural Language Encoders, Bloom Filter, and Graphical Passwords.

▶ Chapter 5 covers Multi-factor Authentication (MFA) which provides a fail-safe feature in case of compromising any authentication factor as users are authenticated utilizing the other existing non-compromised modalities.

▶ Chapter 6 discusses continuous and active authentication approaches including both uni-modal and multi-modal, respectively. The chapter also covers recent research reports on using various mobile device sensors for continuous authentication.

▶ The last chapter (Chapter 7) discusses an adaptive multi-factor authentication (A-MFA), in particular, it shows how the adaptive selection of authentication factors can improve MFA to validate the users (at any given time) by sensing the current operating environment (such as devices, media, and surroundings conditions) so to make the decisions unpredictable and unexploitable by hackers.

This book could not be completed without the continued support of the University of Memphis. While Prof. Dipankar Dasgupta is the lead author of the book, several research students at the Center for Information Assurance (CfIA) contributed directly and indirectly to this book project; Arunava Roy (a postdoctoral fellow) and Abhijit Nag (a doctoral student who graduated in December 2016) contributed as co-authors and helped in putting the chapters together under the guidance of Prof. Dasgupta. Authors would like to acknowledge to all whose seminal works which are referenced and cited in this book. We also acknowledge strong support of our families to complete this book project in a timely manner.

Memphis, USA   Dipankar Dasgupta
                                                        Arunava Roy
                                                         Abhijit Nag

Contents

About the Authors

Dr. Dipankar Dasgupta is a Professor of Computer Science at the University of Memphis. His research interests are broadly in the area of scientific computing, design, and development of intelligent cybersecurity solutions inspired by biological processes. He is one of the founding fathers of the field of artificial immune systems, in which he has established himself with his works on nature-inspired cyber defense. His previous graduate-level textbook on "Immunological Computation", was published by CRC press in 2009. He also edited two books: one on Evolutionary Algorithms in Engineering Applications and the other is entitled "Artificial Immune Systems and Their Applications", published by Springer-Verlag.

Dr. Dasgupta is at the forefront of research in applying bio-inspired and machine learning approaches to cyber defense. Some of his groundbreaking works, like digital immunity, negative authentication, cloud insurance model, and Auth-Spectra put his name in Computer World Magazine and other News media. Prof. Dasgupta is an Advisory Board member of Geospatial Data Center (GDC), Massachusetts Institute of Technology since 2010, and worked on joint research projects with MIT.

Dr.Dasgupta has more than 250 publications with 12,000+ citations and having h-index of 54 as per Google scholar. He received four Best Paper Awards at international conferences (1996, 2006, 2009, and 2012) and two Best Runner-Up Paper Awards (2013 and 2014): one from ACM Information Security Curriculum Development in October 2013, and the other from ACM Cyber and Information Security Research (CISR-9) Conference in April 2014. He is the recipient of 2012 Willard R. Sparks Eminent Faculty Award, the highest distinction and most prestigious honor given to a faculty member by the University of Memphis. Prof. Dasgupta received the 2014 ACM SIGEVO Impact Award, and was also designated as an ACM Distinguished Speaker. In addition to Prof. Dasgupta's research and creative activities, he spearheads the University of Memphis' education, training, and outreach activities on Information Security. He is the founding Director of the Center for Information Assurance, which is a nationally designated Center for Academic Excellence in Information Assurance Education and Research. He developed the University of Memphis' Graduate Certificate Program in Cyber Security and also introduced undergraduate concentration on this topic.

Dr. Arunava Roy did his Ph.D. from the Department of Applied Mathematics, Indian Institute of Technology-ISM, Dhanbad in August 2014 and worked as a postdoctoral researcher in the Dept. of Computer Science, The University of Memphis, TN, USA. Dr. Roy has also worked as a research scientist in the Dept. of Industrial and Systems Engineering at the National University of Singapore (NUS). Presently, he is working as a research fellow in the Corporate Lab at the Singapore University of Technology and Design (SUTD) (Collaborated with Massachusetts Institute of Technology (MIT)). He has published a number of papers in various reputed journals. He is also involved in filing two US Utility patents on collaborative research with Prof. Dipankar Dasgupta. He is also a reviewer of various journals of reputed publishing houses like IEEE, Elsevier, Springer, Wiley, etc. He was a M.Sc. gold medallist and got the prestigious INSPIRE fellowship (Govt. of India).

His areas of interest are web reliability, cyber security, authentication, software reliability, deep learning, statistical forecasting, statistical and mathematical modeling, and fuzzy time series.

Dr. Abhijit Nag obtained his Ph.D. in Computer Science from The University of Memphis. Previously he received his masters in Computer Engineering from The University of Memphis and got his bachelor degree in Computer Science and Engineering from Bangladesh University of Engineering and Technology. His primary research interest includes various authentication approaches, mainly continuous authentication and multi-factor authentication systems. His other research interests include evolutionary algorithms, biometric approaches, cloud computing, computer and network security, bio-inspired/nature-inspired computing, and anomaly detection. He is an inventor of a submitted Utility Patent on Adaptive Multi-factor Authentication System. He serves as a reviewer for many reputable peer-reviewed journals and conferences. He is currently an Assistant Professor of Computer Information Systems department at The Texas A&M-Central Texas, USA.

Authentication Basics

Key to the kingdom – Access a Computing System

©Mikko Lemola/ Shutterstock.com

© Springer International Publishing AG 2017
D. Dasgupta et al., *Advances in User Authentication*, Infosys Science Foundation Series,
DOI 10.1007/978-3-319-58808-7_1

1

This chapter covers the basic protection mechanism against unauthorized access to a computing system, known as Authentication. Authentication mechanism ensures the process of identifying and authorizing entry to computing systems by the legitimate users and denying the use of others. This chapter first discusses systems authentication processes and describes the advantages and challenges of each of them. The second section narrates different authentication categories, which are now widely used; and highlights how a combination of these categories can provide strong authentication in comparison to a single category of authentication factor.

Introduction

Authentication is the mandatory process to verify the identity of a user and restricts illegitimate users to access the system. It varies from identification, which is an initial process to confirm the identity of users through requesting his/her credentials. For example, a user enters an ID (username) to logon to a system or network where the username is the unique identity of a person for that system. Authorization is considered as the second step in the overall authentication process, and it establishes the user's credential with the computing system for a certain period based on the defined usage policy. ◘ Figure 1.1 illustrates different steps to perform authentication process.

In this chapter, the basic authentication process and different types are described in details. Every type has its own advantages as well as disadvantages. Passwords are the most common and extensively used form of authentication approach to date. Biometric is another widely accepted authentication approach for highly secured applications and adapted for bank ATMs and mobile devices. Additionally, swipe card, photo ID, security codes, PINs, passphrases, and smart card are also used in different computing environments. Depending on the system requirement and organization's policy type, one or the other is used more than one authentication type. Due to alarming threats and compromise of user authentication credentials, organizations are moving towards strong authentication to secure users' confidential information from attackers.

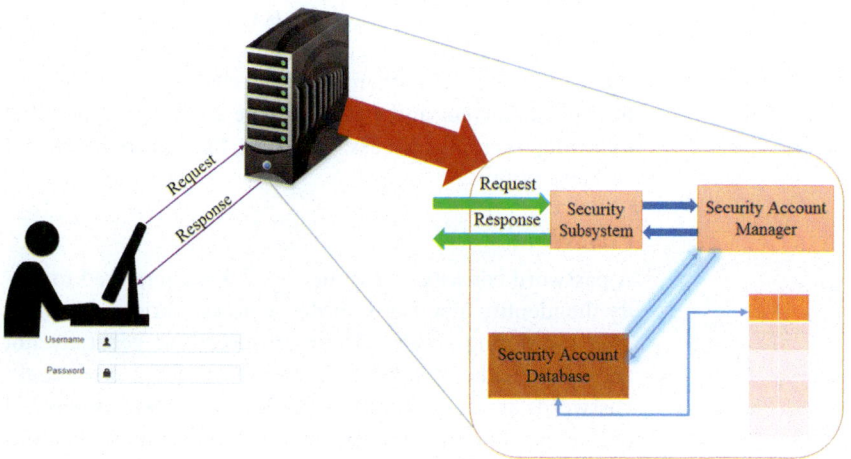

■ **Fig. 1.1** Different steps in authentication process [21]

Types of Authentication Approaches

In general, four authentication types are used for user verification as follows: These types can be combined to provide better and stronger way of authentication. Organizations choose different types based on their number of users, cost for the authentication process, the cost for their employees and customers, a particular type or combination of their security requirements, and the cost of managing the authentication system storing authentication information, user preference, company policy, etc. Four major authentication types:

Type I: What you know—your cognitive information.

Type II: What you have—items that you possess.

Type III: What you are—your physiological and behavioral attributes.

Type IV: Where you are—your location information.

Type I: What You Know?

This type of authentication is most common, and the information used to authenticate should only be known to the person and needs to be kept secret. Passwords, security codes, PINs, passphrases, etc., are the most common examples of this type. The information is sometimes known as cognitive information as only the correct user knows the actual answer to it.

One of the issues with Type I authentication is that it can be shared with other people and can easily impersonate

1

the legal user. Another drawback is that it is easy to uncover using different advancement guessing tools and techniques. Again, this type of authentication needs lots of memorization of characters and hence can be forgotten easily. The following sections discuss in detail the different Type I authentications.

Password

A password consists of a sequence of characters used to verify the identity of a user in order to access various resources in a computing system. These resources are generally not accessible without a valid password or other credentials. Passwords are used nowadays to login into a wide variety of online accounts, access emails, web applications, database systems, bank accounts, tax documents, etc. Most organizations have established their password policy in order to manage and use robust passwords. In general, password policy determines the requirements of minimum length and many other criteria for choosing passwords. Some of these requirements are the use of uppercase and lowercase letters, numbers, special characters, etc. In 2013, Google released a list [1] of most common password types which are easy to guess by the crackers as they can be found on the user's other social profile such as:

- Pet names
- A notable date, such as a wedding anniversary
- A family member's birthday
- Your child's name
- Another family member's name
- Your birthplace
- A favorite holiday
- Something related to your favorite sports team
- The name of a significant other
- The word «Password»

According to a study by splashdata in 2014 [2], it was found that people use the most common words as passwords which can easily be remembered. Here is a list of 24 common passwords (Table 1.1).

▪▪ *Guidelines to Select Good Passwords*

To make a strong password, the following options can be followed [3]:

1. It should be at least eight characters long.
2. It should not contain username, real name, or institution name.

◻ Table 1.1 Worst passwords used by users as reported by splashdata [2]

Rank	Password	Rank	Password	Rank	Password
1	123456	9	Dragon	17	access
2	Password	10	Football	18	shadow
3	12345	11	1234567	19	master
4	12345678	12	Monkey	20	michael
5	Qwerty	13	Letmein	21	superman
6	123456789	14	abc123	22	696969
7	1234	15	111111	23	123123
8	Baseball	16	Mustang	24	batman

3. It should not contain any complete word or dictionary word.
4. It is sufficiently different from previously used password.
5. It contains characters from each of the following four categories:
    (a) Uppercase letters (example: A, B, C)
    (b) Lowercase letters (example: a, b, c)
    (c) Numbers (example: 0–9)
    (d) Symbols found on the keyboard and spaces (example: ` ~ ! @ # $ % ^ & * ( ) _ - + = {} [] \ | :; " ' < > , . ? /)

■■ *How Password Is Stored*

In order to use passwords for authentication, operating system password and other application passwords need to be stored securely. If the storing process is not secure enough, there is a chance of compromising passwords by attackers. National Institute of Standards and Technologies (NIST) provides the guide [4] to store passwords securely with additional security controls. Some of these controls are listed below:

1. Encrypt the files containing passwords. This task can be done by the operating system, any application, or any password management software, which is designed to protect confidentiality of passwords.
2. Incorporate the OS access control features to limit the access to certain files that contain passwords. For instance, a host machine can be configured to allow only administrators and some processes with root privilege to access a password file.

3. Use of one-way cryptographic hash functions for passwords to store. The use of hashed password allows the authentication system to verify whether the correct password is entered without storing the plain text password in the database. The purpose is to keep the password secret even if a cracker gains access to password profile; an intruder cannot gain access to the actual password.

The most commonly used one-way hashing algorithms are MD5, SHA-1, and SHA-256. As the hashed output has the same length irrespective of the password length, hashing of passwords does not reveal any information regarding the size of the password. However, if the same password is chosen by two users, the hashed password will also be the same.

■■ *Salted Passwords*
In order to overcome this weakness, a random number called salt is added in the hash password. However, if two users choose to add salt in their hashed passwords, there will be some benefits:

— It prevents duplicate passwords to be stored in hashed password file or database.
— It increases the difficulty level of offline dictionary attacks or guessing attacks.
— It makes it very hard for the password crackers to find whether a user maintains the same password to authenticate different system or not.

In general, when a user tries to login his account with a password, the password is hashed and then it is compared to the hashed output stored in the password file. If they match with each other, the password management grants the access to the system. In the hash values, the salt is added to the algorithm to create different conceivable hashes for a particular password. Hence, salt is a random string added to the password; it adds another layer of obfuscation to the attackers. Actually the usage of salt does not prevent cracking the passwords, it just prolongs the cracking time of the passwords during the hashing process. The salting process of passwords and the verifying of a user password are shown (adopted from [5]) in ◘ Figs. 1.2 and 1.3.

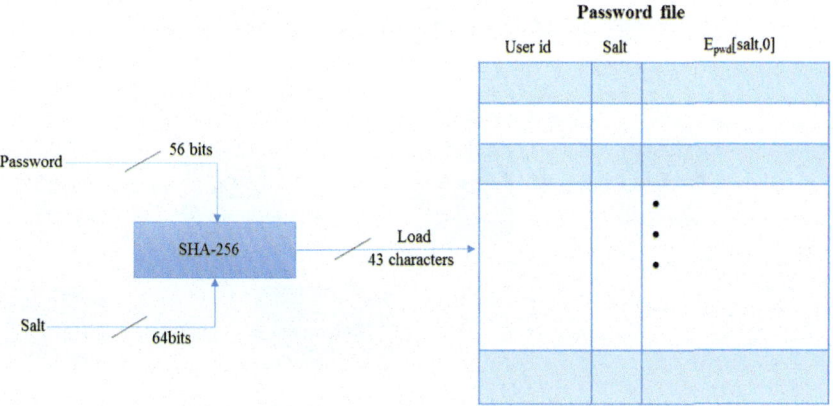

■ **Fig. 1.2** The loading of a new password in the password database

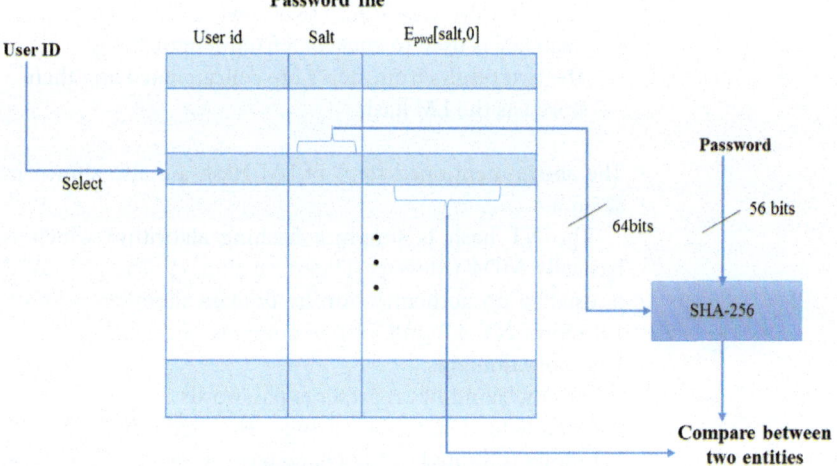

■ **Fig. 1.3** Verify a user's password against a password database

■■ *Passwords Storing in Windows OS*

Windows OS stores passwords for users in different ways. In Windows networking, the password profile is generally stored in two ways—LAN Manager one-way function (LM OWF) and NT OWF (NT one-way function) [6]. The LM OWF algorithm is included in Windows to allow backward compatibility with the software and hardware which cannot use newer algorithms. LM hash is computed in the following way [6]:

1. The password is padded with NULL bytes to make exactly 14 characters. If the password is longer than 14 characters, it is then replaced with 14 NULL bytes for the remaining operations.

1

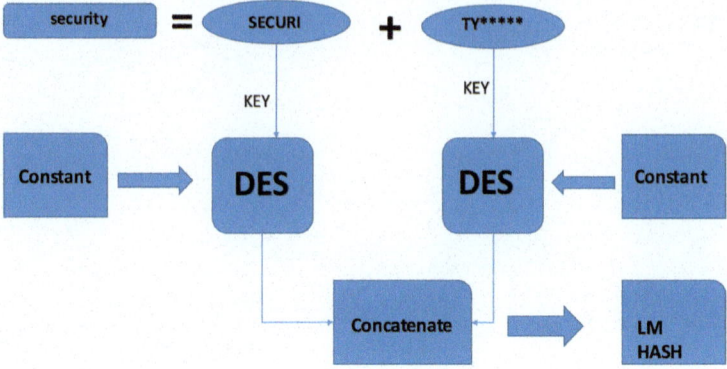

Fig. 1.4   Flow diagram for computation of LM HASH

2. The password is converted to all uppercase form.
3. The password is split into two 7-byte (56-bit) keys.
4. Each key is used to encrypt a fixed-length string.
5. The two results from step 4 are concatenated and then stored as the LM hash.

The above-mentioned steps of LM Hash are also shown in ◘ Fig. 1.4.

The NT hash is simply a hashing algorithm which is basically MD4 (Message Digest version 4). The NT OWF is used to do authentication by domain members in both Windows NT 4.0 and earlier domains and also in Active Directory domains.

Internally, Windows represents passwords in the format of 256-character UNICODE strings. The logon dialog box in Windows is limited to 127 characters.

▪▪ Passwords Storing in Unix OS

In UNIX system, the password file is stored under /etc., and the name of the password file is **passwd**.

This file contains the username, real name, identification information, and basic account information for each user of a UNIX system.

▪▪ Shadow Password File

In UNIX, the 'passwd' file in /etc. directory is generally readable by anyone in that system. Hence, an attacker can login as a regular user and easily retrieve the passwords by running any password cracking software on that file. To avoid this threat, UNIX system uses a shadow password file, which is a system file of encrypted user passwords that is stored to make it not available to the regular users. This file can only be read or used by root process—(the UNIX administrative account) processes which require access to the shadow password file must check whether the user has root access which provides much greater security against password snooping.

▪▪ Password Strength

Strong passwords provide a defense against guessing and cracking of passwords. In general, the password strength is measured by the length of the password and its complexity in terms of the unpredictability of its characters. In most cases, the complex password is designed by choosing characters from at least three of the four categories—lowercase letters, uppercase letters, digits, and symbols. ◘ Table 1.2 (mentioned in [4]) illustrates the effect of password length and its complexity with the help of possible approximate keyspace for passwords. Here *Keyspace* is defined as the total number of possible values that a key (aka password) can have. As the value of the keyspace increases, it takes more time to perform an exhaustive brute-force attack against the password.

◘ Table 1.2 indicates that increase of keyspace also increases the complexity of the passwords as well as with the increase of length of the passwords.

▪▪ Graphical Password

Graphical password [7] is one of the alternatives to the strive-based password approach. The motivation behind this type of password is that humans can remember pictures better than text or number. In addition, if the number of possible pictures is significantly large, the possible password space for graphical password also can exceed that of the strive-based password. An example of the graphical password is shown in ◘ Fig. 1.5. In this a user is asked to select or click a certain number of images from a given set of randomly arranged images generated by a program. Then the user is asked to do the authentication by identifying the preselected images. The primary purpose of introducing graphical passwords is to defeat shoulder surfing attacks. While shoulder surfing is only relevant in a shared office environment, the use of graphical passwords is becoming more useful in mobile devices. Detailed descriptions regarding graphical password are given in ►Chap. 4.

Table 1.2 Possible keyspaces by password length and character set size [4]

Char. set size	Character types letters				Password length				
	Digits	Letters	Symbols	Other	4	8	12	16	20
10	Decimal				1×10^4	1×10^8	1×10^{12}	1×10^{16}	1×10^{20}
16	Hexa-decimal				7×10^4	4×10^9	3×10^{14}	2×10^{19}	1×10^{24}
26		Case-insensitive			5×10^5	2×10^{11}	1×10^{17}	4×10^{22}	2×10^{28}
36	Decimal	Case-insensitive			2×10^6	3×10^{12}	5×10^{18}	8×10^{24}	1×10^{31}
46		Case-insensitive	10 common		4×10^6	2×10^{13}	9×10^{19}	4×10^{26}	2×10^{33}
52		Upper and lower			7×10^6	5×10^{13}	4×10^{20}	3×10^{27}	2×10^{34}
62	Decimal	Upper and lower			1×10^7	2×10^{14}	3×10^{21}	5×10^{28}	7×10^{35}
72		Upper and lower	10 common		3×10^7	7×10^{14}	2×10^{22}	5×10^{29}	1×10^{37}
95	Decimal	Upper and lower	All symbols on standard keyboard		8×10^7	7×10^{15}	5×10^{23}	4×10^{31}	4×10^{39}
222	Decimal	Upper and lower	All symbols on standard keyboard	All other ASCII characters	2×10^9	6×10^{18}	1×10^{28}	3×10^{37}	8×10^{46}

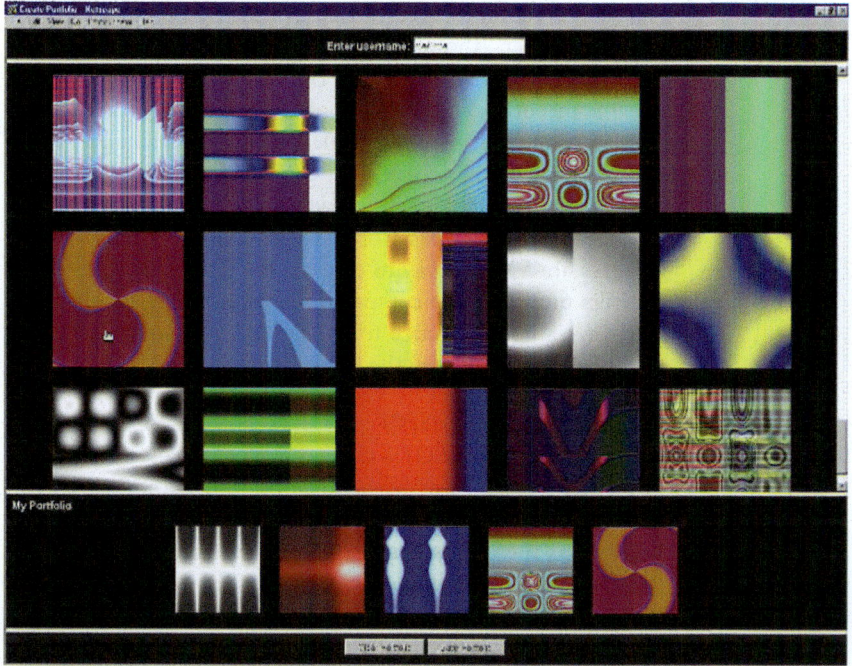

□ **Fig. 1.5**   Random images used by Dhamija and Perrig [22]

■■ *Password Manager*

A password manager is a software to store all user passwords in one location with the intention to keep them
protected but accessible with an easy-to-remember master password. When a password manager is used, it is only
needed to remember the master password to access all the
passwords necessary for different systems.

There are many types of password managers available,
and some of these are listed here.

Desktop password manager is a software which is
installed in a local machine and passwords stored in a hard
drive. In particular, it stores all users' credentials (username
and passwords) who use that machine only. Portable password manager is another type of password manager which
can be used in smartphones and other portable devices.
Online password manager is a website which stores login
credentials of users. This is basically the web-based version
of the desktop password manager. Token-password manager is another type where credentials are protected using
a security token (Type II). The cloud-based password manager is one special type of password manager where the

users' credentials are stored on a service provider's dedicated server. These credentials are handled by password management software which runs on the client system.

Some of the best password managers according to PC Magazine [8] are LastPass 3.0, Dashlane 3, Sticky Password, RoboForm Everywhere, etc. However, on June 15, 2015 LastPass reported their breaches of email addresses and encrypted master passwords. In general [9], password managers are a smart way to escalate the online security of the users until they are compromised by cyber hackers. So it is always a good practice to change the passwords at regular intervals to minimize the risk of theft of passwords.

▪▪ *Single Sign-On (SSO)*

Single Sign-On (SSO) is mechanism or process where a single action of a user authentication and authorization can allow him/her to access all of resources and systems that s/he has the access permission. In this process, there is no need to enter multiple passwords as the user moves from one system to another. Login process in SSO is generally done using Lightweight Directory Access Protocol (LDAP), where LDAP database is kept on different servers.

In general, different applications and services use different authentication mechanisms to support SSO; it is essential that the system internally does the processing and stores the required credentials from different authentication mechanisms beforehand.

The benefits of SSO are as follows:
- Able to apply uniform authentication policies across different enterprise systems.
- Significantly reduce the password resetting helpdesk maintenance costs.
- Reduce password fatigue of the users to enter the username and passwords multiple times for accessing different enterprise resources.
- Reduce the time required to re-enter the passwords and reduce human errors in authentication.

▪▪ *Attacks on Passwords*

With the sophistication of computing processes, different password cracking tools and techniques are becoming available to crack passwords. Some of the most widely used password-based attacks are mentioned below.

▪▪ *Password Guessing*

The most common form of password attacks is the guessing attack. If a simple password is used then attackers can quickly find the correct password through guessing or using the brute-force search. In this process, attackers try every possible combination of characters for a given length of the password. If the length of the password is large, it takes longer for an attacker to run the brute-force approach to finding the actual password. A dictionary-based attack is another approach to do the guessing attacks. This approach is based on trying all the possible strings in an arranged listing of words derived from a dictionary. Unlike brute-force approach, dictionary attack tries those combinations which are dictionary words and hence the chance of succeeding to gain the actual password is increased. People are more likely to use common words or phrases as passwords to remember. This type of attack can be easily reduced by not choosing a common or dictionary word as passwords.

▪▪ *Password Resetting*

Attackers often find it easier to reset users' passwords than to guess them. Some password cracking software are actually password resetters. In some cases, an attacker boots from an external drive, floppy disk, or CD-ROM to get around the typical Windows protections. In Unix/Linux environment, use a bootable version of Linux that can mount NTFS volumes and thereby, locate and reset the administrator's password easily. For security reasons, it is advisable to change the password on a regular basis (at least once in 3–6 months).

▪▪ *Retrieving Forgotten Password*

Users sometimes forget the password and are required to use the resetting options. Web applications generally use the following techniques to help users to retrieve or reset their passwords [10].

- Sending an email with the existing or a temporary password;
- Sending instructions regarding password reset through email;
- Displaying the current password after answering some predefined security questions;
- Providing previously saved password hints or password reminder.

These are common practices though considered as weaker ways to reset passwords. A more secure way recently suggested to reset through a multi-step procedure for forgotten

■ **Table 1.3**   List of hard data to be asked to the user [10]

Hard data
Email address
Last name
Date of birth
Account number
Customer number
Last 4 digits of social security number
Zip code for address on file
Street number for address on file

passwords [10]. The stepwise procedure for password reset is given below:

Step 1: Collect the hard data.

In general, a minimum of three inputs is recommended to ask user to verify their identities. The first one should be the username of the account. The others can be chosen from ■ Table 1.3.

Step 2: Ask personal security questions

If the user cannot answer any of the questions of Step 1, then his request for changing the password will be denied. Upon success, the system can ask some of the user's pre-established personal security questions (for example: mother's maiden name, favorite pet's name, etc.).
Upon Success of Step 2, the user is allowed to reset the password by typing a new password and then confirming the new password by typing it again. If the user successfully completes these two steps, the system will update the new password and shows him the successful message of resetting the password.

PIN Passwords

In general, PIN used to access some physical facilities or Bank ATMs are short in length and commonly used in a number pad (only 0–9 are usable characters); common PIN passwords are also considered weak (■Fig. 1.6).

◘ Fig. 1.6 Keypad is used to
provide PIN passwords

1111: This option is extremely easy to guess without decryption software and wears down the «1» key, making it obvious which key has been pressed many times.

1234: This option is also very easy to guess without software because the numbers are sequential. Hence, it is a bad choice for a pin number.

5947: In this particular case of PIN password, randomization or not following a standard (like sequential order) can make such password the better option.

When the characters of the password are limited, the order of those characters becomes extremely important and can make the difference.

An Incident Regarding PIN-Based Authentication

» A bank in Canada was forced to upgrade their PIN security system after two ninth graders were able to hack into one of their ATM machines during their school lunch hour.

One of the teenagers found an ATM operator's manual online, and that manual showed them how to put an ATM into operator mode. They followed the manual's directions, and once they got it into operator mode, the ATM asked for the password in order to make changes. They were able to guess the password on their first try. They used a standard, default password to gain access. It was not that hard to guess, but the bank staff did not believe them after the first time. They did not bring any proof that the system was vulnerable, so they went back and hacked the ATM a second time.

Once they gained access the second time, they were able to print off the reports of that ATM's activity and gave it to the nearest bank to prove what had occurred and alerted them of the security flaw.

1

■■ *Common Issues with PIN-Based Systems*

(1) Public Manual: Public manuals of ATM machines are available online. Hence, secrets may get out over time. Again, fired/retired employees know things even if manuals are confiscated. Default security procedures are generally written in the manuals. The change of these procedures is needed on a periodic basis.

(2) Operator Mode: Once a user is in operator mode of ATM machine, he/she could see advanced options and the system is considered trusted. In essence, because the ATM was not designed to be an antitrust system towards operators, intruders had a lot of power once authenticated.

(3) Password change: Passwords should be changed as often as turn-over occurs. The credentials of former employees should always be removed to prevent unauthorized access.

Information gets leaked over time. Easily guessed passwords should not be used in ATM machines, and the default configuration settings should be changed to other possible settings at periodic intervals.

Security Questions

Security questions are asked (algorithm password) during the authentication process in order to provide an extra level of security to verify user credentials. The answers are mostly related to the users' personal, professional life, or even places where they grew up. Many websites use the security questions to do password reset or retrieval and verify the user's identity at the time of sign-in. These security questions are also asked by the IT help desk employees to verify identity over phone. In 2008 US Presidential election, one of the presidential candidates had her personal email account hacked by someone answering one of her security questions. The person gained access to her email account through answering one of the simplest security questions: «What year were you born?» The answer to this question was available online through different search engines. However, choosing bad security questions can trigger some undesirable consequences.

— The user may not be able to remember the answer or the recently changed answer;

— The specific question is not appropriate for the user;

- The answer of a question can be guessed by others through social engineering;
- The question is very specific to that user and can be used in multiple websites.

One major challenge is to select the security questions to reset the passwords. In general, «What is your birthdate?», «Where did you go to high school?» are very common and most of the people can find these answer by searching online and social media. Sometime people reveal such information through normal conversation and even put in their public profile. Hence, two different ways for setting the answers of the security questions:

- Real/correct answers are easy to remember but can be discoverable via online searches.
- False answers are hard to remember, but they are safe from search results.

One possible solution can be choosing fuzzy answers to security questions. For example, the response to «What is your favorite color?» may be chosen as «*Redish*» or «*Greenish*» instead of «Red» or «Green». Such a strategy can address the two issues (mentioned above) in a slightly better way.

It is therefore a crucial need to choose a good set of security questions. According to some experts [11], a good security question should meet the following five criteria:

1. Safe: It cannot be easily guessed or researched through web search engines.
2. Stable: The answer to the question does not change over time.
3. Memorable: It can be easily remembered by the user.
4. Simple: The answer should be precise, easy as well as consistent.
5. Many: It can have many possible answers.

It is surprisingly difficult to find security questions that can meet all the criteria mentioned.

Type II: What You Have?

The examples of this type are of medical card (with health record), etc., with different information cards (with a magnetic strip in it), photo ID, smart card (with embedded

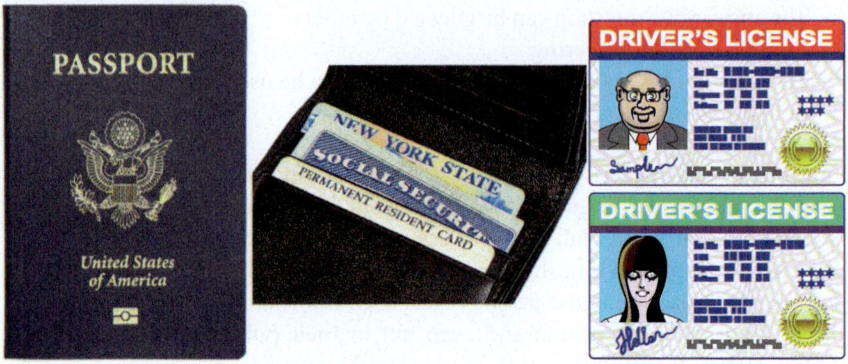

1

Fig. 1.7 Examples of Photo ID as Type II authentication in USA. The *left one* shows passport, *middle picture* shows residence card, and *third picture* (*right*) shows driving license. ©topseller/Shutterstock.com ©M. Unal Ozmen/Shutterstock.com ©VECTORWORKS_ENTERPRISE/ Shutterstock.com

memory chips). Type II authentication is sometimes known as authentication by ownership or in possession (■ Fig. 1.7).

Replacement of the Type II authentication objects is more costly and it is considered as one of the weakness of this type.

» In general, owners of these objects have to make significant effort to keep them safe (either lost or stolen) and require a human or machine to verify.

This means there are inherent weaknesses when using them as standards. For instance, these types of cards are only as good as the personnel tasked with verifying them. Additionally, because these are standardized, they are open to other weaknesses such as manipulation, forgery, and duplication.

Photo ID

Photo IDs are the most common example of Type II authentication. The personal information mentioned in the ID includes bearer's full name, age, date of birth, address, an identification number, photo, and other information to identify the person. In many countries, a driver license is the most common example of Photo ID.

It has been used by a wide variety of organizations including schools, businesses, and many government organizations. Photo IDs are generally checked by some authorities to verify the person's identity. The time taken to verify the identity depends on the information that is being requested. For instance, entering into a movie theater sometimes requires a photo ID, and it takes about 30 s to verify someone. On the other hand, passing through customs for any international flight requires a passport and may take about 2–3 h to

■ **Fig. 1.8** Photo ID verification process. ©Volt Collection/Shutterstock.com ©Rawpixel.com/ Shutterstock.com

finish the checking process. Recently OPT scans are used for automated photo ID verification purpose. ■ Figure 1.8 illustrates some of these identity verification scenarios.

Swipe Card and Smart Card

These are generally used in the automated authentication system. This type differs from photo ID as manual inspection does not require authentication. In general, accessing a secure server with sensitive data using a swipe card can take about 2–4 s, and gaining access to a secured facility or building takes about 15 s.

So the automated systems (using smart or swipe card) are significantly faster in comparison with non-automated systems (photo ID, for instance). However, non-automated systems can notice visual clues which automated systems are usually not programmed to notice (■ Fig. 1.9).

Security Dongles

Security dongles are used in the electronic device-based authentication systems. These can act as random code signal receivers, random code generators, encrypted memory passphrases, or any combination including all or none of the above. Because these are used for specific systems, they can be highly specialized and extremely difficult to duplicate while easy to use. Different types of dongles are available as shown in ■Fig. 1.10.

■■ *Randomized Signal Receiver*

In this type of dongle, one-time passwords (OTP) are generated by the host (server) at authentication requests from users. If the correct receiver (dongle) bounces back the same code and the other credentials match, access is granted to the user.

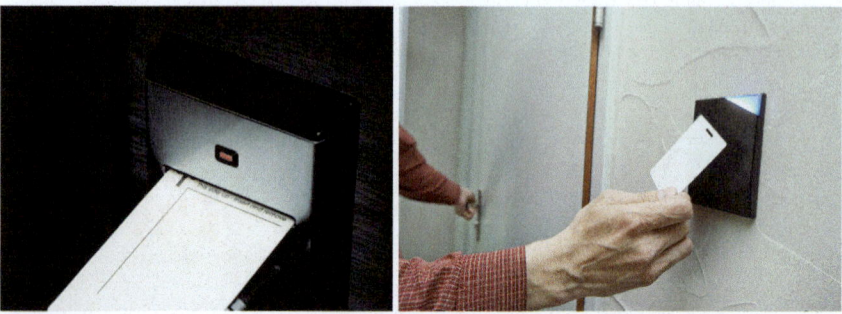

◘ **Fig. 1.9**   The authentication process using swipe card and smart card. ©Zdorov Kirill Vladimirovich/Shutterstock.com ©Dmitry Kalinovsky/Shutterstock.com

◘ **Fig. 1.10**   Different types of security dongles. *Left image* shows dongles as randomized signal receiver. *Middle image* shows dongles as random code generator. *Right image* shows dongles as token storage. ©Morrowind/Shutterstock.com, ©Yevgen Romanenko/Shutterstock. com, ©Charles B. Ming Onn/Shutterstock.com

▪▪ *Random Code Generator*

In this type, a one-time password (OTP) is generated by the user's dongle after the host sends an authentication request back to the user. If the credentials match, access is granted.

▪▪ *Token Storage*

In this type of dongle, **encrypted memory passphrases** are stored on a dongle or memory partition. When the dongle receives a signal from the host, the passphrase is then unlocked and returned like an extra password for verification.

CAPTCHA

CAPTCHA (Completely Automated Public Turing Test to Tell Computers and Humans Apart) is a set of programs that are used to differentiate between human and computer programs acting like humans (◘ Fig. 1.11). This classification is done through generation and grading of tests that are supposed to be solvable by human only. The development of CAPTCHA supports the following properties [12]:

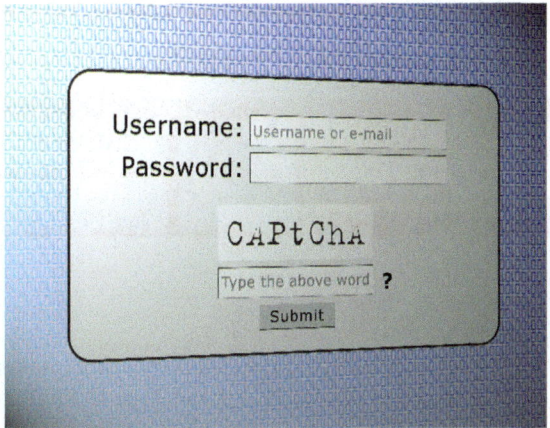

■ **Fig. 1.11** A use of CAPTCHA during the authentication process.
©Maen Zayyad/Shutterstock.com

(a) Automated: Computer programs should be able to generate and grade the generated tests.
(b) Open: The underlying database(s) and algorithm(s) used to generate and grade the tests should be public.
(c) Usable: Humans should easily be able to solve these tests in a reasonable amount of time. The effect of any user's language, physical location, education, and/or perceptual abilities should be minimal.
(d) Secure: The program-generated tests should be difficult for machines to solve by using any existing algorithm.

For CAPTCHA, the system asks the user to respond to an image or recording of certain characters. The user needs to enter the characters in the order shown or heard to prove that he or she is a human. reCAPTCHA is the most commonly used form of CAPTCHA. In this type, characters are shown in a distorted manner in an image. The characters are distorted to prevent Optical Character Recognition (OCR) software from reading the characters (■ Fig. 1.12).

Audio CAPTCHA is used for the visually impaired persons where the characters are pronounced at randomly spaced intervals and background noise is added to make the test more robust against voice-recognizable bots.

Type III: What You Are?

Type III uses humanistic features about a person. Type III authentication tells something unique about the user and is commonly known as biometric authentication. This type of

■ **Fig. 1.12**   Example of reCAPTCHA [23]

authentication is considered to be more robust in comparison with other types. While Type I and Type II authentication information is shareable, Type III is almost impossible to share and its repudiation is very unlikely. As this type of authentication goes with a person, forging or stealing such identity is more difficult in comparison to other types. Moreover, this type of authentication cannot be lost or stolen.

The most common types of biometrics are face recognition, fingerprint, palm scan, retina scan, iris scan, voice recognition, keystroke dynamics, handwriting dynamics, signature, etc. As they are difficult to compromise, in order to access high-security systems and sites, these types of authentication are widely used. Type III authentication does not require human involvement as the process determines automatically the identity of the users. Again, due to automated measurements, the comparison of the captured entity with the stored entity is done in real time. However, this type of authentication system is relatively expensive. In order to verify the identity, a scanner is required to capture the user biometric feature and then verify against the stored entity in the database. Since additional hardware costs are required, it is not used for the mass-consumer authentication process (■ Fig. 1.13).

Biometric authentication is a three-step process:
1. A system has to collect the biometric data with the use of sensors.
2. The system has to convert the input into a digital representation called a template.
3. The system then needs to compare the input template with stored templates in the database. If a match is found then the user is authenticated.

Unlike other type of authentications, here a matching threshold is used for handling inaccuracies in input signals. With Type I and II authentication, the authentication inputs have

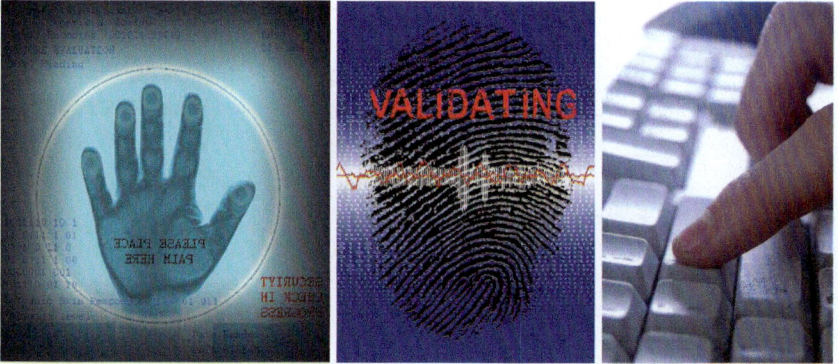

Fig. 1.13 Examples of Type III authentication. ©Bruce Rolff/Shutterstock.com ©Undergroundarts.co.uk/Shutterstock.com ©Dmitri Mihhailov/Shutterstock.com

to exactly match the user profile stored in the database. With biometrics, the information provided does not have to be exactly the same, but must fall within acceptable percentage levels set by the person who administrates the authentication system. The lower the percentage level, the higher the chance that the system will return a false positive. False positive occurs when a user is mistakenly granted the access. On the other hand, setting a higher value of acceptable percentage level may result in more false negatives, i.e., could not recognize the user and the system authentication denied access. In biometrics, such false positive and false negative are formed as false acceptance rate and false rejection rate.

Biometric authentications are categorized in two ways: behavioral attributes and physiological attributes. Voice, keystrokes, signature, mouse movement are the examples of behavioral attributes. Fingerprint, hand geometry, iris, retina, face are the examples of physiological attributes. The details of the biometric authentication modalities are discussed in ►Chap. 2.

Behavioral Attributes

These attributes are those which are related to a person's certain behavior pattern and examples of these attributes are mentioned below:

- The way a person talks (voice)
- The way a person walks (gait)
- The way a person types (keystrokes), etc.

■■ Gait Recognition

Gait Recognition is a visual type of recognition that authenticates someone based on how he/she walks. There

1

■ **Fig. 1.14**   Show people having different gaits that can be used to authenticate.
©Rawpixel.com/Shutterstock.com

are two ways to recognize gait. One way is to record a video of someone walking, measuring the trajectories of the joints and angles of leg movements over time to generate a unique user template. Another way uses radar technology to record the gait cycle of a person during walking and creating a template based on that information for authentication. Several recent researches have shown how to identify gait using different wearable sensors [13, 14] or cameras [15]. ■ Figure 1.14 illustrates the different walking patterns of people which can be used for identification of an individual.

■■ *Keystroke Recognition*

Keystroke recognition is another type of behavioral authentication which compares users' typing patterns. This can be done by observing how long it takes to move between keys, how hard the buttons are pressed, and how long a key is pressed before it is released. Recent works [16–19]

■ **Fig. 1.15** Keystroke-based authentication of users. ©Dmitri Mihhailov/Shutterstock.com ©StepanPopov/Shutterstock.com

on keystroke dynamics provide a better way to authenticate users on mobile devices using this behavioral attribute (■ Fig. 1.15).

■■ *Voice Recognition*

It is an auditory type of authentication that uses the way someone speaks (tone, pitch, speed, etc.) by measuring the sound waves of a person's speech (■ Fig. 1.16). This is also known as speaker recognition (identify who is speaking) rather than speech recognition (recognize what is being spoken). Speaker recognition system generally falls into two categories, namely, text-dependent and text-independent. If the same text is spoken during the enrollment/registration and verification purpose, it is called text-dependent authenticable system. If the text used for enrollment and verification differs, it is known as text-independent systems. Some speaker recognition software include Dragon, Voice Finger, ViaTalk, E-speaking, Tazti, etc.

Physiological Attributes

Physiological attributes are those attributes that are related to human body and cannot be easily altered. These attributes are more reliable in authentication than the behavioral attributes. Examples of this type include fingerprints, facial recognition, retina scans, iris scans, etc. as shown in ■ Fig. 1.17.

■■ *Facial Recognition*

Facial recognition is a visual type of recognition that considers the image of a face and measures the distance between the eyes, the width of the nose, distance between the cheek bones, lips (center position, shape and orientation),

1

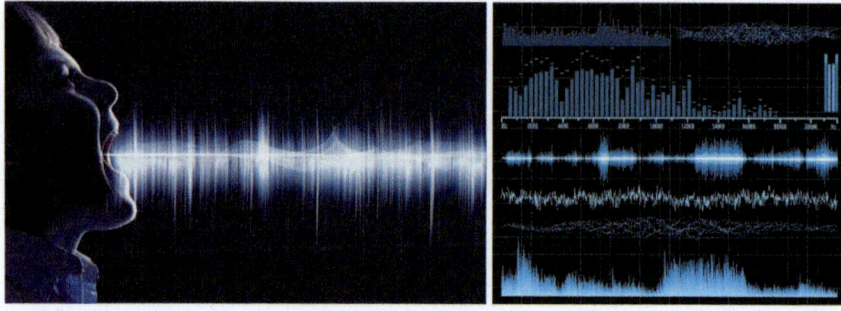

■ **Fig. 1.16**   Voice-based authentication of users. ©Sergey Nivens/Shutterstock.com
©Eliks/Shutterstock.com

■ **Fig. 1.17**   Different types of physiological attributes of Type III authentication. ©Juergen
Faelchle/Shutterstock.com   ©Chaikom/Shutterstock.com   ©Franck Boston/Shutterstock.com

and many other distinctive characteristics (■ Fig. 1.18). It then calculates the same set of features from the facial database of that person and identifies the actual person through comparing those features.

Facial recognition has many challenges, like changes in lighting conditions, camera angles, etc. This technique is used at major foreign embassies and government agencies. Facial recognition also guarantees unique identification of a person in order to prevent identity theft. Many bank ATMs and casinos use this technique to identify users.

■■ Fingerprint Recognition

Fingerprint Recognition is a visual type of authentication that uses the ridges and valleys found on the surface of fingertips (■ Fig. 1.19). A digital image of a finger is taken using a fingerprint scanner. The system then finds unique points on the fingertips and matches the image to one that is found in a database. The parts of the fingerprint used to capture different characteristics of finger geometry are as follows:

▬ Crossover
▬ Core
▬ Bifurcations

■ **Fig. 1.18** Face recognition using different facial features of users

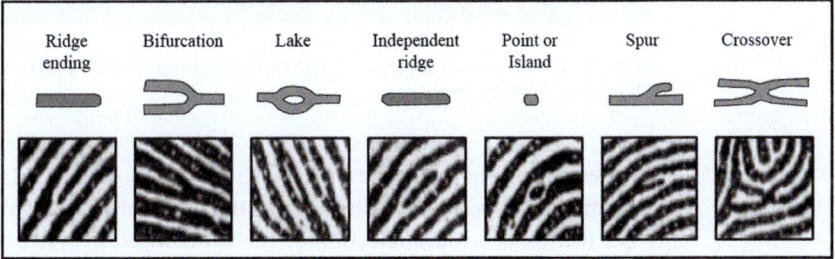

■ **Fig. 1.19** Seven most common minutiae types used to detect in fingerprint sensors [24]

- Ridge ending
- Island
- Delta
- Pore

Some well-known fingerprint scanners are BioMini, Veifi P5100, Thumprint Security USB, U.are.U 4500, FS88, etc.

■■ *Retina Scan*

It is also a computerized recognition system that shines light into the back of the eye. Because the blood vessels absorb light differently than the surrounding tissues, this process allows the computer to get an accurate distribution of the blood vessels, which is unique for each human. Then, just like fingerprint recognition, the computer finds unique points in the picture and measures the distance between them. In addition to blood vessel patterns, retina scan uses vein patterns to find the unique characteristics of

1

🔹 **Fig. 1.20**   Retina scan as a way of authenticating users. ©Left-Handed Photography/Shutterstock.com

user retina. Retina scan requires significant effort than Iris scan, and it is more challenging as the slightest movement of the eye can cause rejection by the authentication system. Also, sophisticated cameras are needed to capture the retina images than face recognition (🔹 Fig. 1.20).

Type IV: Where You Are?

This type of authentication process utilizes location of the user to determine his/her identity. In general, global positioning system (GPS), IP address, cell tower ID, etc., are used for this type of authentication. In general, it is used in connection with other authentication to verify the user's identity. If the user has the correct credentials then the system will look at where he/she is located to see if the user is authorized to access the system from that specific location.

This type of authentication can also keep track of when and from where the person logged in. For example, if a person logged in from one place (New York) but 30 min later he/she logged in from another far away location (Europe), then one of them must be an imposter and the account needs to be locked to prevent suspicious activity.

Most of the implementation of Location-Based Authentication (LBA) requires a location signature (LS) which is generated with the help of location signature sensor (LSS). Location signature describes the physical location of the person and the time that it is requesting access. Once the LS is generated, it is then transmitted to the system that authenticates the user.

Suppose Anna is an employee at a bank. Her employers use a location-based authentication to verify logins. Anna's LSS is stolen, and the thief takes it to another state. From there, the thief tries to access the bank's computers.

Anna's LSS generates a signature and transmits it to the bank. The bank's LSS determines that Anna's LSS is in a different state and is supplying a correct signature. However, Anna is not authorized to access the bank's computer from other states, so the authentication is rejected.

If a thief tries to forge a message indicating that Anna is connecting from inside the state the bank is located, the host's LSS would report that Anna was at a different location and would reject the connection.

The interesting point is that this type of authentication can be done in a continuous manner. The LSS simply intermingles signature data with the transmitted data and the host checks it. If the connection were hijacked, the data from the LSS would be lost.

User Account Types

In a recent survey by Google researchers [20], they found that there exist different types of user accounts and all those accounts may not require strong authentication. For example,

- Users might create *throw-away accounts* for participating in the pseudo-anonymous conversation, doing an online survey, or one-time purchase, etc.
- *Routine accounts* are generally long-lasting which may not be high-valued but used for reading online newspapers, magazines, etc.
- *Spokesperson accounts* are created to participate in political, social, philosophical, professional discussions, blogging; compromising such accounts may defame the user, spread misinformation, embarrassment to extensive cleanup costs.
- *Sensitive accounts* include user's primary email accounts, financial accounts, employment/service accounts, and exposing such accounts may have severe and sometimes unforeseen consequences.
- Very high-value *transaction accounts* are specialized access use for irrevocable actions such as cross-border fund transfer, weapon release, etc.

1

As it is seen that the setting up of these accounts is purpose-driven and needs to be treated separately, linking one type to the other may result in severe consequences. There is an alarming trend of mixing one with the other such as login to social network accounts with an email account or vice versa. Also for usability reasons, users merge official, banking, and social network accounts on a mobile device, which may allow intruders to compromise one and get to the other accounts.

Chapter Summary

This chapter presents different well-used authentication mechanisms that are increasing to prevent unauthorized access to the cyber system as the strong defense against various emerging threats. These mechanisms are categorized into four authentication types; however, new types are also evolving based on emerging sensor technology. Type I uses cognitive information such as passwords, PINs, and security questions are typically used to authorize. Best practices in password selection and password examples, despite their likelihood to compromise, are provided to avoid password selection. Next section presents a brief description of several encryption techniques, such as salted password-based approach along with various password storing procedures in different OS (Windows and shadow password for UNIX systems) for a better understanding of password management. Detailed guidelines are provided for choosing strong passwords. New approaches like Graphical password (non-textual password) and single sign-on are briefly described. Other Type I authentication approaches including PIN and security questions along with their features are presented; password-based attacks including password guessing and resetting are also highlighted for general awareness.

Next, Type II authentication approaches including Photo ID, smart card, swipe cards, and different types of security dongles are presented as examples of authentication strategy by ownership. Additionally, the concept of CAPTCHA is discussed as a part of Type II authentication approach.

Topics on Type III authentication include physiological and behavioral biometrics to allow legitimate access in highly secured applications. Physiological biometric includes face recognition, fingerprint verification, and retina scan, where behavioral biometric contains keystroke, gait, and voice-based recognition. Lastly, location-based authentication (Type IV) is introduced as LS is increasingly being used in mobile device tracking and identification of remote connection.

Descriptive Questions

Question 1:
What are the basic steps in authentication process?

Question 2:
What are different types of authentication? Give at least two examples of each type.

Question 3:
How are the passwords stored? Why is salt added to the hashed passwords? Describe with a diagram.

Question 4:
What are the five major criteria to select secure passwords?

Question 5:
What is the role of password manager? List three recently used password manager tools.

Question 6:
What is SSO? Describe the benefits of choosing SSO over regular authentication process?

Question 7:
What are the most common ways of resetting forgotten passwords?

Question 8:
What are the common issues with PIN-based systems? Describe two such issues.

Question 9:
List two Type II authentication systems. Briefly describe their characteristics.

Question 10:
What are the basic differences between physiological and behavioral biometrics? What are the possible benefits of choosing biometrics over typical passwords?

Multiple-Choice Questions

Question 1:
In terms of time, which group of passwords should be easiest to guess due to commonality?

A. password baseball qwerty dragon 12345789 football
B. shorty ragequit numbers no1knows god123 offby1 iam-2safe 69erfan drinker
C. cellar1door rage1quit drinks1on2me candy1apples computers1are2safe

Question 2:
In terms of time, which group of passwords should be easiest to guess due to shortness and/or low character variation?
A. password baseball qwerty dragon 12345789 football
B. shorty ragequit numbers no1knows god123 offby1 iam-2safe 69erfan drinker
C. cellar1door rage1quit drinks1on2me candy1apples computers1are2safe

Question 3:
You are trying to create a bank account and they ask you to enter a PIN number for the account. What would be a good PIN for you to use?
A. 1111
B. 1234
C. 4571
D. 9876

Question 4:
Mary is creating an account with a bank. They ask her to implement her own security questions as another type of security measure. What would be a good security question for her to come up with?
A. What is the name of your first pet?
B. What is the name of your favorite movie?
C. What is the name of your best friend?
D. What was the name of the first person that you hugged?

Question 5:
Mary is creating an account with a bank. This time when Mary gets to the security question part she can no longer create her own question but is forced to use one that is provided to her. The question that she has to answer is: «What is the name of your first pet?» Which of the following should she choose about how she should answer this question?
A. With the name of her first pet plus the name of her second pet.
B. With the name of her first pet.
C. With a random name that has nothing to do with her pets.

D. With her mother's maiden name.

Question 6:
Which token authentications require personnel to verify? (Select all that apply)
A. Passports
B. Drivers' Licenses
C. Swipe Cards
D. Smart Cards
E. Dongles

Question 7:
Which token authentications are usually automated? (Select all that apply)
A. Passports
B. Drivers' Licenses
C. Swipe Cards
D. Smart Cards
E. Dongles

Question 8:
Which security dongles require the host to produce a one-time password and send it to the user's dongle for authentication?
A. Randomized Signal Receiver
B. Random Code Generator
C. Token Storage

Question 9:
Which security dongle creates a one-time password after the host sends an authentication request to the user's dongle?
A. Randomized Signal Receiver
B. Random Code Generator
C. Token Storage

Question 10:
Which type of token-security stores an encrypted password required for authentication that is unlocked by the host?
A. Randomized Signal Receiver
B. Random Code Generator
C. Token Storage

Question 11:
What are some weaknesses associated with token authentication? (Select all that apply)
A. Tokens can be broken, lost, or stolen.

B. Standardized tokens may be easy to forge.
C. Tokens are difficult to manage and use.
D. Tokens take a long time for authentication.

Question 12:
Who or what are CAPTCHA designed to work against?
A. Administrators
B. Users
C. Hackers
D. Bots

Question 13:
What is the meaning of acceptable percentage level in biometrics?
A. How close an input pattern matches that of a pattern stored in the database?
B. How many patterns are found in the database?
C. How a match is found?

Question 14:
What is it called when a user is wrongly authenticated?
A. False negative
B. False positive
C. True positive
D. True negative

Question 15:
What is considered a Behavioral authentication method?
A. Veins authentication
B. Fingerprint authentication
C. Voice authentication
D. Facial authentication

Question 16:
What are some of the problems with biometric authentication? (Select all that apply)
A. Your biometric data can be stolen.
B. You can be injured.
C. You can easily misplace your biometric data.
D. Your data is easily reproduced.

Question 17:
What is the name of the device that generates the location signature?
A. Location Signature Sensor
B. Location Station

C. Location Base Unit
D. Global Positioning Service

Question 18:
Where is the **passwd** file stored in UNIX system?
A. /root directory
B. /user/var directory
C. /etc. directory
D. /home directory

Question 19:
What is the correct expansion of LDAP?
A. Lightweight Directory Admin Protocol
B. Lightweight Directory Authentication Protocol
C. Lightweight Directory Access Protocol
D. Lightweight Directory Approve Protocol

Question 20:
Which of the following is not a criterion of security questions?
A. Safe
B. Inconsistent
C. Stable
D. Memorable

References

1. Techlicious/Fox Van Allen @techlicious (August 08, 2013). "Google Reveals the 10 Worst Password Ideas | TIME.com". Techland. time.com. Date accessed 01 July 2015. URL: ► http://techland.time.com/2013/08/08/google-reveals-the-10-worst-password ideas/?iid=biz-article-mostpop2
2. ► http://splashdata.com/press/worst-passwords-of-2014.htm. Date accessed 01 July 2015
3. Tips for creating a strong password. Date accessed 15 July 2015. URL: ► http://windows.microsoft.com/en-us/windows-vista/tips-for-creating-a-strong-password
4. Scarfone K, Souppaya M (2009) Guide to enterprise password management: recommendations of the National Institute of Standards and Technology, NIST Special Publication 800-118
5. Stallings W (2010) Cryptography and network security: principles and practice, 5th edn. Prentice Hall Press, Upper Saddle River
6. Passwords technical overview (2 May 2012). Date accessed 15 July 2015. URL: ► https://technet.microsoft.com/en-us/library/hh994558(v=ws.10).aspx
7. Suo X, Zhu Y, Scott Owen G (2005) Graphical passwords: a survey. In: Computer security applications conference, 21st annual, 10 pp. IEEE

8. Rubenking N (2 June, 2015) The best password managers for 2015, PC magazine. Date accessed 15 July 2015. URL: ► http://www.pcmag.com/article2/0,2817,2407168,00.asp

9. Kate V (15 June, 2015) Password manager LastPass hacked, exposing encrypted master passwords, Forbes magazine. Date accessed 15 July 2015. URL: ► http://www.forbes.com/sites/katevinton/2015/06/15/password-manager-lastpass-hacked-exposing-encrypted-master-passwords/

10. Dave F (2010) Best practices for a secure "Forgot Password" feature, fishnet SECURITY. Date accessed 15 July 2015. URL: ► https://www.fishnetsecurity.com/sites/default/files/media/10WP0003_BestPractices_SecureForgotPassword%5B1%5D_0.pdf

11. ► http://goodsecurityquestions.com/. Date accessed 01 July 2015

12. Kluever K, Zanibbi R (2009) Balancing usability and security in a video CAPTCHA. In: Proceedings of the 5th symposium on usable privacy and security (SOUPS), p 14. ACM, Mountain View, CA, USA

13. Gafurov D, Snekkenes E, Bours P (2007) Gait authentication and identification using wearable accelerometer sensor. In: 2007 IEEE workshop on automatic identification advanced technologies. IEEE, pp 220–225

14. Gafurov D, Helkala K, Søndrol T (2006) Biometric gait authentication using accelerometer sensor. J Comput 1(7):51–59

15. Shiraga K, Trung NT, Mitsugami I, Mukaigawa Y, Yagi Y (2012) Gait-based person authentication by wearable cameras. In: 2012 ninth international conference on networked sensing systems (INSS). IEEE, pp 1–7

16. Joyce R, Gupta G (1990) Identity authentication based on keystroke latencies. Commun ACM 33(2):168–176

17. Hwang S-S, Cho S, Park S (2009) Keystroke dynamics-based authentication for mobile devices. Comput Secur 28(1):85–93

18. Nauman M, Ali T (2010) Token: trustable keystroke-based authentication for web-based applications on smartphones. In: Information security and assurance. Springer, Berlin, pp 286–297

19. Zahid S, Shahzad M, Khayam SA, Farooq M (2009) Keystroke-based user identification on smart phones. In: Recent advances in intrusion detection. Springer, Berlin, pp 224–243

20. Eric G, Upadhyay M (2013) Authentication at scale. IEEE Comput Reliab Sci

21. Dasgupta D, Azeem R (2008) An investigation of negative authentication systems. In: Proceedings of 3rd international conference on information warfare and security

22. Dhamija R, Perrig A (2000) Deja Vu: a user study using images for authentication. In: Proceedings of 9th USENIX security symposium

23. iRevolutions (17 June, 2013). How ReCAPTCHA Can Be Used for Disaster Response. Date accessed 1 Jan 2017. URL: ► https://irevolutions.org/2013/06/17/recaptcha-for-disaster-response/

24. Maltoni D, Maio D, Jain AK, Prabhakar S (2009) Handbook of fingerprint recognition. Springer Science & Business Media

Biometric Authentication

Authentication through human characteristics

© Springer International Publishing AG 2017
D. Dasgupta et al., *Advances in User Authentication*, Infosys Science Foundation Series,
DOI 10.1007/978-3-319-58808-7_2

This chapter focuses on the biometric authentication process which helps to prevent unauthorized access to computing resources. This chapter focuses on the biometric authentication process which helps to prevent unauthorized access to computing resources. First, biometric authentication steps are discussed, and then the performance of each biometric modality is illustrated. Next sections provide details of physiological and behavioral biometrics along with the available applications where these authentication techniques are in wide used. The last section of the chapter discusses different known attacks of biometric systems along with possible remedies for each of them.

Introduction

Biometric authentication (sometimes known as biometrics) is a process of recognizing individuals through their physiological and behavioral traits or attributes. The term 'Biometric' is a combination of two Greek words, namely Bio (life) and Metric (to measure), and it refers to a type of authentication «something that you are.» Hence, biometric authentication actually measures as well as analyzes the biological attributes of an individual. In many situations, biometric-based authentications are capable of providing better accuracy in identifying a user than typical password-based systems. ◘ Figure 2.1 shows some most common keywords used in biometric-based authentication process.

The main reasons for using biometric authentication are as follows:

1. The risk of identity theft is significantly low compared with other authentication technologies, since it uses human biometric traits.
2. No need to remember any information while authenticating to use a system.
3. Biometric authentication modalities cannot be easily guessed by the hackers.
4. It is not possible to share the biometric information to any other person.

As security breaches have increased, it is now crucial to build a more secure identification and verification system for accessing online resources.

Some of the disadvantages of biometric authentication systems are:

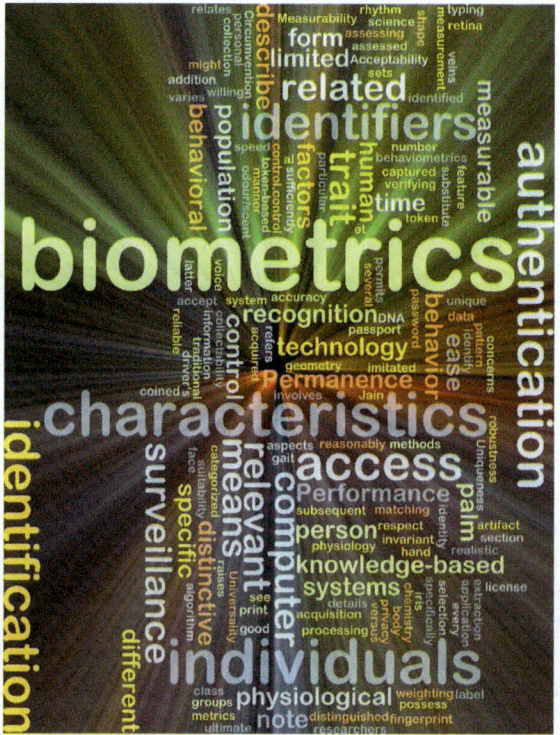

□ **Fig. 2.1** Different keywords used in biometric authentication process. ©Kheng Guan Toh/Shutterstock.com

1. Since the biometric features do not change frequently, in the case of compromise of an authentication system and the biometric data getting leaked, this confidential information may become public and abused.
2. Biometric authentication systems mainly depend on biometrics of all registered users. So disabled people and persons with some defects in required biometric cannot be authenticated using the biometric authentication system. For instance, fingerprint-based authentication is one of the widely used biometric today. If a person has an injury in the thumb, he is not able to authenticate at that time. Hence in airports or banks, there should be an alternate way of authentication other than a specific biometric.
3. One of the most concerning disadvantages of biometrics is privacy. As these authentication data are stored in a central database and are sometimes checked against this stored information from different places for identity veri-

2

fication, there is a risk of losing these personally identifiable information (PII) in the case of compromise of the database. Some people do not want to share their PII and are reluctant to use biometric authentication system.

4. The cost of developing biometric authentication system is relatively high. Also maintaining and updating the authentication database incurs additional cost than other types of authentication systems. For example, an iris scanner is more expensive than a typical keyboard where a user enters his/her password. The size of data that needs to be stored for the biometric authentication is much larger than that of a password database.

The International Biometric society [1] and other groups are continuously exploring new biometric traits which are easy to use, nonintrusive, and provide most accurate results in identification. A paper by Ricanek [2] published in Computer Magazine September 2014 issue titled «Beyond Recognition: The Promise of Biometric Analytics» mentioned that soon biometric analytics will be used as the discovery of potentially interesting information about a person other than verifying identity using biometric signal patterns.

Different Biometric Modalities

Biometric traits can be categorized in different ways based on their features and the way these are used for authentication. In general, those traits are categorized in two types, namely, physiological biometrics and behavioral biometrics [3].

Physiological biometrics are related to human body shape and features. With the change of this body geometry, these biometrics need to be updated to avoid failure of authentication. Some examples of this type are fingerprint, face, hand geometry, iris, retina, etc. In general, the accuracy of physiological biometrics is higher than behavioral biometrics. ◘ Figure 2.2 shows fingerprint and hand geometry as examples of physiological biometrics.

Behavioral biometrics are related to certain kind of behavior of an individual. Hence, this authentication system can prevent a person from accessing a cyber-system, if his current behavior pattern is different from the stored behavioral pattern. Examples of this type of biometrics are—voice,

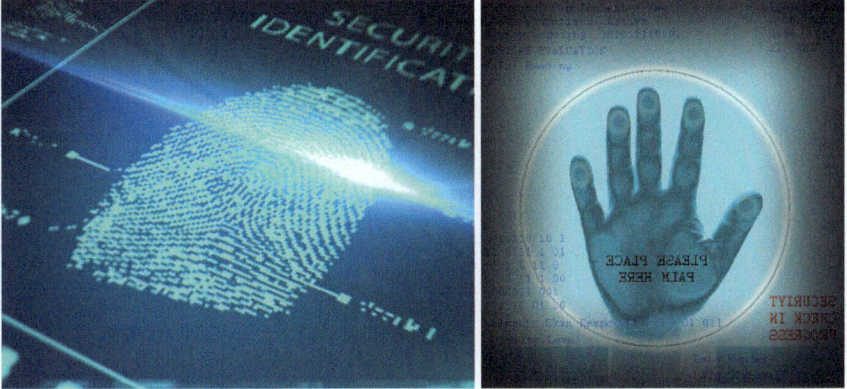

■ **Fig. 2.2**   Examples of how physiological biometrics are used for authentication.
©Peshkova/Shutterstock.com ©Bruce Rolff/Shutterstock.com

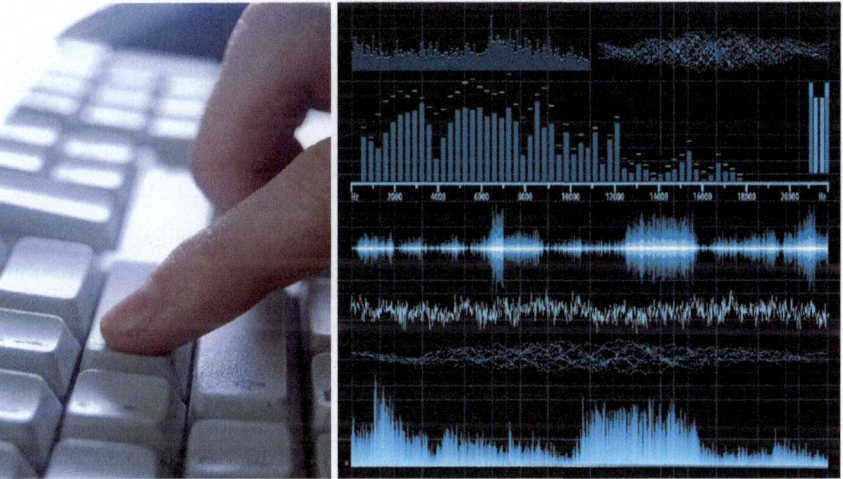

■ **Fig. 2.3**   Illustration of behavioral biometrics use for authentication.
©Dmitri Mihhailov/Shutterstock.com © Eliks/Shutterstock.com

keystroke analysis, mouse dynamics, signature, gesture, etc. ■ Figure 2.3 shows keystrokes- and voice-based biometrics as examples of behavioral biometrics.

Different subcategories of these two categories [4] are shown in ■ Fig. 2.4. According to the figure, physical biometric is categorized in head and hand biometrics. Head biometrics include face, retina, iris and ear, while hand biometrics include fingerprint, palm print, hand geometry.

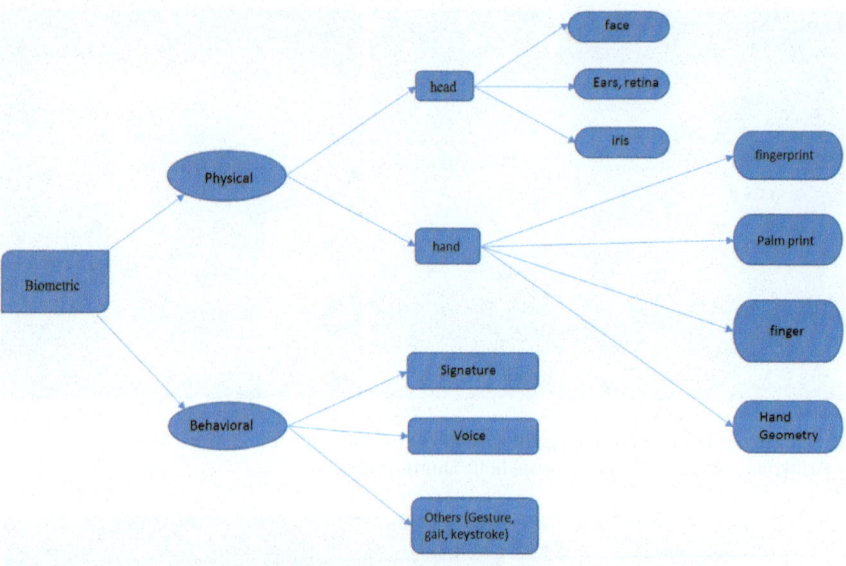

◘ Fig. 2.4 Physiological and behavioral traits used in different biometric authentication [32]

Biometric Authentication Process

A biometric system is generally composed of four major components, namely, enrollment unit, feature extraction unit, template matching unit, and decision-making unit. Enrollment unit takes biometric inputs from an individual user, creates a template, and stores it in the database. During this process, a biometric reader scans person-specific biometric characteristics and then saves it in the digitized form. Feature extraction module takes the digitized input from the enrollment unit and produces a compact representation of the captured biometric data. This compact data is stored in the identity database for future use. The template matching unit does the comparison of the user's identity against the stored biometric in the database and calculates a score. In the case of biometric identification, the matching is performed on many-to-one basis (captured input with stored templates of multiple users). In case of biometric verification where prior identification was done, the matching of captured information is done as one to one with the stored database. The decision-making unit accepts or rejects a user based on the predefined threshold value of score and the calculated scores in the template matching unit. ◘ Figure 2.5 describes the processes involved in biometric-based authentication. The basic steps are:

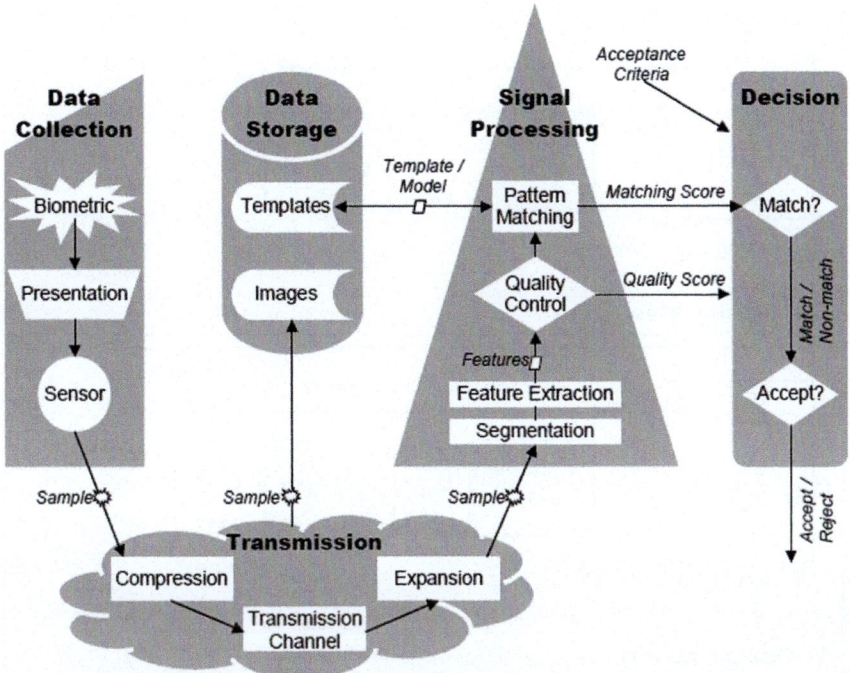

◻ Fig. 2.5 Basic steps in biometric-based authentication including data collection, data storage, processing of biometric data, and decision-making [33]

Data Collection: This step includes the sample collection procedure of biometric data from the sensor.

Transmission: This step uses compression techniques over the captured samples and transmits the output to the data storage unit. It also sends information to the signal processing unit.

Data Storage: This unit gets the input from the transmission unit and stores the biometric images (sensor captured data) into different templates.

Signal Processing: This unit does segmentation and feature extraction steps to get the required features. Then it passes the features through quality control unit and pattern matching unit.

Decision: This unit computes the matching algorithm and gives accept or reject decision based on the quality score.

The enrollment, identification, and verification processes are shown in the flow diagram in ◻ Fig. 2.6.

Enrollment process includes the capture of the biometric signal, extracting of features, the creation of templates, and storing the templates in the database.

2

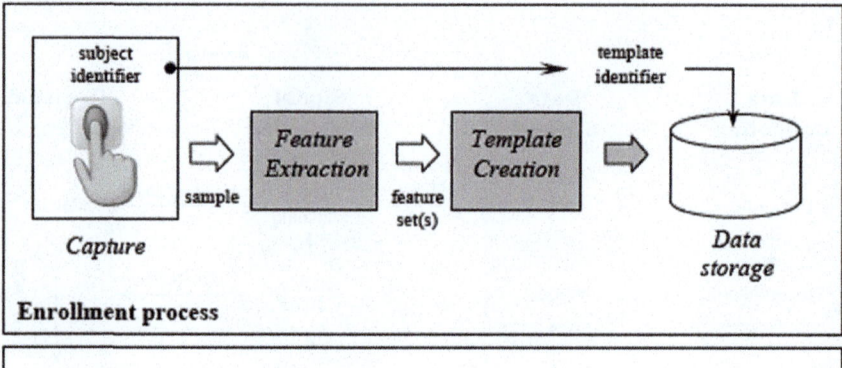

Enrollment process

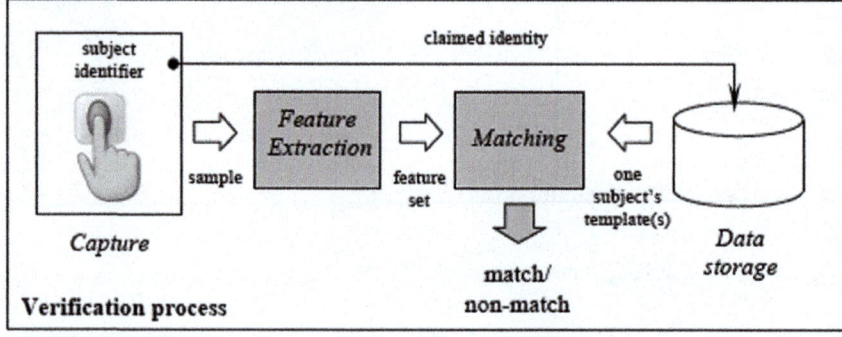

Verification process

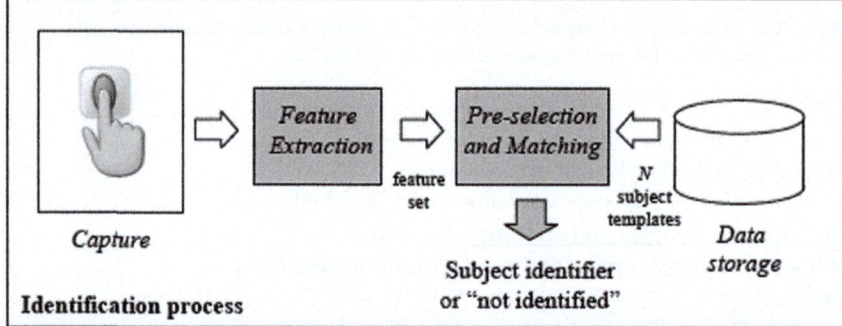

Identification process

■ **Fig. 2.6**   Enrollment, identification, and verification in a biometric system [5]

The *verification process* includes the capture of the biometric signal and extraction of the features from the captured biometric data. Then these sets of features are compared with the stored subject's template. The template is extracted for the claimed identity while capturing through biometric sensors.

The *identification process* also includes the capture of the biometric signal and extraction of the features from the captured biometric data. These extracted set of features are then compared with all the stored subject templates to identify the subject.

Performance Measures of Biometric Systems

The performance of biometric-based authentication systems is generally represented in terms of decision error rates— false acceptance rate and false rejection rate [5].

False Acceptance Rate (FAR)

In biometrics, the instance of incorrectly identifying an unauthorized person is referred to as a type II error or false acceptance. It is considered as the most serious biometric security error as it gives unauthorized users access to systems. Accordingly, the false acceptance rate (FAR) or false match rate (FMR) is the measure of the likelihood that the biometric security system will incorrectly accept an access attempt by an unauthorized user. FAR is calculated using the following formula:

$$\text{FAR}(\mu) = \frac{\text{Number of false sucessful attempts made in authenticating users}}{\text{Total number of attempts made in authenticating users}},$$

where 'μ' is the security level.

False Rejection Rate (FRR)

False rejection is a condition in biometric-based system, where an authorized person cannot be identified by the authentication process. False rejection rate (FRR) is the measure to calculate the chance of the biometric-based system failing to authenticate a legitimate user and is calculated as the probability of the system to fail in finding a match between an input biometric template and the registered biometric stored in the database. Sometimes it measures a ratio of false rejections and a total number of attempts to do authentication. The term 'rejection' refers to the claim of a user in biometric-based authentication system. It is also known as false non-match rate (FNMR).

$$\text{FRR}(\mu) = \frac{\text{Number of false reject made in authenticating genuine users}}{\text{Total number of attempts made in authenticating users}},$$

where 'μ' is the security level.

The lower these FAR and FRR values, the better the biometric trait.

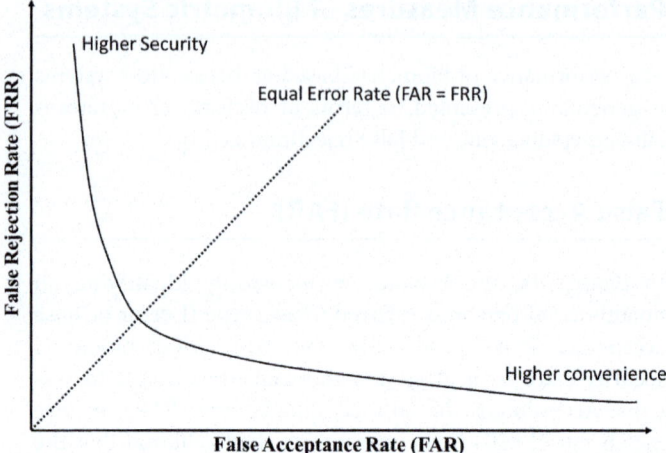

□ Fig. 2.7 The representation of FAR, FRR, and EER in receiver operating characteristic (ROC) curve

Equal Error Rate (EER)

This error rate is defined as the value obtained at some threshold level of a biometric system where the False Acceptance Rate (FAR) and False Rejection Rate (FRR) are the same. It is also called a crossover error rate.

In general, lower the equal error rate, the higher the accuracy of that biometric system. All available biometric systems have the capability of adjusting the sensitivity of these error rates. If the false positive is not desired, the system sensitivity can be set to require (nearly) perfect matches of the enrollment data and the input data of users. If FRR is not desired, the sensitivity value can be readjusted to accept the approximate matches of enrollment data and input data. The following □ Fig. 2.7 (ROC curve) illustrates the threshold level to get the desired ERR value.

Other performance metrics are also used in biometric-based authentication systems such as *Image acquisition errors*. Irrespective of the accuracy of the matching algorithm in a biometric system, the performance of that system gets degraded in case of failing to properly enroll a user. Two examples of this type are failure to enroll rate (FTE) and failure to acquire rate (FTA).

Failure to Enroll Rate (FTE)

This error occurs due to a system's inability to generate the repeatable templates for a particular portion of the population. This can happen if a person does not have that biometric feature, a person cannot produce an image of significant quality during enrollment, or a person cannot reliably match his template in several attempts at the time of verification. FTE also depends on the enrollment policy of the biometric system, and the person can be allowed to enroll at a later date.

Failure to Acquire Rate (FTA)

This error happens for a person when the system is unable to capture or locate the biometric data with a sufficient level of quality. This error rate mainly depends on the threshold level set by the biometric system and can be changed due to the adjustment of the threshold levels of different input biometrics.

Details of Biometric Authentication Modalities

Different physiological biometrics are described in this section. The focus is mainly on the features used to do the authentication and the usability of different biometric traits.

Face Recognition

Facial Recognition is a visual type of recognition that looks at a picture of a face and measures the distance between the eyes, the width of the nose, the distance between the cheekbones, and many other distinctive features as shown in ◘ Fig. 2.8.

Different features of the human face are mentioned below along with their error rates and usability as reported in the literature [6]:

Geometrical Features: Seven categories of features are considered as follows:

Lip: Lip center position (x_c and y_c), lip shape (h_1, h_2 and w), lip orientation (θ). The EER of this feature lies between 5.2 and 6.8%.

Ete: A circle with three parameters (x_0, y_0 and r) for Iris; two parabolic arcs with six parameters ($x_c, y_c, h_1, h_2, w, \theta$) to

2

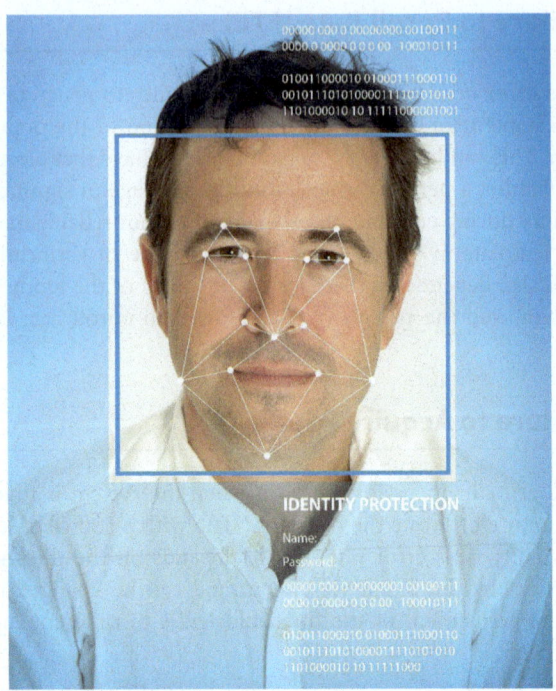

🔲 **Fig. 2.8** Face recognition process, where the facial points are identified to match with the stored template with the captured image. ©Franck Boston/Shutterstock.com

model boundaries of the eye; last parameter differs in closed and open eye position. The false match probability for the iris recognition is in the range of 1 in 10^{13} and the FNMR is very low. The error rate of retina scan is 1 out of 10,000,000 (almost 0%). Again, its FMR and FNMR are 0 and 1%, respectively. The FAR, FRR, and the crossover rates of the retina scan are 0.31, 0.04, and 0.8% respectively.

Brow and Check: Left and right brow: triangular template with three coordinates (x_1, y_1), (x_2, y_2), and (x_3, y_3) is used. Cheek also has six parameters (both up and downward triangle template).

Both brow and cheek templates are tracked using Lucas–Kanade algorithm [7]. Here, reported EER is about 15%.

Textural Features: Textural features will be elicited using Local Ternary Pattern (LTP) and Genetic and Evolutionary Feature Extractor (GEFE) [8]. GEFE uses the LTP feature extractors to elicit the distinctive features from the facial images. Its ERR is not much less, about 10%. As reported,

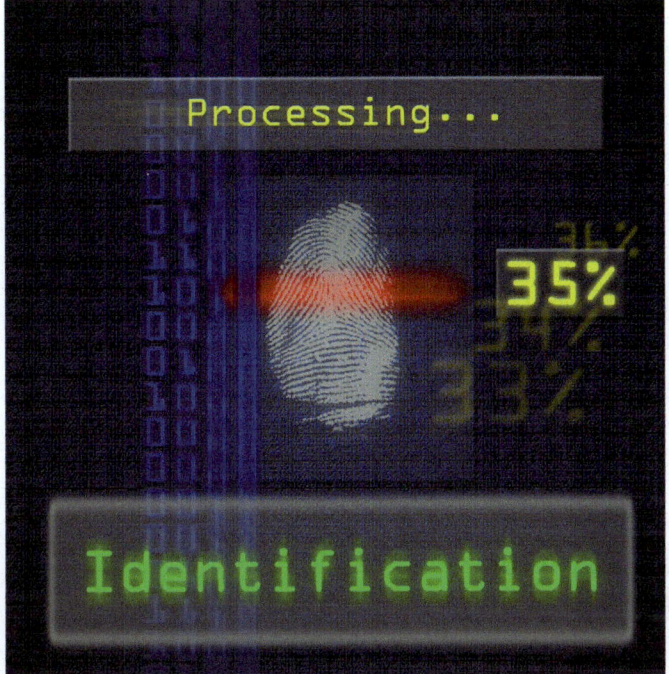

Fig. 2.9 The fingerprint recognition of a person is in progress. ©Ford Prefect/Shutterstock.com

overall FAR and FRR for the face recognition modality are 1 and 20% respectively.

Usability: Under normal lighting conditions, geometric features will be used. Under varying lighting conditions, textural features will be utilized. This modality is not reliable if the facial features are altered through plastic surgery [6, 9]. It is also dependent on the motion of the person at the time of taking the picture.

Fingerprint Recognition

Fingerprint Recognition is a visual type of authentication that uses the ridges and valleys found on the surface of fingertips. A picture is taken with the use of a fingerprint scanner that reads the digital image of a fingerprint (shown in **Fig. 2.9**). The system then finds unique points on the finger and matches the image to one that is stored in the database.

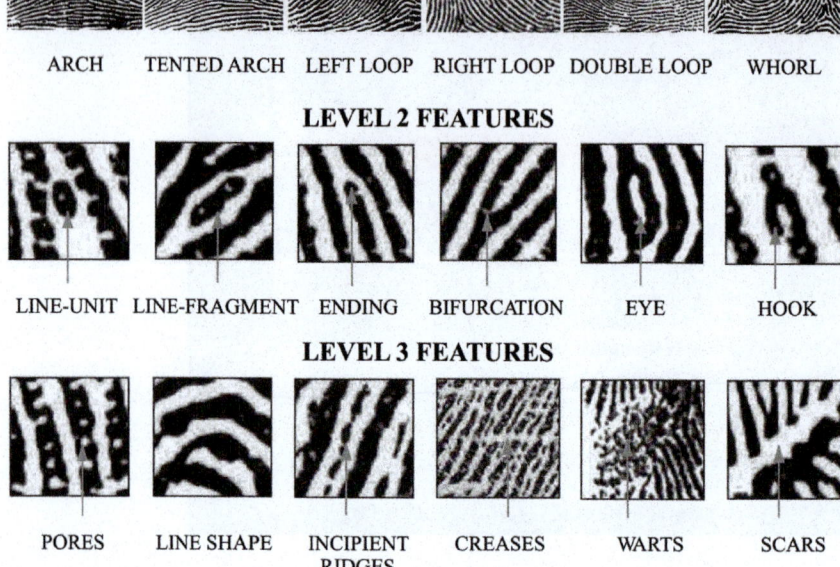

Fig. 2.10 Fingerprint features at level 1 (*top row*), level 2 (*middle row*), and level 3 (*bottom row*) [34–36]

Three levels of fingerprint features are usually considered as mentioned below (Fig. 2.10):

Level 1 features: (Global Fingerprint features): This level features include singular points, ridge orientation map, and ridge frequency map.

Level 2 features: (Minutiae-based features): This level features include ridge ending and ridge bifurcation, minutiae location, and orientation. Each of these is a combination of a number of features in order to make a unique identification of a person.

Level 3 features: (Sweat-pore-based features): Sweat pores are considered to be highly distinctive in terms of their number, position, and shape. The best algorithm for fingerprint verification yields EER less than 2.07% and more than 30% of the existing algorithms yield EER less than 5% [6].

According to a study, the overall FAR, FRR, crossover rate and failure to enrollment rates are 2, 2, 2, and 1% respectively. Figure 2.10 shows details of three levels of features:

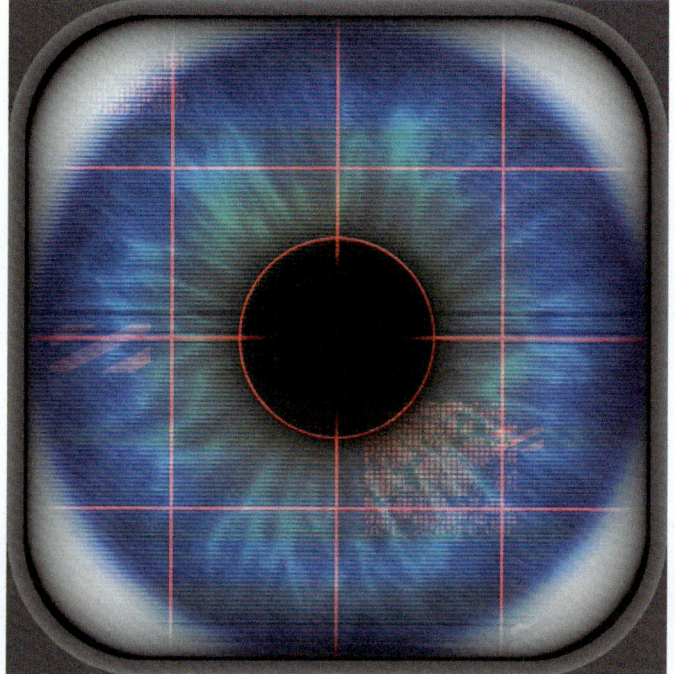

■ **Fig. 2.11** Iris recognition process of an individual. ©Juergen Faelchle/ Shutterstock.com

Usability: Fingerprints can be forged and altered by transplant surgery [6, 10, 11] and in many other ways.

Iris Recognition

Iris Recognition is another type of biometric recognition that uses a camera to capture an image of the iris. This biometric technology acquires an image of the eye, analyzes image data to generate pattern data, and tries to find a match in the stored data. ■ Figure 2.11 shows the image of an iris of an individual.

To calculate the features of an iris, localization features are used. These include iris and pupil boundary localization and pupil center detection. Also, shape, edge, and region information are fused to create a list of features for iris recognition.

Capturing the image of an iris needs a high quality digital camera. Now, commercially available sophisticated cameras use infrared light to illuminate the iris which is supposed to cause no harm or discomfort to the participants.

Usability: This modality is resistant to false matching and is ideal for high-security applications. The technology is suitable for access control, border control, and large-scale identification and is currently being used by the Immigration Services of the United Kingdom, Canada and many other countries for passport control. Iris recognition technology is used to identify an individual from a crowd and is accurate 90–99% of the time according to the National Institute of Standards & Technology (NIST) [12]. However, the difficulty of operation leads to increased false non-match rates and capture error rates; also user discomfort with eye-based technologies prevent its use than fingerprint.

Retina Recognition

Retina Recognition is a visual type of recognition that shines light into the back of the eye. Because the blood vessels absorb light differently than the surrounding tissue, this process allows the computer to get an accurate picture of the blood vessels. Retina scan requires significantly more effort to use than Iris scan, and it is more challenging as the slightest movement of the eye can cause rejection by the recognition system. It also needs more sophisticated cameras to capture the retina images than Iris scan. ◘ Figure 2.12 shows different retinal veins of an individual which are used to identify the person.

The advantages and disadvantages of retina-based recognition are mentioned below

Advantages:
- Fewer occurrences of false positive rate
- Close to zero in false negative rate
- Very reliable due to the fact that no two persons have the same pattern
- Verification of user's identity is quick.

Disadvantages:
- Accuracy can be affected by some eye diseases
- Scanning process can be considered as invasive
- It is not considered as a user-friendly authentication process
- The higher cost of equipment and the higher cost of deploying this authentication approach.

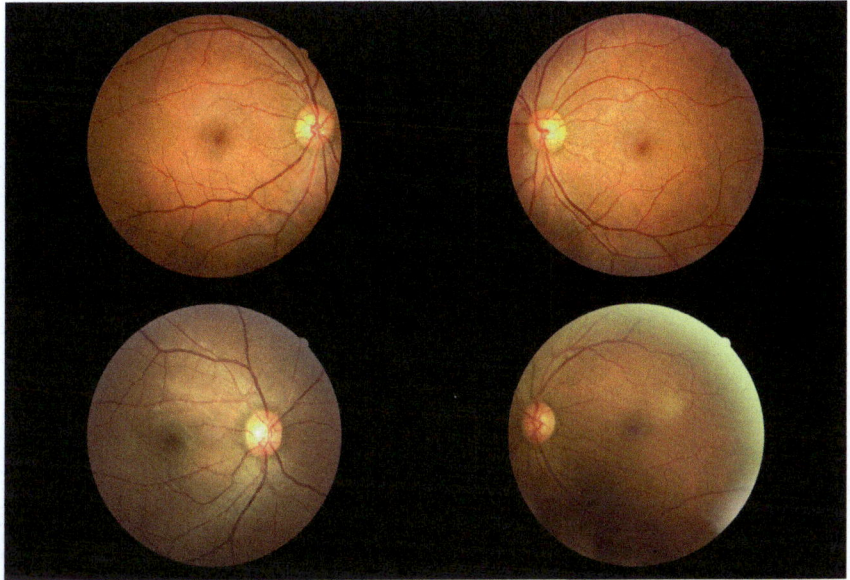

Usability:

Retina scanning is considered as the most accurate and invulnerable biometric trait to authenticate users in high-security environments. It has been utilized by different government agencies including FBI, CIA, and NASA.

◘ Figure 2.13 shows iris recognition and retina recognition which can go together to identify an individual.

Hand Geometry

Biometric hand recognition system measures and evaluates the structure, shape, and proportions of the hand. Different parts of the hand include length, width, and thickness of hand, fingers, and joints. In addition to these features, different characteristics of hand skin surface such as creases and ridges are also considered for hand geometry.

A user puts his palm on a reader's surface and properly aligns his hand to fit the fingers in the appropriate distance. The scanning device then captures hand geometry and extracts the features. These features are then compared with the stored record of the user in the database. This process of

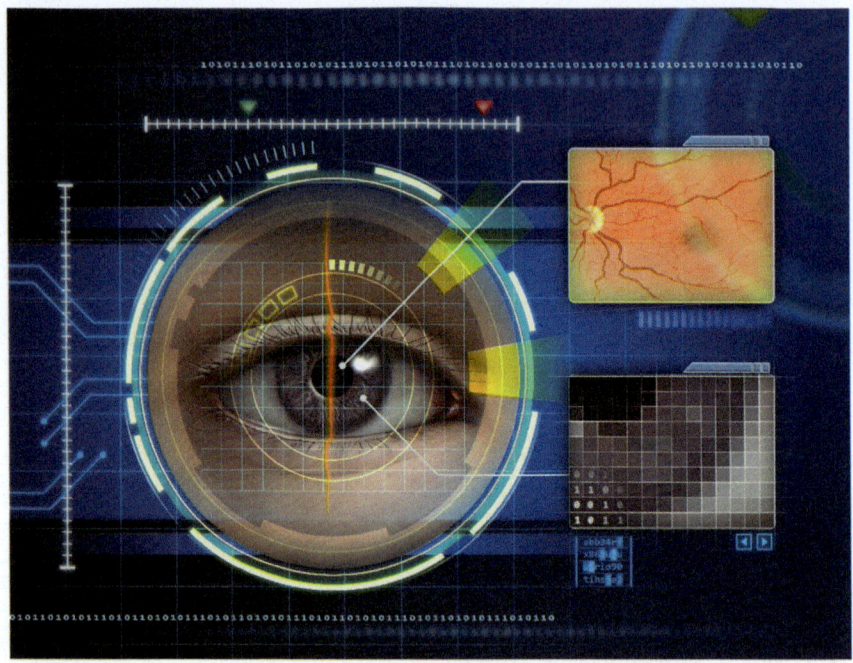

Fig. 2.13   Iris recognition and retina recognition in progress. If both identities match, access is granted to an individual. ©Andrea Danti/Shutterstock.com

authentication generally takes a couple of seconds to verify individuals. Hand-based biometric depends mainly on hand and finger geometry and hence, this biometric trait can also work with dirty hands.

Usability:

Possible applications of hand geometry-based authentication systems are

- Cash vault applications
- Point of Scale applications
- Interactive kiosks
- Parking lot entrance

Figure 2.14 shows the hand geometry scanning is done along with swipe card and face recognition techniques to identify users. Generally, the hand geometry scanner starts scanning from tip of the finger to the end of the palm.

Next, different behavioral biometrics are illustrated with their respective advantages and disadvantages and their usability issues.

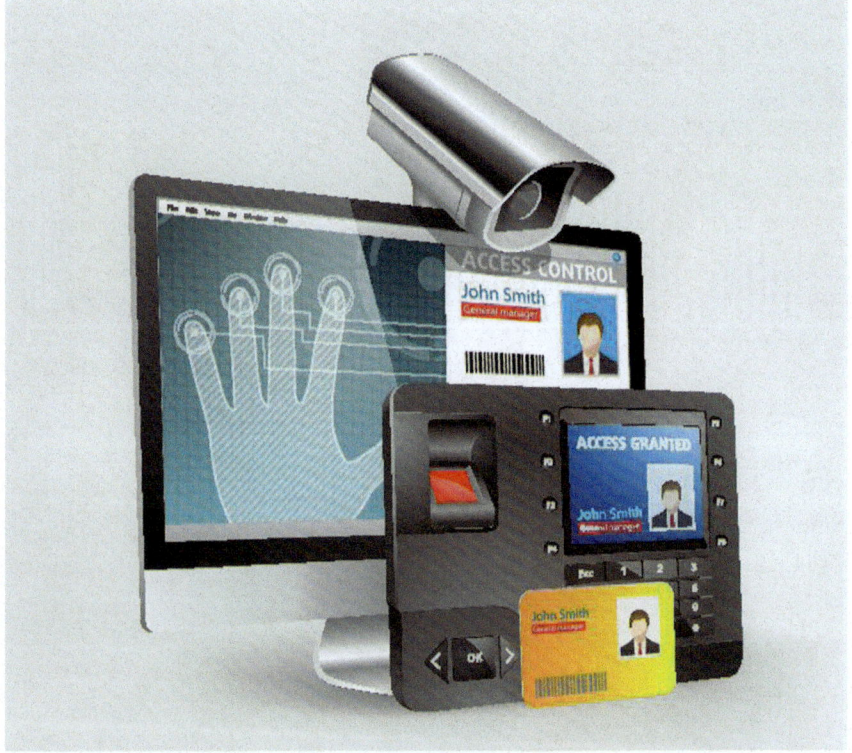

◼ **Fig. 2.14** Hand geometry is used along with swipe card to identify users. ©Black Jack/Shutterstock.com

Voice (Speaker) Recognition

Voice recognition is an auditory type of authentication that uses the way a person speaks (tone, pitch, speed, etc.) to authenticate a user. This type of recognition measures the sound waves of a person's speech. This is also known as speaker recognition (identify who is speaking) rather than speech recognition (recognize what is being spoken). There are two forms of speaker recognition: text-dependent and text-independent mode. In the first case, the person is given a fixed phrase to say and it records different attempts of the same user to track the variability of his/her voice. A text-independent system is more flexible in terms of presenter's phrasing, and it considers different situational variation of the speech of the user. It is relatively difficult to identify users in text-independent system.

2

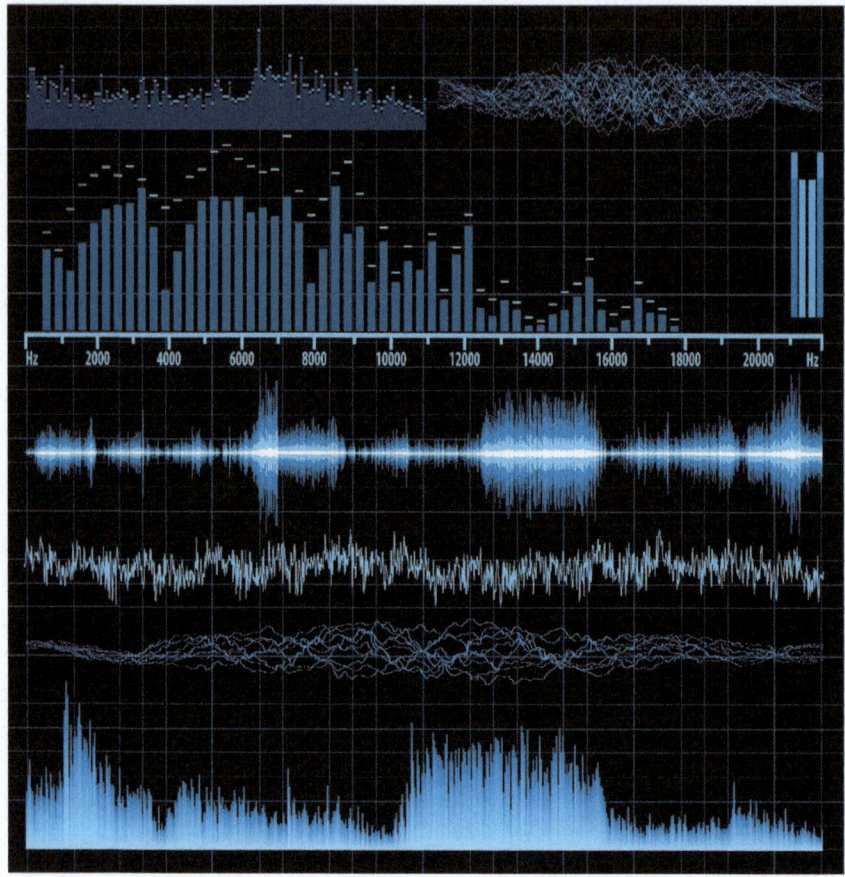

■ **Fig. 2.15**   Illustration of voice or speech recognition for identification of users. © Eliks/
Shutterstock.com

Voice recognition [6] uses pitch and different formant features. Recent evaluations based on more than 59,000 trials from 291 speakers in identification mode have reported an accuracy in the range of 1% false recognition rate at a 5% miss rate (i.e., a 1% FIR plus a 4% non-identification rate), for the most accurate among the 16 participating systems [6]. The FAR, FRR, and the crossover rates for the voice biometric are 2, 10, and 6%, respectively.

Speech samples are composed of waveforms where time is shown in the horizontal axis and loudness on the vertical axis (■ Fig. 2.15). Speaker recognition process analyzes the frequency of the voice signal and derives characteristics such as quality, duration, intensity, and pitch of the signal [11].

Usability: Some well-known speaker recognition software includes Dragon, Voice Finger, Via Talk, E-speaking, Tazti, etc.

Keystroke Recognition

Keystroke recognition is a behavioral biometric trait which captures the unique way a person types in order to correctly verify the identity of the individual. Typing patterns are generally extracted from computer keyboards, phone's virtual keyboard, etc. In order to extract features for keystroke recognition, it is needed to consider the time taken to move between two keys, how hard the buttons are pressed, and how long a key is pressed before it is released. In general, there are five elements that are captured to do keystroke recognition [13]. These are:

1. Key-down: This event occurs while a person presses a key down. The speed at which this key-down event happens is dependent on the users.
2. Key-up: This event occurs when a currently depressed key is subsequently released by the user.
3. Keystroke: This event is a combination of initial key-down event and corresponding key-up event. These two events together make keystroke event.
4. Hold time: The duration between a key-down event and key-up event happens. It is sometimes known as dwell time. It varies with the users.
5. Delay: The length of time between two successive keystroke events is called delay. The value of this length can be positive or negative (in case of overlapping keystrokes). This term is sometimes called latency.

The features of keystroke recognition are derived from the previously mentioned five elements and first-order statistics are used to calculate the features of keystrokes. Some example of first-order statistics are—minimum, maximum, mean, median, and standard deviation of hold times and latencies [14].

The overall FAR, FRR, and the crossover rates for the keystroke modality are 7, 0.1, and 1.8%, respectively [6]. ◘ Figure 2.16 shows the keystroke events by a user on a laptop, while ◘ Fig. 2.17 shows the finger positions of a person while navigating the keys of a keyboard and mouse.

Usability: Keystroke-based recognition is widely used in continuously authenticating users in different environments.

Gait-Based Recognition

Gait Recognition is a visual type of recognition that authenticates someone based on how they walk. Gait is

2

◘ **Fig. 2.16**   A user is typing in a laptop keyboard. This process captures the keystroke patterns of the users. ©Piotr Adamowicz/Shutterstock.com

◘ **Fig. 2.17**   The finger orientation for navigating keyboards and mouse of a person. ©wrangler/Shutterstock.com

generally considered as a complex locomotion pattern that comprises of synchronized movements of different body parts, joints, and also how they interact with them [15].

Fig. 2.18 Irregular walking patterns can prevent a person from authenticating a system.
©Rawpixel.com/Shutterstock.com

Every person has his/her own walking pattern. Hence, it can be considered as a unique feature for identifying individuals. Figure 2.18 shows the walking pattern of the different individuals that are used to identify them.

The advantages of this biometric trait are [16]:

- Free from background noise
- Does not depend on other person's appearance or camera viewpoint.
- Derive dynamic gait features from the model parameters.

However, this trait requires many parameters to compute from extracted gait features and hence, computational time, data storage, and maintenance costs are relatively higher than other behavioral biometrics.

2

☐ **Fig. 2.19** Walking patterns of individuals that are used to identify them. ©alphaspirit/Shutterstock.com

There are two ways to recognize this biometric trait. One way is to record a video of someone walking, looking at the trajectories of the joints and angles over time to create a mathematical model which is converted into a template. The other way uses radar to record the gait cycle that a person creates when walking and creates a template based on that. Both methods then compare the template created with the templates that are stored in the database for authentication.

Recent work reported [6] an improvement in EER from 8.6 to 7.3% by using Bayesian probability rather than Euclidean distance as a comparison metric. These results were obtained based on relatively large databases, consisting of 1079 video sequences from 115 persons. When this technology is used, irregular walking patterns can prevent a person from authenticating as shown in ☐Fig. 2.19.

Usability:
Gait-based recognition is mostly used in mobile-based authentication systems where users authenticate themselves in their smartphones using accelerometer and gyroscope sensors.

Passthoughts

Recent research focuses on authenticating humans to their computing systems using brainwave signals, which in other terms are called passthoughts [7]. These brainwaves are captured through consumer-grade wireless headsets and wearable devices having EEG sensors. Passthoughts can be treated as two-factor based authentication approach as they incorporate both knowledge factor and the inherence factor. Knowledge factor consists of a person's mental thought, which only that person knows. Inherence factor consists of EEG signal that comes from a person's brain. In addition, this biometric approach can be treated as one-step two-factor authentication as both factors can be presented at the same time. This provides a great benefit in terms of usability perspective to the participants as they do not need to undergo extra attempt to authenticate themselves.

Brainprint

A biometric Brainprint system [8] is designed to identify individuals with 100% accuracy measuring their EEG signals that represent their brain activities. In general, every individual's brain reacts in a different manner to a same set of images. Hence, the brain activities through EEG signals are able to identify every brainprint (in other words, individuals) with absolute accuracy.

A brainprint of an individual is recorded by having a user look at an image while hooked up to an electroencephalograph that captures his/her brain activity in response to the stimulus. This step is basically the registration phase for the biometric modality. During authentication phase, the user's identity would be verified by exposing her to the stimulus again, recording her current response, and using different pattern recognition algorithms to compare the registered EEG data with authentication EEG data. This biometric-based system has the great potential to be deployed in high-security based system, where 100% accuracy is an absolute need.

Applications of Biometrics

There are many applications of different biometric technologies and their use is being increasingly expanded as new biometrics are becoming available; also more authentication systems are adopting biometrics to authenticate users. ◘ Figure 2.20 shows sample application domains of different biometric modalities.

The International Biometric Group (IBG) [1], a technology-neutral, vendor-independent consulting, and inte-

Fig. 2.20 Illustrates the application of different biometric modalities. ©Kheng Guan Toh/Shutterstock.com

gration firm founded in 1996, provides objective analysis and independent services on biometrics to public and private sector clients. According to IBG, the following (shown in Fig. 2.21) are some important determining factors for applying different biometrics; while this Zephyr analysis is relatively old, this study provides a guideline in selecting different biometric technologies based on intrusiveness, distinctiveness, cost, and effort required by the users.

In Fig. 2.21, the further away the biometric characteristic is from the center, the better is the biometric technique. So for instance, keystroke scan and signature scan are low cost, require very little effort, and are not intrusive at all; however, they are not distinctive. On the other end of the spectrum, retina scan and iris scan provide very high distinctiveness; however, they are both expensive and intrusive.

In general, biometric applications fall into three different categories—law enforcement, governments, and commercial.

Law enforcement: Biometric-based authentication is mostly used for law enforcement purposes. Police departments use fingerprint identification systems to identify criminals or fugitives. Forensic applications also use biometric to expedite the investigation process.

Governments: Different branches of the government or federal agencies use biometrics to identify the citizens and foreign nationals. Some of the examples are driving license, national ID card, social security card, border control, passport, US-Visit program, etc.

Business: Commercial sectors increasingly use biometrics to secure their authentication process. ATM booth, credit card, electronic data access, network login are some examples of this type.

Some specific biometric application areas are discussed next:

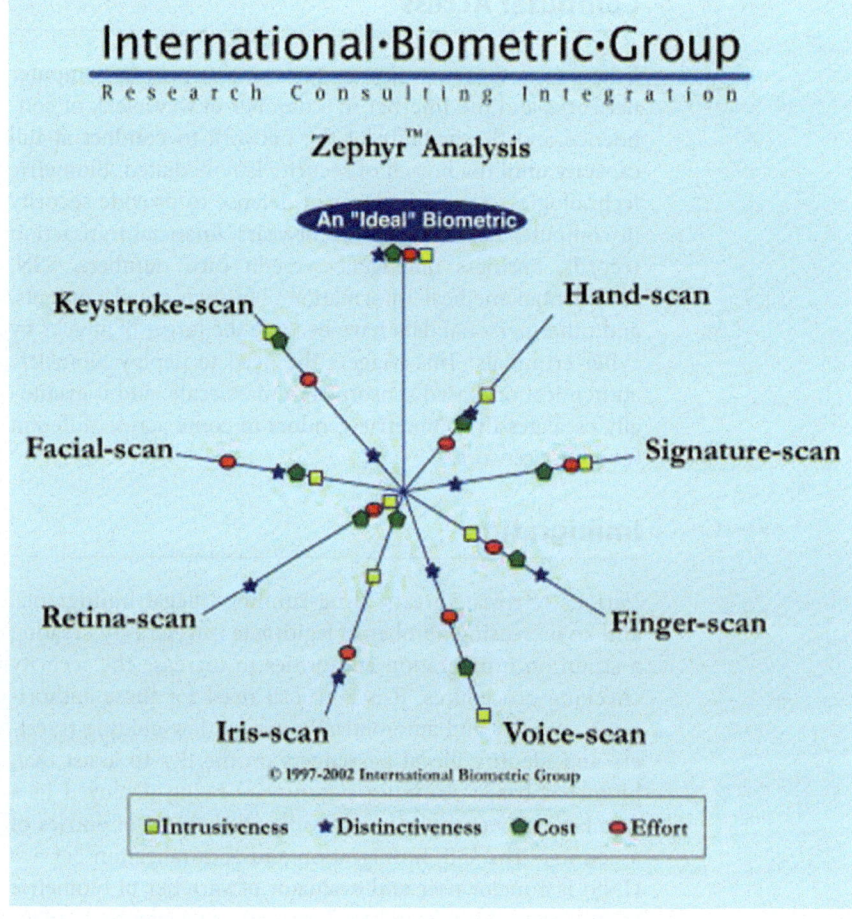

□ **Fig. 2.21** Shows the characteristics of each biometrics and their applicability criteria

Banking

Financial institutions like banks have been using biometrics for many years. Automated teller machines (ATMs) and monetary transaction banks are evaluating a range of biometric technologies for many years. Financial fraud and breaches of security are needed to be controlled for providing better service to the customers. ATMs and transactions at the point of sale (POS) are the most vulnerable to fraud, and they can easily be secured through adopting biometrics technologies. Other emerging sectors such as telephone banking and Internet banking require utmost security for bankers and customers to secure their financial transactions.

Computer Access

Fake access to computing systems affects private computer networks and the Internet in a number of ways: lack of confidence and the inability of the network to conduct at full capacity until the breach of security is remediated. Biometric technologies are a great layer of defense to provide security in computer networks. In recent years, financial transaction records, business intelligence, credit card numbers, SSN, confidential medical information, sensitive email contents, and other personal data have become the target of attacks by cyber criminals. This triggers the need to deploy biometric authentication-based sensors on a mass scale, and it eventually escalates the biometric vendors to come across different biometric sensors.

Immigration

Various terrorist threats, drug-running, illegal immigrants, and an increasing number of legitimate travelers are creating a strain on immigration authorities to increase the security checking procedures. It is a crucial need for these authorities to quickly and automatically process law-abiding travelers and identify illegal passengers on the fly. To assist that, biometric-based scanning systems are being deployed in a number of diverse places including all the port of entries of a country. The U.S. Immigration and Naturalization Service (INS) is a major user and evaluator of varieties of biometric technologies. Biometric-based systems are currently in place throughout the U.S. to automate the identification and verification of legitimate travelers and prevent illegal immigrants entering the country. For this purpose, the false acceptance rate (FAR) of the systems should be as low as possible.

National Identity

Biometrics is now widely used to assist government officials to do their everyday jobs. Their tasks include a recording of growth of population, identifying citizens, and preventing fraudulent activities during local and national elections. In order to facilitate that a biometric template is stored on a card, which in turn, acts as a national identity document for a person. Fingerprint scanning is particularly used in these scenarios to provide higher accuracy rate in person identification. Other than fingerprints, face recognition is also used to identify citizens.

Prisons

Prisons, as opposed to law enforcement, use biometric systems not only to catch criminals, but to make sure that they are securely detained in the given facility. There are many evidences where a good number of prisoners walk out of prison gates before they are officially released by the legitimate authorities. A wide range of biometrics are now being deployed worldwide to provide more secure prison access and police detention areas, enforce home confinement orders, and control the movements of probationers and parolees.

Telecommunications

With the immense growth of global communication systems, cellular networks, dial inward system access (DISA), and a range of telecommunication services are under cyber attacks from cyber criminals and terrorists. Cellular companies are the most vulnerable to cloning (a new phone set is built using information of stolen codes) and new subscription fraud (a phone number is obtained using a false identity of a person). Subsequently, DISA, which allows authorized individuals to communicate with a central exchange unit and to make free calls, is being targeted by phone scammers. To mitigate these threats, biometric-based authentication approaches are being used. Speaker ID in the phone device is a good viable option for the users to verify the numbers and then check the authenticity of the numbers in case of any doubt.

Time and Attendance

Recording and monitoring the movement of employees as they come to the workplace, take breaks for lunch, and leave for the day was traditionally accomplished by 'clocking-in' machines. However, manual checking systems can be circumvented by someone "punching in" for another person. This process really creates confusion in the tracking of the employees of a company. This also interrupts time management and costs companies millions of dollars in a year.

Replacing the manual checking process with biometric-based scanners prevents all types of system abuses.

Limitations of Biometric Systems

Biometric systems that operate using any single biometric characteristic have the following limitations [17, 18]:

1. **Noise in the captured data**: The data captured by the scanner might be noisy or distorted. A fingerprint with a scar, or a voice altered by catching cold or throat infection are examples of noisy data. Noisy data can be captured also due to the defective or improperly adjusted sensors (e.g., accumulation of dirt on a fingerprint device) or unfavorable ambient conditions (e.g., poor illumination of light on a user's face in a face recognition system). Noisy biometric data can be incorrectly matched with the stored templates in the database which results in a legitimate user being incorrectly rejected.

2. **Intra-class variations**: The biometric data acquired from an individual during the phase of authentication may be largely different from the data that was used to generate the template during enrollment. This can affect the matching process significantly. This variation is typically caused due to incorrect interaction with the sensor, or when sensor characteristics are changed during the verification phase. For example, the varying psychological makeup of a legitimate individual might result in vastly different behavioral traits at various verification events.

3. **Distinctiveness**: A biometric trait is expected to vary significantly across individuals. In some cases, it is possible to have large inter-class similarities in the feature sets used to represent these biometric traits. Golfarelli et al. [19] have shown that the information content (number of distinguishable patterns) in two of the most commonly used representations of hand geometry and face are only of the order of 10^5 and 10^3, respectively. Therefore, every biometric trait has some theoretical upper bound that limits its discrimination capability.

4. **Non-universality**: While all registered users are expected to possess the biometric trait being acquired, it is possible for a subset of the users in real scenario who do not possess a particular biometric trait. A fingerprint-based biometric system may be unable to extract

the set of features for certain individuals, due to the poor quality of the ridges in his/her hand. Thus, there is a failure to enroll (FTE) rate associated with using every single biometric trait. It has been empirically estimated that as much as 4% of the population may have poor quality fingerprint ridges that are difficult to capture through fingerprint sensors and that results in increasing FTE errors. Den Os et al. [20] report the FTE problem in a speaker recognition system.

5. **Spoofing attacks**: An imposter may make different attempts to spoof the biometric trait of a legitimate user in order to circumvent the security of the system. This type of attack is especially relevant when behavioral traits such as signature [21] and voice [22] are used. However, physical traits of an individual are also susceptible to spoofing attacks. For example, it can be possible (although difficult and cumbersome and requires the help of a legitimate user) to construct artificial fingers/fingerprints in a reasonable amount of time to circumvent a fingerprint verification system [23].

New Biometric Challenges

With the development of sophisticated skin-like facemasks for use in movies and other recreational purposes, such masks can also be used to defeat the face or fingerprint recognition system for illegal access; ◘ Fig. 2.22 shows wearing such a face mask.

A research on identity science [2] shows that it is possible to alter biometric features of a person through different medical procedures, which are listed below.

▪▪ Fingerprint Modification:

The fingerprint is a widely used biometric specifically in border control and immigration. As fingerprints cover a little area at the tips of fingers, it is possible to obscure this particular area using various methods such as chewing off the skin or applying acid to scar the fingerprint permanently. Criminals sometimes can modify their fingertips surgically by replacing the skin with that of a donor or some other part of their own body.

▪▪ Plastic Surgery:

Plastic surgery poses significant recognition challenges for face-based biometric systems because they rely on various

2

■ **Fig. 2.22**   Illustration of the use of mask in obfuscating face recognition. ©aastock/
Shutterstock.com

facial characteristics, such as the relative position, shape,
and size of the eyes, nose, jaw, and cheekbones. The alteration of the face can be done locally (reshaping and restruc-
turing facial features) or globally (skin peels and facelifts).
This type of modification of face presents a big challenge
for face recognition algorithms to accurately identify a person in real scenario.

■ ■ **Hormone Replacement Therapy:**
An innovative challenge to face recognition-based systems
is the medical alteration of levels of sex hormones to transform gender. Transgender hormone replacement therapy
(HRT) impacts fat distribution in the region of the face
which changes the overall shape and texture of the face biometric. Hence, the current face recognition algorithm will
fail to detect that person from his/her previously registered
data. Figure 2.23 shows the significant changes in male
faces after undergoing hormone replacement therapy.

◘ Fig. 2.23   Example face images of four male-to-female transgendered women before (*top row*) and after (*bottom row*) hormone replacement therapy [2]

Attacks on Biometric-Based Authentication Systems

Biometric systems work accurately if the verifier of the system can validate the following things:
- The biometric data comes from a legitimate person at the time of verification.
- The captured biometric matches with the stored template in the database.

In general, there are eight points in a biometric system which are vulnerable to attack [24]. The details are mentioned below (see ◘ Fig. 2.24) [25].

■■ Attack the Biometric Sensor:
In this particular type of attack, a fake biometric is presented at the sensor to verify. The fake instance can be false image or fingerprint.

■■ Resubmit a Previously Stored Digitized Biometric Signal:
In this mode of attack a previously recorded biometric signal is replayed to the system through bypassing the biometric sensor.

■■ Override the Biometric Feature Extractor:
The feature extractor of the biometric system is forced to generate a set of features chosen by the attackers rather than actual values generated from the captured data of the sensors.

2

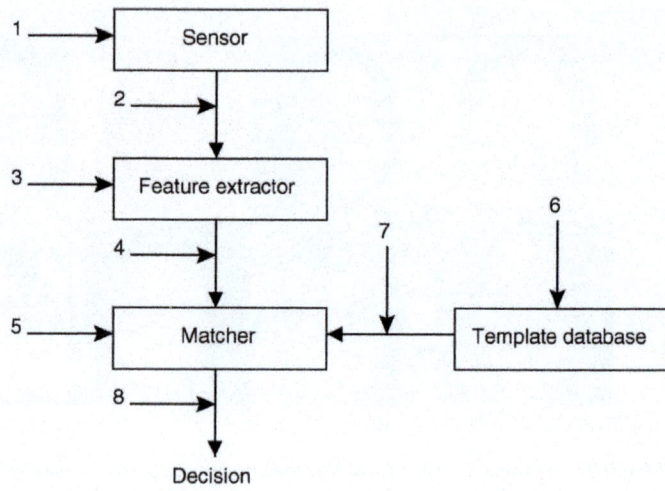

◘ Fig. 2.24   Possible different attack points in a generic biometrics-based system [27]

■■ Tamper with the Biometric Feature Representation:
The features extracted using the data obtained from the sensor are replaced with a different fraudulent feature set.

■■ Corrupt the Biometric Matcher:
The biometric matcher can be attacked in order to produce a predefined match score irrespective of the input feature captured by the biometric sensor.

■■ Tamper with the Stored Biometric Templates:
The attackers can alter any numbers of stored templates in order to allow a fraud to authenticate into the system. In that case an unauthorized person can do the regular biometric scanning and allow access to the sensitive data of a system. Any smart card where the template is stored is equally vulnerable to this type of attack.

■■ Attack the Communicating Channel between the Stored Template and the Biometric Matcher:
The attack can happen in the communication channel between the stored template in the database and the biometric matcher logic module. In this type of attack, any communicating data can be altered to change the final decision of matching.

■■ Override the Final Decision of Matching:
The final decision regarding the matching of biometric can even be hacked or overridden by the attacker, which allows the intruders to gain access to a biometric system.

Mitigating Different Attacks on Biometric Systems

In order to resist the previously mentioned attacks, several approaches are taken to ensure the integrity of the each component of a biometric authentication system. Some of these mechanisms are described below [25, 26].

Liveness Detection Mechanisms

Liveness detection in a biometric system guarantees that only 'real' biometric data are used to generate templates for enrollment, verification, and identification. Detection of liveness can be done using software or hardware implementation. The first attack point can be mitigated with the incorporation of the liveness detection in the biometric sensor.

One should incorporate extra hardware to capture life signs like temperature, pulse detection, etc., for fingerprint detection and movement of face for face recognition. The drawback of using extra hardware is that the system will be expensive and difficult to deploy for the mass population.

Facial recognition systems can use multiple cameras to obtain 3D properties of the head. This will avoid an attack if a photo is used as authentication. Text-prompted voice systems can ask the person to say a random phrase. Again a combination of face and voice recognition can verify the lip movement.

Liveness detection can also be done through challenge-response based approaches, where a small impulse current is passed into the finger and then response from the finger is captured. Liveness detection through perspiration patterns of a fingerprint image can be done. Moreover, procedural techniques like supervision are greatly effective to detect the liveness of the biometric.

Steganographic and Watermarking Techniques

Steganographic and Watermarking techniques are used to resist attacks at the previously mentioned attack points 2 and 7 (Channel between the sensor and feature extractor and also the channel between the stored template and the matcher). Steganography means secret communication and it involves hiding critical information through unsuspected carrier data. Steganography-based techniques are suitable to transfer critical biometric information from a client

2

application to a secured server. In the following section, two application scenarios are illustrated where hiding process of data is similar, but they (scenarios) vary in the characteristics of the embedded data, host image, and medium of data transfer [27].

The first scenario shown in ◘ Fig. 2.25 involves an application based on steganography. The required biometric data (for example, fingerprint minutiae) to transmit over a communication channel is hidden in a host image. Sometimes it is called carrier image due to its purpose of carrying the data only. The carrier image can be a synthetic fingerprint image, a face image, and any random image. The usage of this synthetic image to carry actual fingerprint data provides more security towards preventing interception of actual fingerprint image. The security of the transmission process can further be increased through encryption technique applied on the stego image before transmitting it.

The second scenario mentioned in ◘ Fig. 2.26 focuses on hiding the facial information (different facial points to identify a person) into a fingerprint image to increase the security of the fingerprint biometrics, and this information is stored on a smart card of a user. At the time of authentication, the fingerprint of a person is compared with the stored information on the smart card. Then the facial information embedded in the fingerprint biometric is recovered. This additional information is a good source to verify the authenticity either in an automated manner or by a human in a supervised biometric application.

Challenge-Response Systems

Challenge-response systems are a better way to prevent replay attacks mentioned in the attack 2 and attack 7 in the previous section. One approach can be to use image-based challenge response method. Using this method, a challenge is presented to the biometric sensor and the output response is computed depending on the asked challenge and the content of the acquired input image [28].

In another approach [29], the smart card containing the biometric data (used to do verification) is protected using a cryptographic checksum. This value is calculated within a security module capable of resisting tamper and integrated with a biometric sensor.

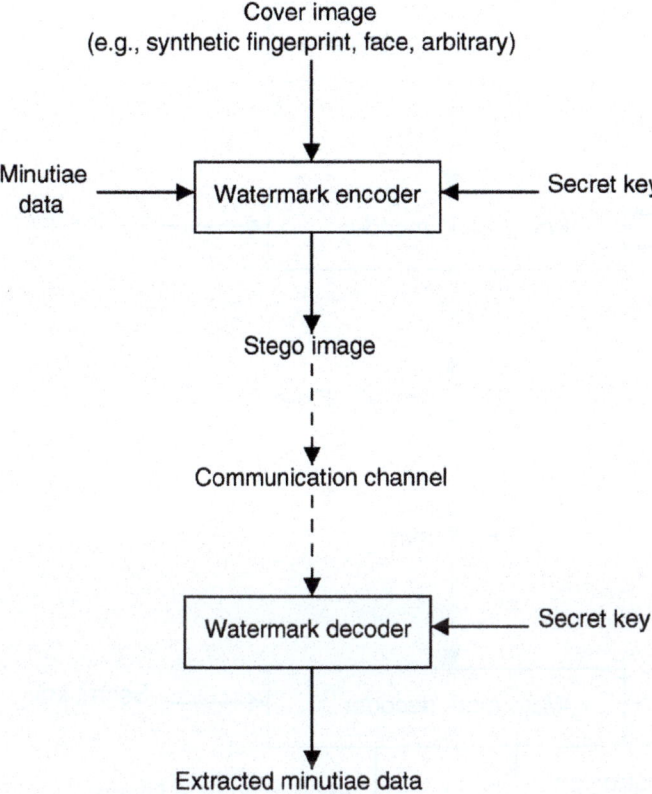

Fig. 2.25 Diagrams of steganographic techniques to prevent attacks [27]

Multimodal Biometric Systems

Unimodal-based biometric systems suffer from a variety of problems including noisy data, intra-class variations, restricted degree of freedom, non-universality, spoof attacks, and unacceptable error rates [30]. Some of these limitations imposed by the unimodal biometric system can be mitigated with the inclusion of multiple sources of biometric information to verify user's identity. Multimodal biometric systems are expected to be more reliable because of the multiple independent pieces of information about the same identity. They address the issue of non-universality by ensuring multiple traits with significant population coverage. They also prevent spoofing since it will be really difficult for an impostor to spoof multiple biometric traits of a legitimate user simultaneously. Above all, they can facilitate a challenge-response

2

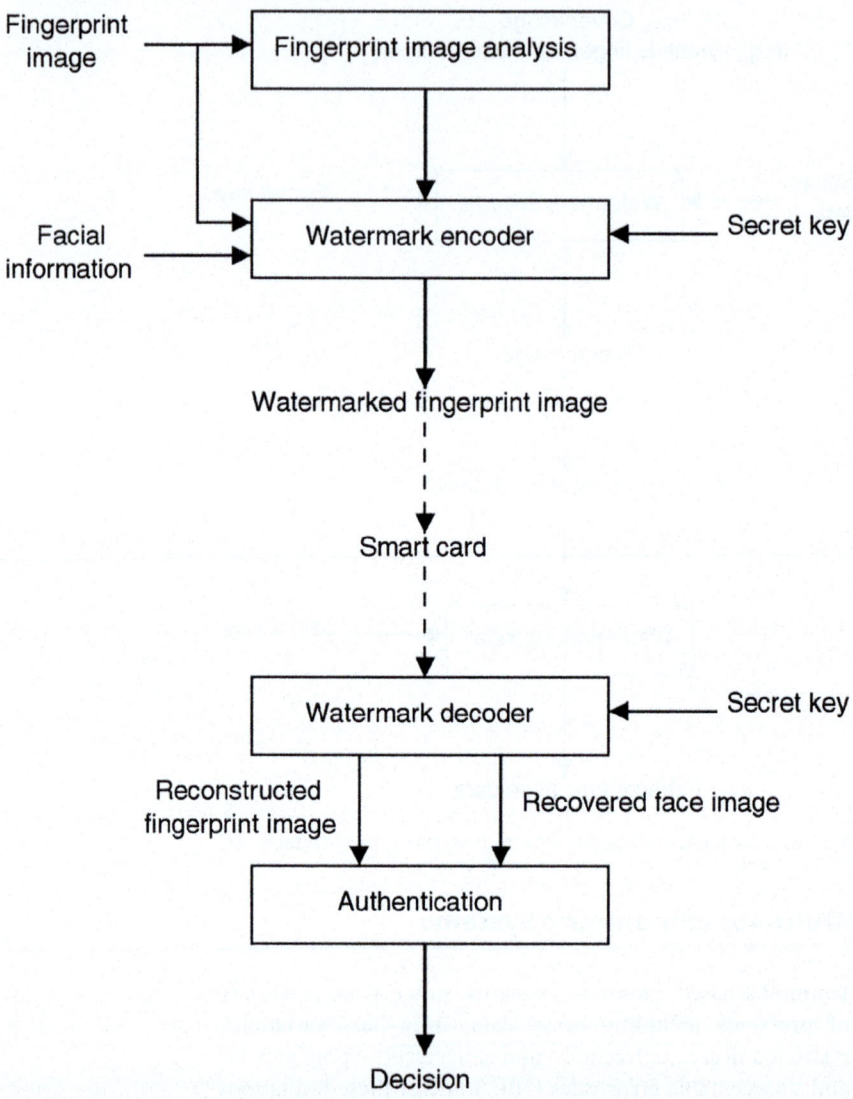

□ Fig. 2.26 Diagram of watermarking techniques to prevent attacks [27]

type of mechanism by requesting the user to present a random subset of biometric traits (authentication modalities) which ensure the liveness of a user at the point of data acquisition. The choice and the number of biometric traits are determined by the type of application using the required computational resources and the cost required to do the authentication.

The combination of different biometric traits can be achieved at the feature extraction level, matching score level, or decision level. At the feature extraction level, features from different biometric traits are combined to make the new sets of features and this new set is going to be used in matching an identity. While in the case of matching level, the individual scores produced by every biometric trait are integrated to generate a new score. The generation step depends on the system administrator regarding what metric should follow to combine them. The new score is then compared with the new threshold value and makes the decision regarding allow or deny the identity. For the case of decision level, each biometric system decides its individual decision regarding the entity and then a majority voting scheme is used to come to the final decision. In practice, the fusion of matching score of the different biometric system is more preferred as it gives more flexibility to choose the importance of one biometric trait over the others.

The issues of noisy data can easily be solved using the fusion of the biometric traits as we can adjust the weight (different degrees of influence) based on the existing operating conditions. The performance of the multi-modal-biometric system depends on computational time and cost. Hence, the cost versus performance measurement should be put into consideration before deploying these systems.

Soft Biometrics

Soft biometrics traits are some characteristics of humans based on their physical, behavioral, or adhered attributes. Skin color, eye color, height, weight, etc., are examples of soft biometrics. Biometrics can be used to defend attacks at points 1 and 8 (mentioned in the previous section). These traits lack the distinctiveness to clearly distinguish two individuals. In general, many biometric systems collect auxiliary information about the registered users during enrollment time. This auxiliary information is stored either in the central database or in the smart card of the user. The soft biometric traits can help to filter out a large biometric database to a significantly smaller number of templates to match. It thus helps the speed and efficiency of the biometric authentication system.

Soft biometric nowadays are used to tune the parameters of the biometric systems as the human biometric traits are going to change over time. So, the adjustment of the threshold value for the matching score or the threshold value to weight different biometric traits can be done with the help of soft biometrics. In addition to that, soft biometrics will decrease the FAR and FRR values of biometric-based authentication that help in preventing spoofing.

Cancelable Biometrics

Cancelable biometrics requires some intentional and repeatable distortion of a biometric trait based on some non-invertible transform [28]. This is done to protect sensitive user credentials. In the case of compromise of the cancelable biometric feature, the distortion process can be changed and the same set of biometric traits can be mapped to a new template and can be used for authentication purpose. This approach is considered as one of the major ways for protecting biometric templates from being compromised.

In order to design a cancelable biometric; four criteria below should be fulfilled:

(a)   **Diversity**: The set of cancelable features can be used in only one application. Other application should use a different set of cancelable features. Hence, a large number of templates need to be created for the same biometric feature (or trait).

(b)   **Reusability**: In the event of compromise or data breach, it should be able to revoke the existing templates and reissue a new set of templates.

(c)   **Non-invertibility**: This property ensures that the template creation process is non-invertible and hence it prevents the recovery of original biometric data from the template.

(d)   **Performance**: Due to some extra processing of original biometric data, the performance of recognition should not deteriorate.

The transformation steps of biometric inputs for different applications are shown in ◘ Fig. 2.27. In general, at enrollment phase, different non-invertible transforms are applied to biometric inputs based on application-dependent parameters. During authentication, captured biometric inputs are transformed and then a comparison with the stored transformed template is done.

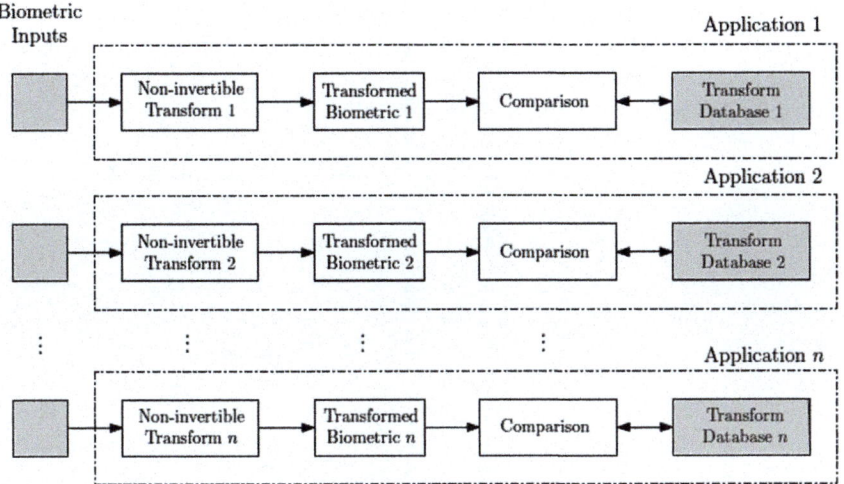

Biometric
Inputs

O Fig. 2.27 The basic concept of cancelable biometrics based on non-invertible transforms [37]

Comparative Evaluation

As there are a good number of biometric traits available, some metrics are necessary to make a comparative analysis of all these traits. In general, a biometric trait has to follow some requirements to be applied in identifying persons [31]:

Universality: Every person should possess the biometric trait.

Distinctiveness: Any two persons should be significantly different in terms of their traits or characteristics.

Permanence: Biometric trait should perform in the same way irrespective of the matching criterion.

Collectability: A biometric trait is able to be measured quantitatively.

In a practical biometric system, other metrics should be considered for choosing the biometric trait. They are:

Performance: Measured in terms of recognition accuracy, computational time, error rates (FAR, FER), etc.

Acceptability: Which set of users can accept to go for the biometric sensor in everyday life.

Circumvention: Measured in terms of how easy to bypass the biometric authentication system.

The following table **O** (Table 2.1) shows a comparison of various biometric traits with the above-mentioned characteristics.

Table 2.1 Comparison of commonly used biometric traits. Entries in the table are based on the perception of the authors [31]. High, Medium, and Low are denoted by H, M, and L, respectively

Biometric identifier	Universality	Distinctiveness	Permanence	Collectability	Performance	Acceptability	Circumvention
Face	H	L	M	H	L	H	H
Fingerprint	M	H	H	M	H	M	M
Hand geometry	M	M	M	H	M	M	M
Hand/finger vein	M	M	M	M	M	M	L
Iris	H	H	H	M	H	L	L
Signature	L	L	L	H	L	H	H
Voice	M	L	L	M	L	H	H

Table 2.2 Advantages and drawbacks of the different protection techniques [25]		
Techniques	Advantages	Drawbacks
Liveness detection	Resists spoofing attacks	Increased cost for the extra hardware and software, user inconvenience and increased acquisition time
Watermarking	Prevents replay attacks and provides integrity of the stored templates	Problem of image degradation and lack of algorithms to deal with it
Soft biometrics	Provides improved performance through filtering and tuning of parameters	Lack of techniques for automatic extraction of soft biometric techniques
Multi-modal biometrics	Improves performance, resists spoofing and replay attacks, and provides high population coverage	Increased system complexity, computational demands and costs

Table 2.2 highlights the advantages and drawbacks of different protection schemes of biometric authentication techniques.

Chapter Summary

This chapter covers different biometric authentication methods and provides details on various aspects of each biometric modality. It also covers the limitations and error rate of each biometrics which can be used as a performance measure while adopting in different applications. While Face Recognition, Fingerprint are in wide use, there is continued search for the Holy Grail, the perfect biometric, and many other biometric modalities are emerging. Some of these are called soft biometrics and others are behavior biometrics which are gradually in adaptation. Recent reports predict that biometric analytics may soon be used to discover potentially interesting information about a person other than verifying identity using biometric signal patterns. These include person's emotional state, longevity, aliveness, continuous authentication, ethnicity, gender, age (demographics, in general), honesty, concentration, mood, attitude, and even frustration in some situations.

Review Questions

Descriptive Questions

Q1: State four reasons to choose biometric-based authentication system.

Q2: Describe the taxonomy of biometric traits.

Q3: Difference between physiological and behavioral biometrics.

Q4: Describe the biometric authentication process.

Q5: What are different performance metrics used for biometric authentication? What is the most dominant error rate in biometrics?

Q6: Describe a few applications that use biometric-based authentication.

Q7: What are the drawbacks of a biometric system? Describe any two of them.

Q8: What are the possible attack points of a biometric authentication system? Illustrate with diagram.

Q9: What is liveness? Describe a process of liveness detection mechanism for biometric systems.

Q10: Distinguish between the following:
(a) Behavioral and Physiological biometric
(b) Soft biometrics and actual biometric
(c) Liveness detection and watermarking technique
(d) Unimodal biometric and multimodal biometric.

Multiple Choice Questions

Question 1:
A company is looking into adding biometric scanners to their building for added security. Which option would NOT be a good idea?
A. Facial recognition
B. Weight recognition
C. Gait recognition
D. None of the above

Question 2:
Which of the following does not use behavioral characteristics of users for authentication?

A. Voice
B. Signature
C. Veins
D. Keystrokes

Question 3:
Jason is the type of person who does not like to give out his personal information and is overly suspicious of other people. What would be the best authentication type for Jason?
A. Cognitive-based Authentication
B. Token-based Authentication
C. Biometric-based Authentication
D. Any of the above

Question 4:
What is true for equal error rate?
A. Lower the error rate, higher the accuracy
B. Higher false positive make lower equal error rate.
C. Lower false negative make higher equal error rate.
D. None of the above.

Question 5:
In which of the following biometrics, will the most sophisticated camera be used to capture the biometrics?
A. Face recognition
B. Fingerprint recognition.
C. Iris recognition
D. Retina recognition

Question 6:
Which biometric has higher universality?
A. Face recognition
B. Hand Geometry
C. Signature
D. Voice

Question 7:
Which biometric has lower distinctiveness?
A. Face
B. Hand Geometry
C. Iris
D. Fingerprint

Question 8:
Which biometric has higher performance?
A. Face
B. Hand Geometry

C. Iris

D. Signature

Question 9:

Which biometric has higher acceptability?

A. Face

B. Fingerprint

C. Iris

D. Hand Geometry

Question 10:

Which biometric has lower circumvention?

A. Iris

B. Face

C. Fingerprint

D. Signature

References

1. The International Biometric Society website. ► http://www.biometricsociety.org/
2. Ricanek K (2013) The next biometric challenge: medical alterations. Computer 46:94–96. doi:10.1109/MC.2013.329
3. Ashbourn J (2014) Biometrics: advanced identity verification: the complete guide. Springer, Berlin
4. Jain AK (2008) Biometric authentication. Scholarpedia 3(6):3716
5. Maltoni D, Maio D, Jain AK, Prabhakar S (2009) Handbook of fingerprint recognition. Springer Science & Business Media, Berlin
6. Vielhauer C (2005) Biometric user authentication for IT security from fundamentals to handwriting, vol 18. Springer, Berlin (2005)
7. Johnson B, Maillart T, Chuang J (2014) My thoughts are not your thoughts. In: proceedings of the 2014 ACM international joint conference on pervasive and ubiquitous computing, Adjunct Publication, ACM, pp 1329–1338
8. Ruiz-Blondet MV, Jin Z, Laszlo S (2016) CEREBRE: a novel method for very high accuracy event-related potential biometric identification. IEEE Trans Inf Foren Secur 11(7):1618–1629
9. Lucas BD, Kanade T (1981) An iterative image registration technique with an application to stereo vision. IJCAI 81:674–679
10. Bowyer KW, Chang KI, Yan P, Flynn PJ, Hansley E, Sarkar S (2006) Multi-modal biometrics: an overview. In second workshop on multimodal user authentication, vol 105
11. Woodward JD, Orlans NM, Higgins PT (2003) Biometrics. McGraw-Hill/Osborne ISBN: 0-07-223030-4
12. Lipowicz A (2012) NIST tests accuracy in iris recognition for identification. Date accessed 1 Jan 2017. Url: ► https://fcw.com/articles/2012/04/23/nist-iris-recognition.aspx
13. Bartlow N (2009) Keystroke recognition. In: Encyclopedia of biometrics. Springer, US, pp 877–882
14. Bartlow N, Cukic B (2006) Evaluating the reliability of credential hardening through keystroke dynamics. In: null, IEEE, pp 117–126

15. BenAbdelkader C, Cutler R, Nanda H, Davis L (2001) Eigengait: motion-based recognition of people using image self-similarity. In: audio-and video-based biometric person authentication, Springer, Berlin, Heidelberg, pp 284–294
16. Ng H, Ton HL, Tan WH, Yap TTV, Chong PF, Abdullah J (2011) Human identification based on extracted gait features. Int J New Comput Architect Appl (IJNCAA) 1(2) 358–370
17. Blum RS, Liu Z (eds) (2005) Multi-sensor image fusion and its applications. CRC press
18. Jain AK, Ross A, Prabhakar S (2004) An introduction to biometric recognition. IEEE Trans Circ Syst Video Technol 14(1):4–20
19. Golfarelli M, Maio D, Malton D (1997) On the error-reject trade-off in biometric verification systems. IEEE Trans Pattern Anal Mach Intell 19(7):786–796
20. den Os E, Jongebloed H, Stijsiger A, Boves (1999) Speaker verification as a user-friendly access for the visually impaired. In: EUROSPEECH
21. Harrison WR (1981) Suspect documents, their scientific examination. Nelson-Hall, Chicago, IL
22. Eriksson A, Wretling P (1997) How flexible is the human voice?–a case study of mimicry. Target 30(43.20):29–90
23. Matsumoto T, Matsumoto H, Yamada K, Hoshino S (2002) Impact of artificial gummy fingers on fingerprint systems. In: Electronic imaging. International Society for Optics and Photonics, pp 275–289
24. Ratha NK, Connell JH, Bolle RM (2001) An analysis of minutiae matching strength. In audio-and video-based biometric person authentication, Springer, Berlin, Heidelbergpp, pp 223–228.
25. Ambalakat P (2005) Security of biometric authentication systems. In: 21st computer science seminar, SA1-T1, pp 1–7
26. Matyáš V, Říha Z (2010) Security of biometric authentication systems. In: international conference on computer information systems and industrial management applications (CISIM), IEEE, pp 19–28
27. Jain AK, Uludag U (2003) Hiding biometric data. IEEE Trans Pattern Anal Machine Intell 25(11):1494–1498
28. Ratha NK, Connell JH, Bolle RM (2001) Enhancing security and privacy in biometrics-based authentication systems. IBM Syst J 40(3):614–634
29. Waldmann U, Scheuermann D, Eckert C (2004) Protected transmission of biometric user authentication data for oncard-matching. In: Proceedings of the ACM symposium on applied computing, ACM, pp 425–430
30. Ross A, Jain AK (2004) Multimodal biometrics: an overview. In: 2004 12th European Conference on Signal Processing, IEEE, pp 1221–1224
31. Maltoni D, Maio D, Jain AK, Prabhakar S (2009) Handbook of fingerprint recognition. Springer Science & Business Media, Berlin
32. Zhang DD (2013) Automated biometrics: technologies and systems, vol 7. Springer Science & Business Media, Berlin
33. Mansfield AJ, Wayman JL (2002) Best practices in testing and reporting performance of biometric devices. Centre for Mathematics and Scientific Computing, National Physical Laboratory, Teddington, Middlesex, UK

34. Jain AK, Chen Y, Demirkus M (2007) Pores and ridges: High-resolution fingerprint matching using level 3 features. IEEE Trans Pattern Anal Mach Intell 29(1):15–27
35. The thin blue line. ► http://www.policensw.com/info/fingerprints/finger06.html. Oct 2006
36. Nieuwendijk HVD (2006) Fingerprints. ► http://www.xs4all.nl/~dacty/minu.htm. Oct 2006
37. Rathgeb C, Uhl A (2011) A survey on biometric cryptosystems and cancelable biometrics. EURASIP J Inform Secur 1:1–25

Negative Authentication Systems

Who am I? "I am not this; no, nor am I this, nor this," then which remains is the identity of I

© Springer International Publishing AG 2017
D. Dasgupta et al., *Advances in User Authentication*, Infosys Science Foundation Series,
DOI 10.1007/978-3-319-58808-7_3

Password-based authentication systems are the oldest and most popular among all authentication methods.

Password-based authentication systems are the oldest and most popular among all authentication methods. Specifically, more than 80% of current authentication systems are password-based though these are prone to direct and indirect cracking via guessing or side-channel attacks. In recent years, different password-based approaches have been introduced to strengthen authentication systems by obfuscating user password credentials (positive identification) allowing access through different (derived) complementary authentication approaches. That means, user identity is confirmed by the negative information or credentials that are not valid, as opposed to valid or positive information. This chapter and the following chapter describe several password obfuscation and complementary approaches.

Introduction

As discussed in ▶ Chap. 1, password-based authentication system stores the user profile data in a password file in an authentication server. Every access is authenticated by comparing the access request with the password profile. One of the basic problems of password authentication is the risk of malicious access to the password data. Various guessing attacks and password cracking methods have been reported to be successful in cracking hashed passwords [1]. Any person who has stolen files of hashed passwords can often use brute-force methods to find out a password 'p' whose hash value $H(p)$ is equal to the hash value stored for a given user's password, thus allowing him/her to impersonate the legitimate user. Also a side-channel attack of stealing password profiles is a severe threat against secure authentication and access control. Weaknesses in Operating Systems are exploited to steal password profiles.

Password cracking was also instrumental in a cyber espionage campaign against the New York Times [2]. A recent study shows that 63% of confirmed data breaches involved leveraging weak, default, or stolen passwords [3]. Due to the widespread reuse of passwords across multiple websites [4], an emerging attack model is to compromise accounts by a guessing attack against a low-security website and then attempt to reuse the credentials at critical websites [5]. In addition, cyberattacks can exploit weak passwords. There are many cases that some anonymous person impersonates common passwords chosen by users which can be avoided by requiring users to use uncommon passwords [6, 7].

This chapter describes a complementary approach of passwords, namely, Negative Authentication System (NAS), its design, and implementation details. In negative authentication, a user authenticity is checked if it is not an invalid request rather than verifying it to be valid.

Concept of Negative Authentication

The idea of NAS is based on the Negative Selection Algorithm (NSA) [8–11]. This algorithm is inspired by the biological immune system which has a censoring process of T cell maturation for distinguishing entities within the body as «self» or «non-self». The Negative Selection Algorithm (NSA) is an abstract computational model, which has been widely used in many real-world applications.

The generic form of Negative Selection Algorithm (NSA) is summarized as follows [12]:

1. Define the set S (Self) as the normal pattern of activity of a system which needs to be monitored.
2. Generate a set D (Detector), those must not match any element from S.
3. The set S is continually monitored for changes. The detector set D is adjusted if any change occurred in S.

This suggests that a set of data elements can be represented by its complement set (i.e., negative database [13]). Such a representation has interesting information-hiding properties when privacy is a concern and seems to be very appropriate in hiding password database. Studies show that the negative information can be represented efficiently, even though the negative space may typically be much larger than the positive dataset [11].

In NAS, authentication data for valid users is termed as password profile or self-region (positive profile); any element other than the self-region is defined as non-self-region in the same representative space. The anti-password detectors are generated which covers most of the non-self-region. Some uncovered regions are left in the non-self-region for inducing uncertainty to the attackers. Briefly, user identities are confirmed not by using valid or positive information, rather by using negative information. A conceptual visualization of the positive, negative (anti-password), or non-self-regions

3

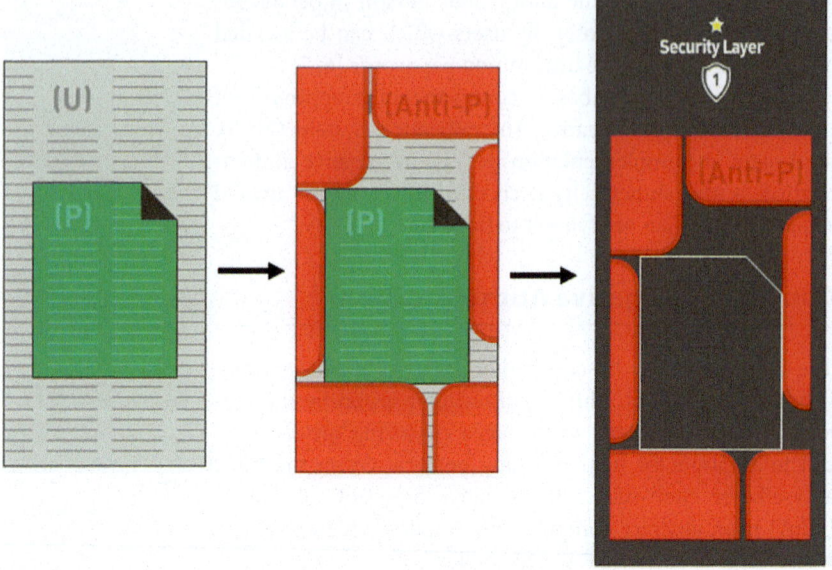

◻ **Fig. 3.1**   A conceptual view of password space (self-region) shown in *green* and anti-password space (non-self-space) shown in *red* in the total username-password space

is shown in ◻ Fig. 3.1. Here, the green region is the positive region or the password space, which contains the actual user passwords hidden from the users using the anti-password (Non-self-regions shown in the red portions) and the obfuscation (the gray portions). Some elements or clusters are created using negative selection algorithm (NSA) to cover the non-self-regions. These elements are called detectors, anti-passwords, and collectively called as anti-password region. The obfuscation has been added to increase the complexity for attackers in guessing the password. It is desirable to generate detectors to cover most of the non-self-region and thereby increase the probability of detecting invalid access requests (◻ Fig. 3.1).

While using NAS, it is appropriate to keep the password profile and anti-password (detectors) in separate servers for security enhancement of the authentication system. The negative detectors that check for guessing attacks are stored on the first server that handles all incoming requests. Every access request arrives at the first server and is checked with the detectors. If an access request information is matched with any detector, that request is marked as an invalid request (◻ Fig. 3.2). Otherwise, that request is sent to the next server that contains information regarding

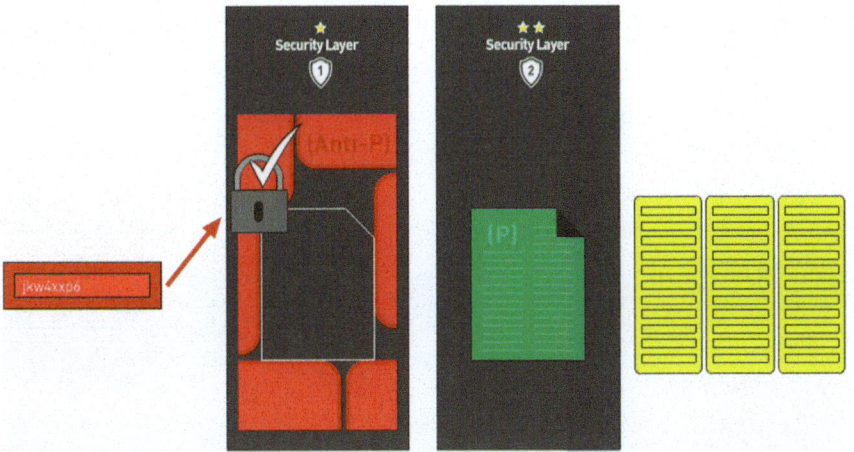

▣ Fig. 3.2 Illustration of two layers of NAS: a detector is matched in negative filtering layer (layer 1) indicating an invalid password entered for login

password-based positive authentication. This server checks the request and if it finds a match, then it marks it as a valid request (▣ Fig. 3.3). In this way, there is no direct communication among the access requests and the positive authentication server (second server), while the communication between the first server and the second server is transparent to the users.

This profile represents the negative abstraction of valid credentials and actually is the problem-specific realization of the negative database described by [13]. This negative authentication for a system login application has been tested successfully workable [14]. This negative approach has some benefits over the traditional positive authentication approach. The negative authentication module is supposed to detect and filter out most of the invalid requests; therefore, such guessing requests have a low probability of accessing the positive authentication module. Furthermore, negative detectors are placed in the front security perimeter. Also, well-crafted negative detectors can reveal very little information of password profile due to the ambiguity introduced intentionally in their (detectors) generation. Hence, NAS can reduce the risk of side-channel attacks of passwords. Also, the password file that contains the authentication data resides behind a highly controlled firewall in the second server and not accessible to users directly ensuring that the positive profile is protected from being

3

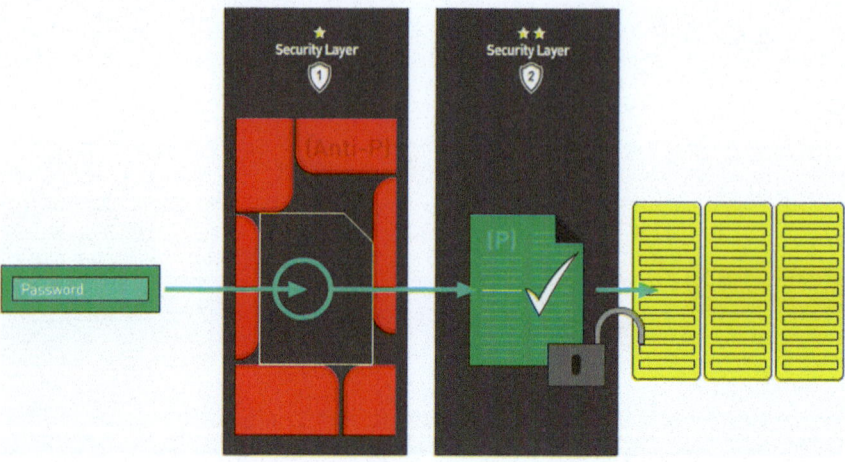

Fig. 3.3 Illustration of two layers of NAS: here a valid password entry did not match negative filtering layer and was also verified by the password checking layer 2

compromised. ◘ Figure 3.4 shows a logical view of the two-layer authentication system.

Types of NAS

There exist different approaches to Negative Authentication Systems (NAS), which are primarily based on the representation of the passwords. The main differences of these approaches are how to generate detectors covering most of the non-self-region while increasing the detection rate of invalid access requests. Each type of NAS maps the information in different representation spaces such as Binary-valued, Real-valued, and Grid-based approaches, which are discussed in the following subsections.

Binary-Valued NAS (B-NAS)

The binary-valued space model is a hypercube-based non-deterministic model of NAS. This model is adopted from the original V-Detector algorithm [15]. The binary-valued model uses N-dimensional binary space (0 or 1), where N is the length of the representation string (encrypted or hashed password). Total number of elements in binary-valued password representation space is 2^N, where N is the dimension of the space as illustrated in ◘ Fig. 3.5.

To create detectors, points are chosen from the non-self-region of the binary space that is not covered

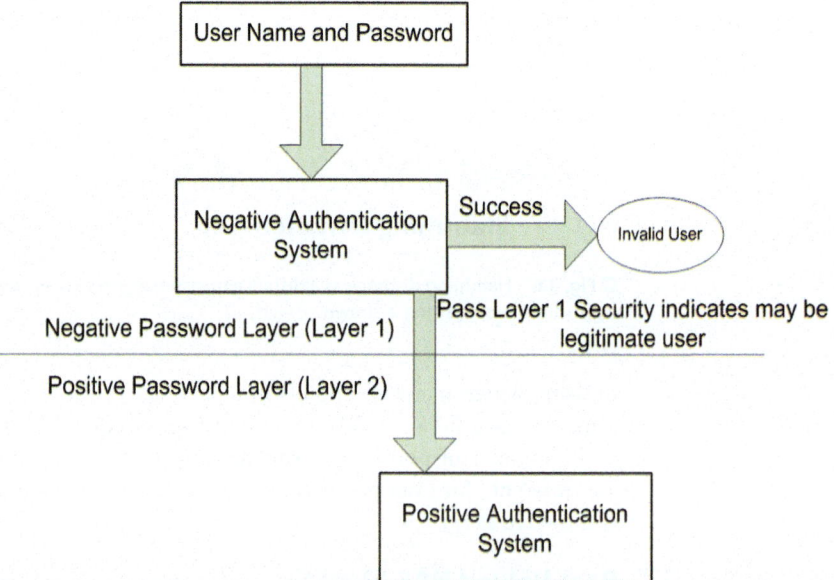

Fig. 3.4 Logical view of negative (anti-password space) and positive (password space) authentication NAS layers

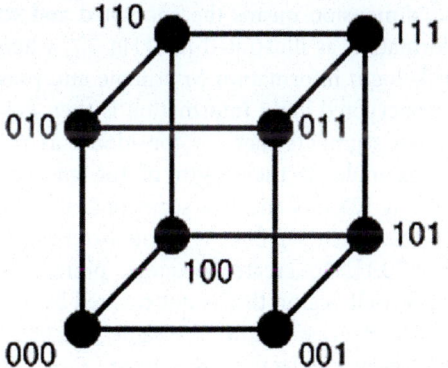

Fig. 3.5 Binary-valued NAS in three-dimensional space (i.e., $N = 3$)

by the password points and the existing set of detectors. Accordingly, the radius of a detector is chosen in such a way that the hypercube representing that detector will not overlap with the existing self-region (containing all self-points). These detectors cover some portion of the non-self-region, so partial overlap among the detectors may be allowed. This B-NAS model uses Hamming distance as a distance

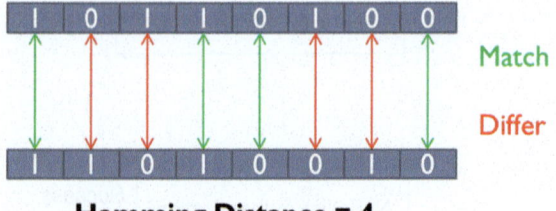

Hamming Distance = 4.

◘ Fig. 3.6   Hamming distance as a distance measure in B-NAS. Here an eight-dimensional string is shown

measure between any two points in binary space. The Hamming distance between two detector strings of equal length is the number of positions at which the corresponding bits are different. An example of Hamming distance calculation is shown in ◘ Fig. 3.6.

Real-Valued NAS (R-NAS)

The real-valued NAS model uses n-dimensional string to represent self-region and detectors. Both detectors and self-regions are some hyperspheres of $n \in \mathbb{Z}^+$ dimensions [15, 16]. Here, dimension means the encrypted and segmented login information as illustrated in ◘ Fig. 3.7, where hashed (encrypted) login information (username and password) is used. The encrypted login information is then divided into n real-valued segments that are considered as the dimension. For example, if the length of the encrypted login information is 128 bits and it is segmented into four 32-bit segments, then each segment is normalized in real-valued space over [0,1]. To create detectors, points are chosen from the non-self-region that is not covered by any existing detectors. The process follows the same concept described in general Negative Selection Algorithm (NSA). The radius of the detector is chosen in such a way that the hypersphere may not overlap with the self-region, whereas partial overlap may be allowed among the detectors. ◘ Figure 3.8 shows a *two*-dimensional representation of R-NAS with self-points (self-region) and negative detectors. However, the four-dimensional real-valued NAS implementation uses generalized Euclidean distance for calculation of distance between any two points in the space.

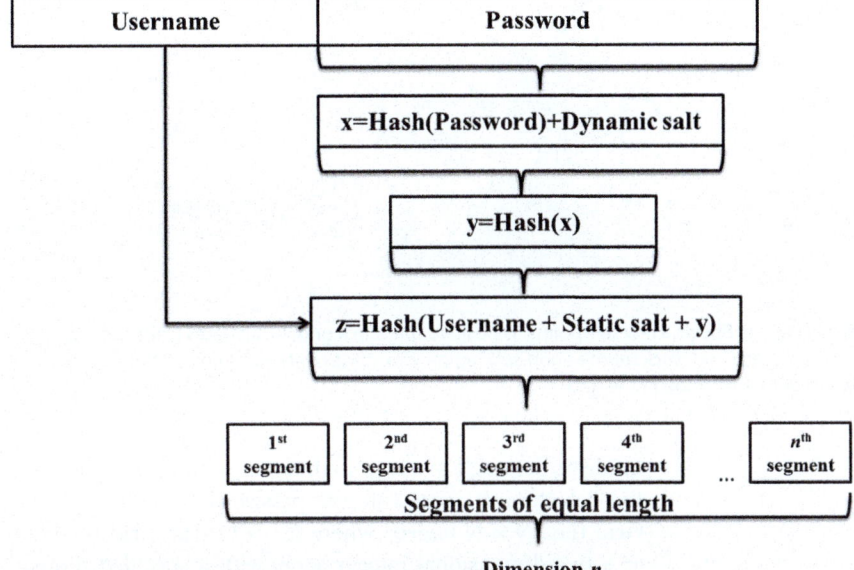

Fig. 3.7 The concept of dimension used in R-NAS implementation

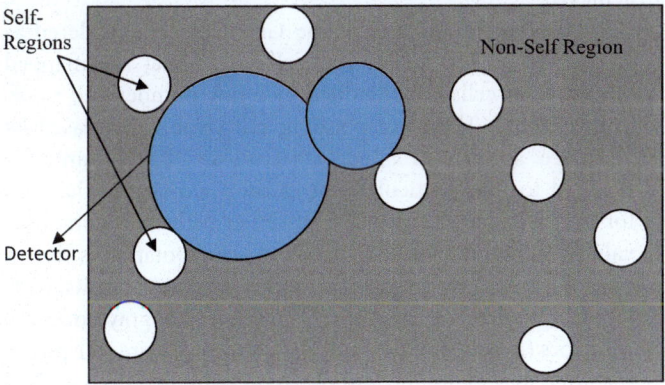

Fig. 3.8 Two-dimensional projection of self-regions and detectors in real-valued space. For higher dimensional representation, the shape will differ from the shown figure

Grid-Based NAS (G-NAS)

Grid-based implementation is a deterministic model of NAS [17], where the representation scheme is adopted from the work by Williams et al. 2004 [18]. This Grid-based NAS

3

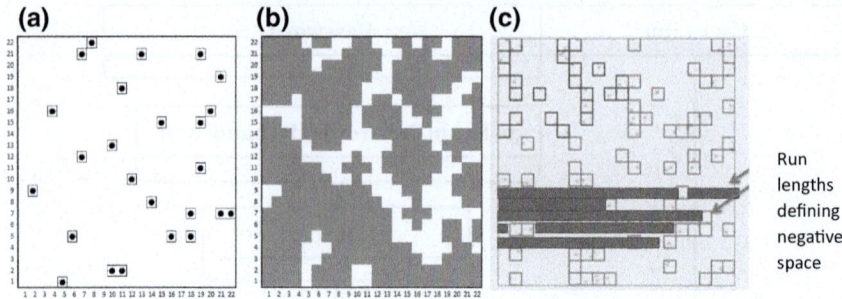

(a) **(b)** **(c)**

Run
lengths
defining
negative
space

◼ Fig. 3.9   a Shows a few self-points as positive space in a grid; **b** illustrates *white space* as positive space and *black space* as non-self-region; **c** displays continuous non-self-region defining as detector region in the grid

model is designed in a way so that there is no overlap among the detectors. It represents the password space in a square Grid (i.e., $N \times N$ *matrix*, where $N \in Z^+$). The Grid is stored as a two-dimensional integer array with a specified dimension, where the optimum Grid size can be chosen to divide the space according to the obfuscation level desired to be achieved.

For example, the size of the Grid can be considered as 512×512 or 256×256, etc. Self-points that are mapped to particular cells are identified by a row number and a column number. The corresponding cell value is increased by 1. Hence, any cell as value $M \geq 1$ means M self-points are mapped to that particular cell in the Grid space. The cells for non-self-points are easily identified with their contained value 0 (zero). A sample Grid-based representation is shown in ◼ Fig. 3.9. ◼ Figure 3.9 shows the password points, self-region, and detector region in Grid space. The gray boxes are detectors (negative space) and the white boxes are self-points (positive space) in the sample representation of the Grid space (◼ Fig. 3.9a, b, c).

The (x, y) coordinates of the cell containing each username–password for the positive space are stored. For N username–passwords, total N coordinates and the Grid size are stored once for the entire space. The complexity of the algorithm to generate the negative detectors is $O(N)$ or less.

To get a detector in Grid space, it is started from a positive cell (where $M > 0$) and is defined a run connecting series of cells (say along the x-axis) until reaching another positive cell. The run length of cells is encoded by the beginning and end cell ids. Thus, the storage required to define the complete negative space will be no greater than $2 \times N+$ some small constant brought in by runs that end on boundaries of the space. If x and y coordinate runs are combined, it is possible to get a larger compression. Thus, even though the negative space is much larger than the positive space, it can be defined in terms of the N cells of the positive space.

Detectors are created by scanning the Grid once and keeping track of all cells that match the criteria, $M = 0$. Consecutive cells of corresponding "$M = 0$" in a row are considered as one detector and its run length is calculated with the number of consecutive cells. In the Grid-based implementation, it is possible to increase or decrease the detectors' coverage by changing the Grid dimension.

Detectors generated from this implementation are stored with their run length and the starting index. In G-NAS, the detectors are saved in row-wise format and in each row, the detector is stored with the starting column index and run length. The benefit of using row-wise format is to find a particular point in the Grid space, the row of that point can be determined easily from the hashed form of the password and the search space is then reduced to that specific row to check whether it falls in a detector cell or self-point cell. This approach also compresses the space required to store the total detector region, and the region can be reconstructed easily from the saved information of the detector space.

The details of the Grid space and generation of positive Grids and detector Grids are discussed in later sections.

Implementation of Negative Authentication System (NAS)

Three NAS approaches introduced in the previous sections are explained in details in this section. The self-regions in these implementations consist of username–password points known as self-points. Some of the common terms that are applicable for all three approaches are mentioned below.

Hashed String Generation

The username and password pairs are converted to the corresponding self-point of the representing space model. At first, the password is hashed with the specified hashing algorithm.

3

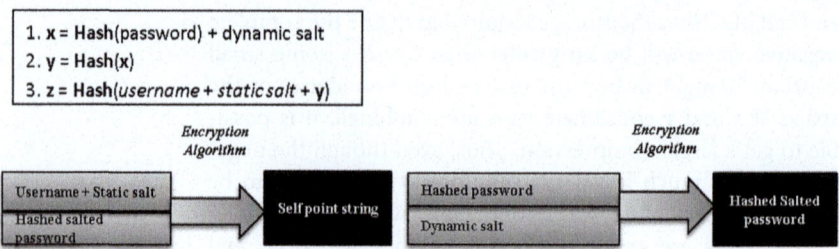

1. x = **Hash**(password) + dynamic salt
2. y = **Hash**(x)
3. z = **Hash**(*username + static salt* + y)

Encryption Algorithm

Encryption Algorithm

Username + Static salt

Hashed salted password

Self point string

Hashed password

Dynamic salt

Hashed Salted password

Fig. 3.10 Creation of hash string from the username–password pair. '+' sign indicates concatenation operation among strings and 'z' represents the derived hashed string. For hashing SHA-128, SHA-256, or SHA-512 can be used

Unique (dynamic) salts are added with hashed passwords and the output string is hashed again. The corresponding username is then added with last output and the result is hashed for the last time after adding a fixed (static) salt. The complete process is shown in **Fig. 3.10**. Self-points are created from these hashed string output using the implementation-specific approaches.

Confusion Parameter

The generated self-points are usually a single point in the model space. To induce some level of obfuscation, a portion of its neighboring region surrounds each self-point. For real- and binary-valued space models, every self-point region forms a hypersphere having a radius specified during implementations and a center, which is the self-point itself. The higher value of confusion parameter increases the self-region but the total number of the detectors covering the non-self-region does not decrease significantly. Total number of detectors depends on the interleaving gaps among hyperspheres, rather than by the total area of non-self-region, as long as this total area does not become too small. As the randomness of hashed self-points is uniformly distributed within the required n-dimensional space (because of the one-way hash function property), the confusion parameter appears to have little effect on the detector set size [19]. For Grid-based space model, every self-point is mapped to a Grid cell and the whole cell is defined as self-cell or self-Grid; the size of the self-Grids (depends on the Grid size) is considered as the confusion parameter for the Grid-based space model.

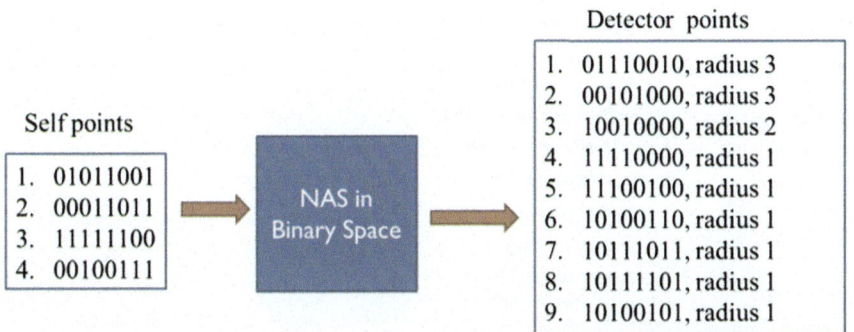

□ Fig. 3.11 Sample detector space generated using B-NAS. The configurations are Dimension $= 8$, total number of self-points $= 4$, Detectors' desired coverage $= 0.60$ and confusion parameter $= 1$. With these configurations, nine detectors are created with different radiuses to cover the non-self-space

Detector Coverage

Detector coverage can be defined as the volume of the non-self-space covered by the detectors. In the NAS approaches, the detector coverage has been normalized between 0 and 1. The detectors' coverage has been calculated using the Monte Carlo simulation technique [20]. Using this technique, 1 million invalid requests are passed through the first layer containing negative detectors, and the percentage of the requests blocked by the set of detectors determines the overall coverage of them for that specific implementation of NAS model.

B-NAS Implementation

In binary-valued space implementation of NAS, no two points will map to a single point in binary space. Based on the password points (self-points), which are hashed output of username and password pair, the implemented algorithm calculates the non-self-detectors. Each detector consists of a center a ($\in$ non-self-region) and radius r, where all the points from a with r hamming distance are the members of that detector. These generated detectors cover some portion of the non-self-space (negative space). In this implementation, it is assumed that both self-points and detectors are hypercubes. Detectors can overlap with each other but they cannot overlap with the volume of self-points. Because of this radius (aka confusion parameter), a self-point or a detector is not a single point in the space; in fact, it is composed of many points in the space. A sample scenario of self-point and detectors in reference to the B-NAS are given in □ Fig. 3.11. Accordingly, four self-points are mentioned with

Binary - valued DetectorSet()

Input :

S : set of self samples

n : Dimension of the space

r_s : Confusion parameter

c_0 : Minimum Coverage

F_0 : Maxmimum Failed Attempt

Output :

A set of V - detectors

1 : $D \leftarrow \varnothing$

2 : $d_0 \leftarrow 0$

3. : $f_0 \leftarrow 0$

4 : Repeat

5. if $f_0 = F_0$, and D.size > S.size, exit

6 : $\quad r \leftarrow n$

7 : $\quad x \leftarrow$ random sample point with n dimension in Binary value in each dimension

8 : $\quad$ Repeat for every d_i in $D = \{d_i . i = 1, 2, ...\}$

9 : $\quad\quad d_d \leftarrow$ Hamming distance between $x(d_i)$ and x, where $x(d_i)$ is the location of d_i

10 : $\quad\quad$ if $d_d \le r(d_i)$ then, where $r(d_i)$ is the radius of detector d_i

11 : $\quad\quad\quad f_0 \leftarrow f_0 + 1$

12 : $\quad\quad\quad$ go to 6 :

13 : $\quad$ Repeat for every s_i in S

14 : $\quad\quad d \leftarrow$ Haming distance between s_i and x

15 : $\quad\quad$ if $d - r_s - 1 < r$ then $r \leftarrow d - r_s - 1$

16 : $\quad\quad$ if $r <= 0$ then

17 : $\quad\quad\quad f_0 \leftarrow f_0 + 1$

18 : $\quad\quad\quad$ go to 6 :

19 : $\quad$ if $r > 0$ then

20 : $\quad$ Repeat for every d_i in $D = \{d_i . i = 1, 2, ...\}$

21 $\quad\quad d_{diff} \leftarrow |d_i - r|$

22 : $\quad\quad$ dsum $\leftarrow d_i + r$

23 : $\quad\quad$ dcenter $\leftarrow$ distance between centerpoint of d_i and x

24 : $\quad\quad$ if dcenter $\le d_{diff}$ or dcenter $\le$ dsum/2

25 : $\quad\quad\quad f_0 \leftarrow f_0 + 1$

26 : $\quad\quad\quad$ go to 6 :

27 : $f_0 \leftarrow 0$

28 : $D \leftarrow D \cup \{< x, r >\}$, where $< x, r >$ is a detector with location x and radius r

29 : $\quad$ Calculate detector coverage and add it to d_0.

30 : $\quad$ if $d_0 >= c_0$, exit

31 : return D

◘ **Fig. 3.12** Pseudocode for detector generation algorithm in B-NAS model

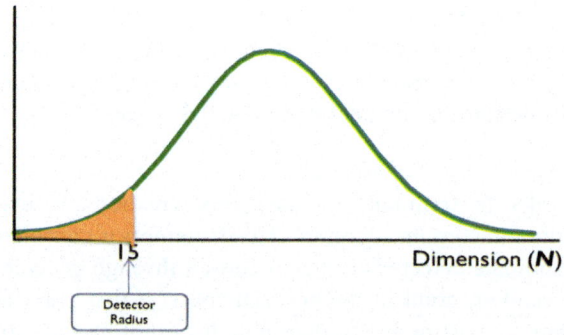

Fig. 3.13 The *orange (dark)* area shows the volume of a detector with a sample radius 15

their radius 1. If the detectors' coverage is set to 60%, total nine detectors are created with different radiuses to cover the non-self-space. The pseudocode for B-NAS algorithm is shown in ■ Fig. 3.12.

The input and output of the detector generation algorithm are mentioned in the pseudocode (■ Fig. 3.12). Code block in line 8–12 checks whether candidate detector, 'x', falls in the existing list of detectors. If the candidate detector falls, the failed count will be increased to keep track of total failed attempts. Code block in line 13–18 computes the minimum distance between all self-points and the candidate detector, 'x'. Code block in line 20–26 checks the overlap with the candidate detector, x, with other existing detectors and decides whether it can be included in the existing detector set. Code block in line 28–31 adds the candidate detector, 'x' to the detector set and checks the termination criteria to exit or not. The algorithm then returns the final set of detectors upon meeting the termination criteria.

Some important terms of B-NAS are discussed in the following subsections:

■ ■ Confusion Parameter for B-NAS

This term defines the radius of a self-point in the binary space. In the binary-valued space model, confusion parameter is set to 5 (very small value in comparison with the total dimension of the space).

■ ■ Expected Coverage of Non-self-Space

In binary-valued space, detector coverage is approximated as a normal distribution [21]. For this considered distribution, the mean is $N/2$ and the standard deviation is $\sqrt{N/2}$,

3

where N is the dimension of the space. The coverage of a detector of radius 'r' ($r \ll N$) is shown pictorially in ▪ Fig. 3.13. The gray area denotes the coverage (or volume) of the detector in the total representative space.

▪▪ *Maximum Failed Attempts*

In order to terminate the detectors' creation process, a count of maximum successive failed attempts is considered. A candidate detector's center is chosen through generation of a random point in the non-self-space. If that point falls within an existing list of detectors, it cannot be the candidate point and the attempt is considered as a failed attempt to create a detector point. If successive failed attempts reach the limit of ‹*maximum failed attempts*›, the detector creation is stopped and the algorithm returns with all the generated detectors. Reaching that limit signifies that many detectors are already created in the non-self-space and the detectors' coverage is sufficient to cover the non-self-space. This criterion reduces the algorithm to create very small detectors (in terms of the radius of the detector) and to run the code for a longer time without much improvement in the total detector coverage. This value is tuned with the dimension of the space and the total number of self-points existing in that space.

▪▪ *Allowed Percentage of Overlap Among Two Detectors*

As detectors are generated, there are some overlaps with the existing set of detectors. Generally, the more overlap among the detectors, the less actual coverage the detectors have achieved. Hence, the actual coverage calculation using the volume of the detectors is not significant with the increase of the overlap. In order to reduce the overlap with the existing detectors' set and a new candidate detector, some criteria are followed. At first, the distance (hamming distance) between existing detectors and the candidate detectors is calculated. This calculation ensures that the candidate detector should not fall into any of the existing detectors.

Suppose the distance between the center point of an existing detector, $D1$, and a candidate detector is d, and their radii are $R1$ and $R2$, respectively. The following three cases can happen. These conditions are checked in the following orders:

(i)   If $d \leq |R1 - R2|$ is satisfied, it means that the candidate detector lies within $D1$ and it can be ignored.

(ii)   If $d \geq (R1 + R2)$ is satisfied, it means there is no overlap between $D1$ and the candidate detector. It is the best possible scenario in creation of detectors.

(iii) If $d > |R1 - R2|$ and $(R1 + R2) \geq d \geq (R1 + R2)/2$ are satisfied, the candidate detector is allowed with the given radius $R2$. In a typical implemented approach of B-NAS, up to 50% overlap between two detectors is allowed. This value can be adjusted to generate the desired number of detectors in order to achieve the required expected coverage. Various values of overlap threshold are tested, and it is found that maximum 50% overlap between two detectors is good enough to optimize both the detection rate and the execution time of the algorithm.

■■ *Detector Radius Adjustment*

The radius of the candidate detector in detector generation phase is adjusted in such a way that there is no overlap in the detector region and the existing self-region. This will be ensured in the following way.

Suppose the minimum distance between the candidate center of the detector and all the existing self-points is d. The confusion parameter (radius of self-points) is r. Then the radius of the candidate detector$= (d - r - 1)$.

Experimental Results for B-NAS Implementation

■■ *Experiment Setup*

In order to measure the performance of B-NAS implementation, a dataset of 33 million user passwords is used. This dataset is extracted from SkullSecurity [22] website that represents exposed user password of RockYou site [23]. As these passwords belonged to a real authentication system, it is used as password base for this experiment. To generate the usernames, another dataset of most common surnames of U.S. is used [24]. This dataset includes all surnames with over 0.001% frequency in the US population during the 1990 census. The performance of B-NAS implementation is indicated by the detector coverage of the configurations. Detector coverage is the percentage of spaces covered by the proposed implementation. In other sense, it is the performance of a specific implementation of NAS to detecting an unauthorized user. It is also noted as detection rate. The detection rate of B-NAS is calculated from the average of 20 runs by applying

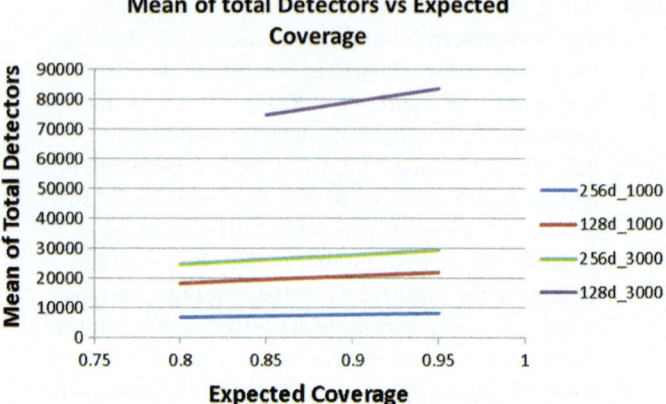

Fig. 3.14 Scatter diagram for total number of detectors with different expected coverage values.256d_1000 means 1000 self-points with 256-dimensional binary space

Monte Carlo Simulation [20] with 100,000 authentication requests. The coverage (detection rate) by the detectors is calculated by using the following ratio:

$$\frac{\text{total number of trials detected by the detectors}}{\text{total number of trials}}.$$

The false rate can also be calculated by subtracting the coverage from 1 (one).

▪▪ B-NAS Results

For binary space implementation, the average count of detectors is compared with various detector coverage values. The experiment is conducted with 1000 and 3000 self-points and two different hashing algorithms; results are shown in ◘ Fig. 3.14. It is clear from the figure that, with the increase of the password points (self-points), the total number of detectors increases. Also it is noted that with the increase of the minimum coverage for a fixed number of self-points, the total number of detectors is increased. Additionally, the dimension of binary space plays a role in the total detector count. It is derived from the figure that, for 256-dimension binary space, the total number of detectors is lower than that of 128-dimension binary space.

The reason is as follows. As the number of self-points and confusion parameter is fixed, for changing 256-dimension space from 128-dimension space, the total number

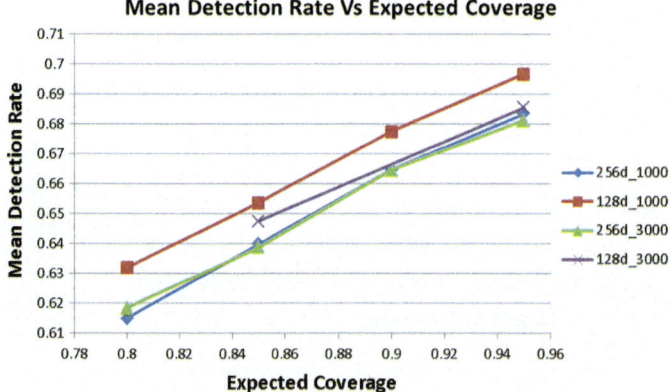

Fig. 3.15 Scatter diagram for mean detection rate with different expected coverages. 256d_1000 means 1000 self-points with 256-dimensional binary space

of points covering the non-self-space is increased exponentially. Hence, the detectors created in 256 dimensions have a higher radius that covers more space in comparison with that of 128 dimensions. So the total numbers of detectors are going down in 256 dimensions to cover the same minimum coverage. One further observation is with the increase of self-points; the difference among two different encryption algorithms in terms of a total number of detectors is larger.

Mean detection rate for B-NAS implementation is compared with different values of minimum coverage.

The results are shown in ◨ Fig. 3.15, where the detection rate increases linearly with the increase of minimum coverage. However, the rate of increase is not constant in different scenarios. This is obvious in the sense that with the increase of dimension and number of self-points, the total points covered are increased not in a fixed ratio.

For a fixed binary space dimension, the mean detection rate is higher for self-points with the smaller region (small confusion parameter value). With the increase of the number of self-points, the total non-self-space is reduced and hence, the generated detectors have more overlap for higher self-points which causes the detection rate to go down. For a fixed number of self-points, the mean detection rate is higher for 128-dimension than 256-dimension space. This result can be explained with the help of total detectors' count. As the total number of detectors is high for 128 dimensions, the proportion of not covered

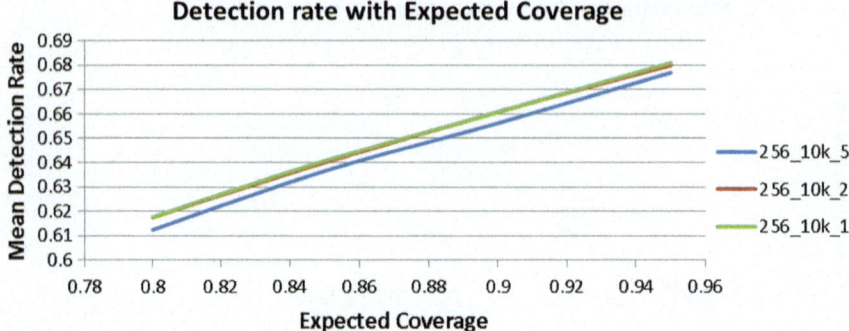

■ **Fig. 3.16**   Scatter diagram for mean detection rate with different minimum coverage for 10k self-points and 256-dimensional space.256_10k_5 means 10k self-points with confusion parameter 5 and 256-dimensional binary space

non-self-space is reduced for this scenario and hence the detection rate goes high.

The effect of different values of confusion parameter for a fixed set of self-points is shown in ■ Fig. 3.16. According to this figure, for 10k password points and 256-dimensional space, the confusion parameter is varied from 1 to 5. It is observed from ■ Fig. 3.21 that detection rate of the detectors is higher with the smaller values of confusion parameter. With the linear decrease of the confusion parameter, the volume of the self-region reduces exponentially and more space is available for detectors to cover. Hence the detectors with higher radius can cover the extra spaces with less overlap. This triggers the detection rate to increase.

Among all different configurations of B-NAS, it is found that to achieve a good detection rate, a higher dimensional binary space (256 or 512 bit) and a smaller value of confusion parameter will serve the purpose to generate an optimal set of detectors.

■■ *Password Cracking Test for B-NAS Approach*

In order to test the efficacy of the B-NAS approach, the password cracking test is designed to simulate the guessing attacks. 100k different access requests are made to test binary-valued NAS approach, and the number of times these requests fall into the self-region is counted. Generally, the first layer detects and blocks all the invalid requests (those which fall into the anti-password space or negative space). The results of password cracking test for B-NAS are shown in ■ Table 3.1.

According to ■ Table 3.1, no invalid authentication requests fall into the self-region. In general, if the self-radius (confusion parameter) is significantly small, the

■ **Table 3.1** Result of password cracking test for B-NAS implementation

Hash dimension	Hashing algorithm	Total number of delf-points (passwords)	Confusion parameter or self-radius	Actual coverage	Number of attempts fall on self-region
256	SHA-256	1000	5	0.612402	0
256	SHA-256	1000	5	0.636626	0
256	SHA-256	1000	5	0.656294	0
256	SHA-256	1000	5	0.676946	0
128	MD5	1000	5	0.624925	0
128	MD5	1000	5	0.647056	0
128	MD5	1000	5	0.667413	0
128	MD5	1000	5	0.687952	0

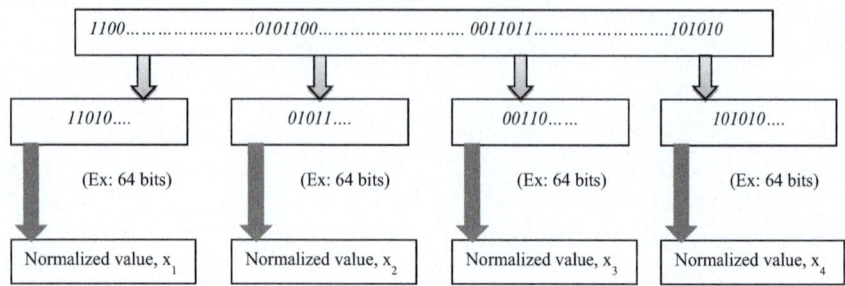

◘ Fig. 3.17 Formulation of a self-point in four-dimensional real space from a 256-bit hashed self-point string

possibility of mapping of the invalid requests fall into self-points' region is very less. Hence, with the proper tuning of the confusion parameter for B-NAS, the password guessing attack can be fully prevented in the proposed approach.

R-NAS Implementation

In the real-valued implementation of NAS, n-dimensional real space model is considered to map the total password space. Though the real-valued NAS is a continuous space model, but in its implementation, only predefined places after the decimal point are considered (typically 3–5 places). As mentioned before, self-region is defined by self-points that are hashed output of username and password. Every hashed string is divided into four segments ($n = 4$) to be mapped into the four-dimensional coordinates of the representational space. The coordinate of a self-point is calculated by normalizing each part of the hashed strings from 0 to 1 (◘ Fig. 3.17). Each self-point forms a hypersphere with a small radius. This radius is defined as confusion parameter. The non-self-region is created by the implemented R-NAS algorithm. This algorithm generates the detectors to cover the non-self-region. Each detector consists of a center and a radius and so defines a hypersphere of four dimensions. So both self-points and detectors form hyperspheres. The radius of a self-point is defined as confusion parameter. The center and radius of detectors are generated in such a way that detectors can overlap with each other but they cannot overlap with the volume of any self-points.

The pseudocode for the real space implementation algorithm is shown in ◘ Fig. 3.18. The input and output of the detector generation algorithm are mentioned in the pseudocode. Code block in line 8–12 checks whether candidate detector, 'x', falls in the existing list of detectors. If that is true, the failed count will be increased to keep track of total

Real Space DetectorSet()

Input :

S : Set of self points

n : Dimension of the space

r_s : Confusion parameter

F : Maxmimum Failed Attempt

Output :

A set of V - detectors

1 : $D \leftarrow \emptyset$

2 : $d \leftarrow 0$

3 : $f \leftarrow 0$

4 : Repeat

5 : if f = F, exit

6 : $r \leftarrow$ dimention of the space

7 : $x \leftarrow$ random sample point in Real Space with n dimension

8 :   Repeat for every d_i in $D = \{d_1, d_2, d_3, ...\}$

9 : $d_e \leftarrow$ Euclidian distance between $x(d_i)$ and x, where $x(d_i)$ is the location of d_i

10 :   if $d_e \leq r(d_i)$ then, where $r(d_i)$ is the radius of detector d_i

11 : $f \leftarrow f + 1$

12 :   go to 6 :

13 :   Repeat for every s_i in S

14 : $d \leftarrow$ Euclidian distance between s_i and x

15 :   if $d - r_s < r$ then $r \leftarrow d - r_s$

16 :   if $r \leq 0$ then

17 : $f \leftarrow f + 1$

18 :   go to 6 :

19 :   if r > 0 then

20 :   Repeat for every d_i in $D = \{d_1, d_2, d_3, ...\}$

21 : $d_{diff} \leftarrow |d_i - r|$

22 : $d_{center} \leftarrow$ distance between center point of d_i and x

23 :   if $d_{center} \leq d_{diff}$

24 : $f \leftarrow f + 1$

25 :   go to 6 :

26 :   overlap $\leftarrow |r - d_{diff}|$

27 :   if(overlap > $0.5 * \min(r_{d_i}, r)$)

28 : $r \leftarrow r - (overlap - 0.5 * \min(r_{d_i}, r))$

29. $f \leftarrow 0$

30 : $D \leftarrow D \cup \{<x, r>\}$, where $<x, r>$ is a detector with location x and radius r

31 :   if $d \geq c$, exit

32 : return D

◼ **Fig. 3.18** Pseudocode for detector generation algorithm in NAS using real-valued space model

failed attempts. Code block in line 13–18 computes the minimum distance among all the self-points and the candidate detector, x. Code block in line 20–28 checks the overlap of the candidate detector, x, with other existing detectors and decides whether it can be included in the existing detector set. Code block in line 29–32 adds the candidate detector, x, to the detector set and checks the termination criteria to exit or not. The final set of detectors is then returned.

■■ *Confusion Parameter for R-NAS*

This term defines the radius of self-hyperspheres created using self-points in the real-valued space. Accordingly, the value of the confusion parameter controls the volume covered by a self-point. Usually, a small value is chosen for the confusion parameter so that the self-region remains negligible in comparison to the total space [19]. In the real space implementation, confusion parameter is set as the minimum value possible (for example, if four places after the decimal point are considered, then confusion parameter is set as 0.0001).

■■ *Expected Coverage of Non-self-Space*

Expected coverage is the term that controls the portion of non-self-region covered by the detectors and it can be defined as the following ratio:

$$\frac{\text{Sum of detectors' regions coverage excluding overlap}}{1 - \text{sum of all self-regions coverage}}.$$

In NAS, some regions from the non-self-region are not covered by negative detectors to induce obfuscation for the attackers. Generally, 98% detectors' coverage is targeted while generating the detectors. As the detectors are created with partial overlap, the actual coverage is smaller than 98%.

Experimental Results for R-NAS Implementation

Same datasets are used for R-NAS implementation similar to B-NAS to test the performance of real-valued NAS implementation.

In this case, experiments are done with a various number of self-points generated from the mentioned dataset. The first set of experiments is conducted to show the relationship between self-point radius and the detector coverage. This experiment is carried out with 1000, 3000, and 5000 self-points. The result is summarized in ◘ Table 3.2 and is plotted in ◘ Fig. 3.19. From ◘ Fig. 3.19, it is found that detectors'

▪ Table 3.2 Change in detector coverage with different settings of the self-point radius

Self-point radius	Detector coverage (%)		
	With 1000 self-points	With 3000 self-points	With 5000 self-points
0.001	79.11	70.25	65.04
0.002	79.23	70.13	64.04
0.004	78.33	67.97	62.93

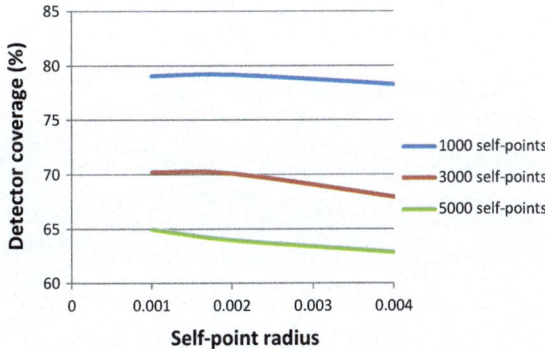

▪ Fig. 3.19 Scatter diagram for coverage by detectors with different self-point radiuses

coverage decreases slightly with the increase of the self-point radius. The reason behind this result is as follows:

Self-point radius with larger value will occupy more spaces leaving less free spaces, causing detectors to have a slightly smaller radius. So the detector coverage decreases slightly with the increase of self-point radius.

Next result shows the relationship between the actual detector coverage achieved with the expected detector coverage by the R-NAS algorithm. The expected coverage was varied from 85 to 98% for 1000, 3000, and 5000 self-points. The result is summarized in ▪ Table 3.3 and graphically represented in ▪ Fig. 3.20. Here the lower detector coverage value results as many detectors overlap one another and the overlapping region covers the same space. But the coverage calculation shows only overlaps with two detectors

3

◼ **Table 3.3** Different values of achieved detector coverage with different settings of expected detector coverage

Expected detector coverage (%)	Achieved detector coverage (%)		
	With 1000 self-points	With 3000 self-points	With 5000 self-points
85	38.12	31.15	30.08
90	52.08	48.47	47.98
95	67.15	60.32	63.18
98	79.11	70.25	65.04

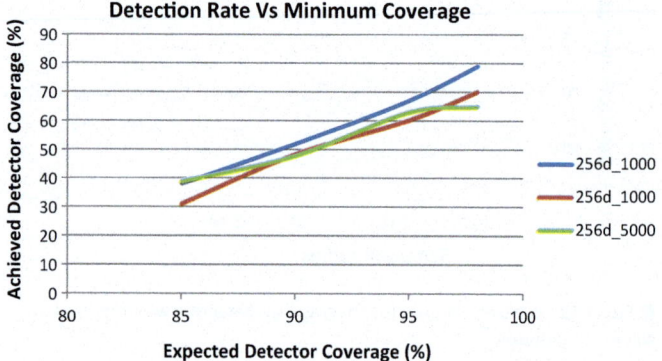

◼ **Fig. 3.20** Scatter diagram for achieved detector coverage by the generated detectors with different expected detector coverages

as measured. Hence, the actual detector coverage value is lower in comparison with the expected coverage value.

Another set of experiments is reported in [19, 25] for real-valued implementation of NAS. These experiments use twofold objectives of R-NAS approach. These decrease the number of Anti-Passwords (detectors) and achieve a good detection rate. Three parameters are involved to show the correlation: number of valid passwords, confusion parameter, and Anti-Password (detector) coverage.

Time and space complexity of negative authentication is a factor of the number of Anti-Passwords. Both the generation time of Anti-Passwords and detection time of an invalid

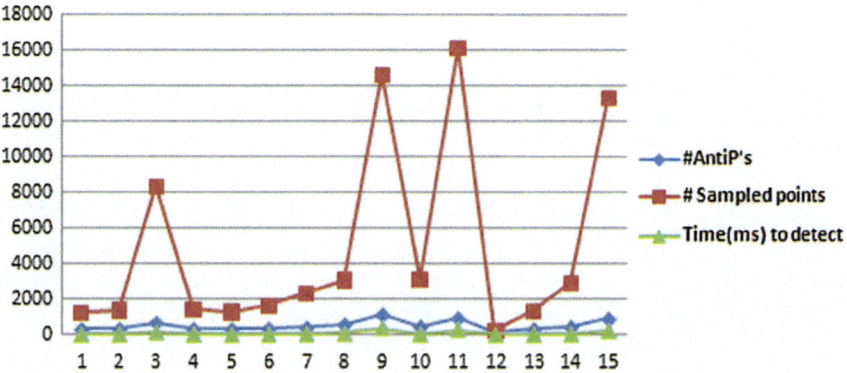

Correlation among #Anti-Passwords, sampled points for Anti-Password generation, and required time to detect invalid requests (ms) in a 2.4 GHz Quad core machine

entity are related to the number of Anti-Passwords as illustrated in ◘ Fig. 3.21. The number of samples used to generate enough Anti-Passwords indicates the generation time of
Anti-Passwords. This number, quite expectedly, has been
found to have a very high correlation (around 0.92) with
the number of Anti-Passwords across different experiments.
Similarly, the detection time is also highly correlated to the
number of Anti-Passwords, as each requested username–
password has to be checked against the set of Anti-Passwords.

The desired detection rate, the second objective, is subject to the choice of the designer. But it should not be too
high or too low. A very high detection rate essentially means
the same thing as in positive authentication, because the positive profile is highly guessable from this complete negative
profile. On the other hand, a very low detection rate adds no
value to the negative detection module. 90% detection rate
has been considered as a good choice in this experiment.

◘ Table 3.4 shows the summary of a number of Anti-
Passwords and detection rate performances for different settings of parameters. All the results have been computed by
averaging 20 runs of the same experiment. As seen from the
table, good results are found for limited sized password files.
However, for very large password files, performance gets
poorer. For example, for a password file of 50,000 entries,
it needs a significantly high number of Anti-Passwords to
achieve even an 86% detection rate. Hence, for the rest of
the experiments, password file size will be restricted to 3000
entities.

Table 3.4 The table summarizes the results of experiments with the following variation: size of password files, Anti-P coverage, and confusion parameters

Anti-P Coverage	Confusion Parameter	#Passwords = 500		#Passwords = 1000		#Passwords = 5000		#Passwords = 10000		#Passwords = 50000	
		#Anti-P	DR	#Anti-P	DR	#Anti-P	DR	#Anti-P	DR	#Anti-P	DR
0.9	0.001	463	0.61	805	0.56	2980	0.46	4530	0.37	14694	0.26
	0.05	421	0.57	728	0.55	713	0.18	781	0.12	201	0.00
	0.1	210	0.42	185	0.22	130	0.03	137	0.02	146	0.00
0.95	0.001	757	0.79	1345	0.76	5640	0.70	11681	0.72	56708	0.70
	0.05	693	0.78	1122	0.73	2827	0.58	2108	0.31	333	0.01
	0.1	317	0.57	286	0.34	164	0.03	178	0.02	512	0.01
0.99	0.001	1500	0.96	2697	0.95	12819	0.92	23655	0.94	88908	0.87
	0.05	1389	0.95	2282	0.94	9283	0.83	6794	0.56	712	0.02
	0.1	811	0.85	814	0.67	312	0.09	203	0.02	505	0.01

The shaded results indicate that for a smaller password file, a reasonable size of Anti-Ps and good detection rates are found in around 99% of coverage with small confusion parameter values. However, very large password files lead to a poor result

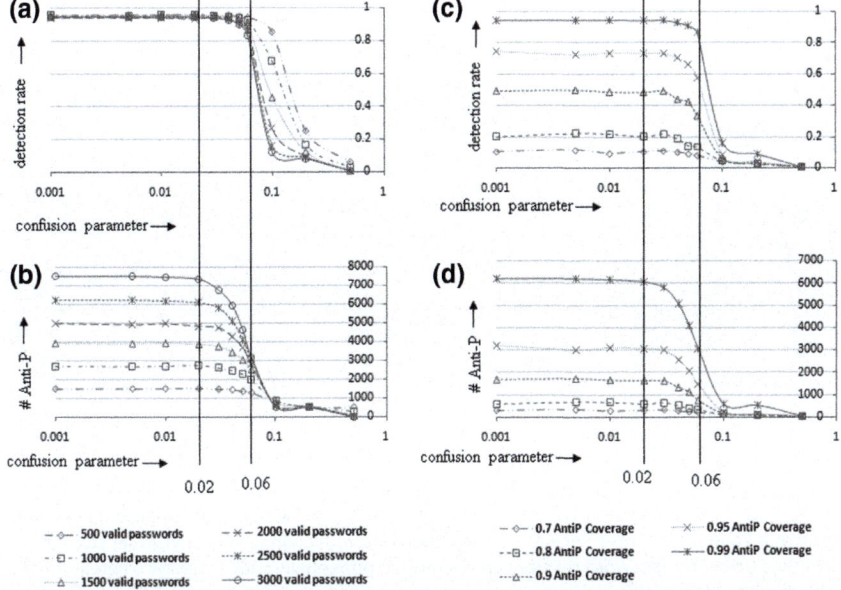

● **Fig. 3.22** Figure shows the changes of the detection rate and the number of Anti-Passwords with variations in the confusion parameter. In particular, **a** exhibits a detection rate with varying sizes of password files while setting the Anti-Password coverage to 0.99. It is to be noted that in the interval 0.02–0.06 of the confusion parameter value, while the detection rate does not change significantly (**a**), the number of Anti-Passwords decreases rapidly (**b**). For example, with $\#P = 3000$ and confusion parameter $= 0.05$, the detection rate is 0.9 that can be achieved with 4625 Anti-Passwords, compared to 0.94 detection rate with 7446 Anti-Passwords as in the case of a confusion parameter of 0.01. It is also observed that the same interval (0.02–0.06) of the confusion parameter produces good results in terms of the number of required Anti-Passwords for the desired coverage as in **c, d**, using a fixed size password file (3000)

■■ Selecting Confusion Parameter

● Table 3.4 indicates very little change in the number of the Anti-Passwords and detection rates for a wide range of confusion parameters (0.001–0.05). However, for relatively higher values (0.1), this change becomes significant. The change of the number of Anti-Passwords and detection rate is shown in ● Fig. 3.22.

The higher confusion parameter shrinks the non-self-space. But interestingly, the number of Anti-Passwords does not decrease that much for an increase in the confusion parameter value up to some level. This is probably due to the sparse nature of the valid password points. Anti-Password set size probably is dictated by the interleaving gaps between corresponding hyperspheres, rather than by the total area of Anti-Password space, as long as this total

(a)   **(b)**

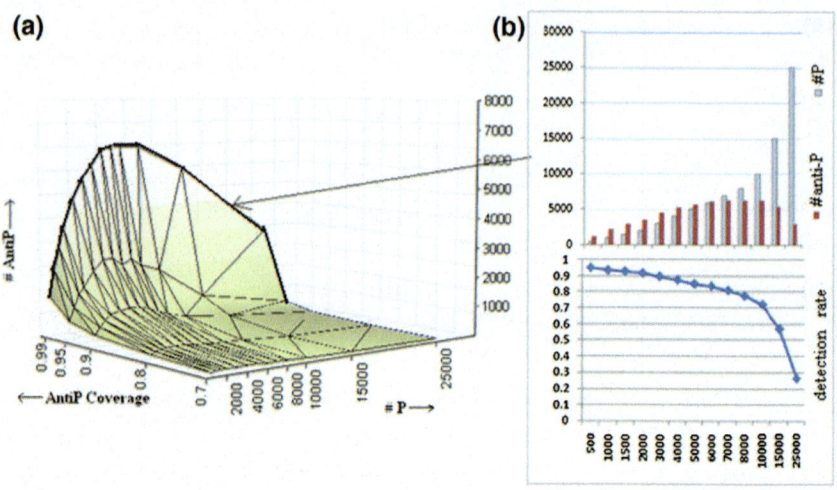

🔲 **Fig. 3.23**   **a** Variation of Anti-P set size with change in Anti-Password coverage and password set size. **b** Variation of Anti-P set size and detection rate with change in password set size for coverage 0.99. Anti-Password set size varies almost exponentially with coverage whereas it has a limit on the increase for varying password set size

area does not become too small. In the scattered space of valid password points, the confusion parameter appears to have little effect on the Anti-Password set size up to values around 0.01–0.02 of the confusion parameter.

After 0.02 the number of Anti-Ps falls rapidly, whereas the detection rate falls after a higher threshold. As shown in 🔲Fig. 3.23, for password files of up to 3000 entities, the range 0.02–0.06 is a good choice for confusion parameters. In this range, the number of Anti-Passwords suffers a rapid fall whereas a good detection rate is almost retained. For small compromise in detection rate, a reasonable decrease of Anti-Password set size is achieved in this range. The same result follows for different Anti-Passwords coverage too. Therefore, the value 0.05 has been chosen for confusion parameter in the next experiments.

▪▪ Effect of Number of Passwords and Anti-P Coverage

The number of Anti-Passwords increases quite exponentially for an increase in Anti-Password coverage, see 🔲 Fig. 3.23. Hence, estimated coverage is to be chosen carefully to maintain a reasonable time and space complexity. However, as seen from the figure, with the increase in password file size, the number of Anti-Passwords increases to some limit and then decreases. This is expected because many hyperspheres in a limited space leave many holes

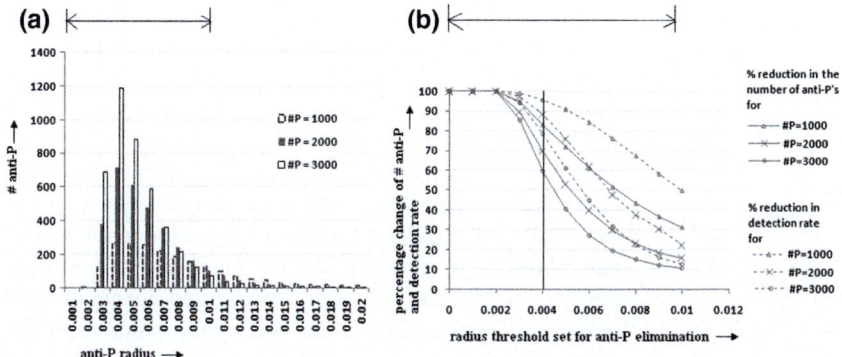

■ **Fig. 3.24** **a** The variation of sizes in a pool of Anti-Passwords produced for password file sizes of 1000, 2000, and 3000. **b** Each pair of the same colored lines represents the variation of Anti-Ps and detection rates for a specific password file size. It is clear from the figure that if smaller sized Anti-Passwords are eliminated, then the detection rate decreases, but the number of Anti-Passwords decreases faster than detection rates. For example, if in the case of 1000 valid password, the Anti-Password sizes of 0.04 or smaller are eliminated, then the # Anti-Password counts reduce to 83% of the original number, whereas detection rate falls only to 95%

which necessitate many Anti-Passwords. However, for a very large number of hyperspheres, the entire space is nearly covered that leaves very little room for Anti-Passwords to cover. Although the size of Anti-P set decreases for very high password sets, it adversely affects the detection rate. To get rid of this problem, a low confusion parameter can be chosen; or the space of the universe set can be expanded by increasing the dimension from four to a higher number. Both of these schemes arrange for more space in the non-self-space. The trend of Anti-Passwords set size has been shown effectively for the coverage of 0.99.

■■ Elimination of Smaller Anti-Ps

It turns out that many small Anti-Passwords are generated to fulfill the desired coverage. Some of them may not contribute that much in detecting invalid passwords. So eliminating such Anti-Ps may lead to better performance.

■ Figure 3.24a shows the distribution of Anti-Password size in terms of radius. As seen from the figure, many small-sized Anti-Passwords are produced. If smaller Anti-Passwords are eliminated, the impact on the Anti-Password set size and detection rate is shown in ■ Fig. 3.24b. Both detection rate and the number of Anti-Passwords decrease, but the decrease of Anti-Passwords is faster. This presents an opportunity to eliminate some small Anti-Ps so that reduction of Anti-Passwords can be achieved at the cost of comparatively less effect on the detection rate. The amount of compromise in the detection rate, or tolerance, is a design choice.

■■ *Password Cracking Test for R-NAS*

Like B-NAS implementation, R-NAS implementation is also tested for password cracking test to simulate the guessing attacks. 100k different access requests are made to test real-valued NAS approach, and the number of times these requests fall into the self-region is counted. Generally, the first layer detects and blocks all the invalid requests (those which fall into the anti-password space or negative space). The results of password cracking test for R-NAS are shown in ◘ Table 3.5.

For R-NAS scenario, for 100k test points, only one test point passed through the second layer and falls into the self-region. It means one invalid request is considered as a valid request. This happens if the dimension of the space is small (here 3 points) but with the increase of the space dimension, no invalid request has fallen into the self-region.

Modification of the Self-region for B-NAS and R-NAS

To remove some existing self-points in binary and real-valued space implementations, the marked-to-be-removed self-points are simply removed from the existing self-point set. If the number of removed self-points exceeds a predefined threshold, then the detector set is generated again to cover up the empty spaces and to create larger detector spheres. The pseudocode is shown in ◘ Fig. 3.25 for removing a self-point. According to pseudocode, if the total number of self-points after the removal is less than a given threshold value, the detector set will be generated again with the given list of self-points. Otherwise, existing detector set can be considered as current detector set.

The case is slightly different if some new self-points are added in the self-region. All detectors are removed that collide with the new self-points and then detectors are generated again with the new self-point set and the existing detector set to cover the vacant non-self-region. When the number of added self-points crosses a threshold limit, the detector set is generated again from scratch to get rid of the smaller detectors and merge them in larger ones. This idea is shown in the given pseudocode in ◘ Fig. 3.26.

■ **Table 3.5** Result of password cracking test for R-NAS implementation

Hash dimension	Hashing algorithm	Points considered after decimal	Number of self-points (passwords)	Confusion parameter or self-radius	Actual coverage
256	SHA-256	3	1000	0.001	76.87
256	SHA-256	4	1000	0.0001	75.734
256	SHA-256	5	1000	0.00001	76.58
128	MD5	3	1000	0.001	76.476
128	MD5	4	1000	0.0001	77.24
128	MD5	5	1000	0.00001	76.982

```
Input:
e <- entry to be deleted
d <- existing set of detectors
n <- no. of self-points from last time NAS is run
s <- set of existing self-points
p <- predefined reset percentage
Output:
d<- set of detectors
Pseudo-code:
s = s − {e} // remove e from s
if |s| < p * n
   then d = NAS(s) // run NAS  with empty detector set and all the self-points, s
return d
```

■ **Fig. 3.25** Pseudocode for deletion of existing self-points from the set of self-points in B-NAS and R-NAS

```
Input
e <- entry to be added
d <- set of detectors
s <- set of self-points
p <- predefined reset percentage
n0 <- no. of self-points when NAS (not using existing entities) is run last time
n <- no. of self-points added after NAS (not using existingentities) is run last time
Output:
d<- set of detectors
Pseudo-code:
if n > p * n0
   d <- φ
     s = s ∪ {e}   // add e to s
     d = NAS(s)  // run NAS with s
else
   for all detector in d
     if e is in the detector set
     d = d − {e} //remove detector                          //from d
     s = s ∪ {e}   // add e to s
     n = n+1     //increase n by 1
     d = NAS(s, d) // run NAS with                           //existing s and d
return d
```

■ **Fig. 3.26** Pseudocode for adding of existing self-points to the set of self-points in B-NAS and R-NAS

Input:

e_{old} <- old entry to be updated

e_{new} <- updated new entry

d <- set of detectors

s <- set of self-points

n <- no. of update operation after NAS is run last time

p <- predefined reset percentage

Output:

d<- set of detectors

Pseudo-code:

remove e_{old} from s

if (n > p * s)

   s = s ∪ {e_{new}}   //add e_{new} to s

   d <- φ   // empty set of detectors

   d =NAS(s)   // run NAS

else

      for all detector in d

        if e_{new} is in detector

           d = d – { e_{new}}   // remove detector from d

           s = s ∪ { e_{new} } //add e_{new} to s

           n = n+1   // increase n by 1

           d = NAS(s, d)   //run NAS with existing s and d

return d

Fig. 3.27 Pseudocode for updating of existing self-points in B-NAS and R-NAS

To update some existing self-points, the old self-point entries are removed and the updated self-point entries are added in the self-region. Corresponding steps for removal and addition of self-points are taken as described previously. Again, the total detector set is regenerated when the number of self-point update crosses a threshold to minimize the number of smaller detectors. This concept is shown in the given pseudocode in ■ Fig. 3.27.

G-NAS Implementation

As discussed in the previous section, Grid approach uses a two-dimensional $N \times N$ Grid to model the total space. At first, the self-region is mapped to the Grid space and later the detectors are formed from the unoccupied non-self-region.

■■ Represent the Space in Grid

The Grid is stored as a two-dimensional array of integers with a specified dimension. Self-points that are mapped to a particular Grid (Grids are identified by row number and column number) increase the occupancy value of that Grid by 1. Hence, any Grid of i-th row and j-th column having occupancy value $(M_{i,j})$ greater than 1 means the Grid is not empty and a number of $M_{i,j}$ self-points are mapped to that particular Grid. The non-self-region Grids are easily identified when their occupancy value is 0, indicating no self-points being mapped to that Grid.

■■ Confusion Parameter in G-NAS

In Grid space implementation, space size for a two-dimensional $N \times N$ grid is N^2 grids, but the total possible space size is $2n$, where n is the length of the representation string of a self-point. Therefore, every Grid is mapped by $(2n/N^2)$ possible points. For every Grid containing m self-points, $((2n/N^2) - m) \approx (2n/N^2)$ points are considered as the confusion parameter.

■■ *Detection of Non-self-Region in G-NAS*

The Grids that are not covered by self-points are considered as the non-self-region. Hence, the non-self-region is created in constant time after getting the Grids occupied by self-points. This is one of the advantages of Grid-based implementation over other implementations of NAS. Detectors can be generated by scanning the Grid once and having consecutive Grids in a row with occupancy value 0 considered as one detector. Run length of each detector is determined by the count of consecutive Grids in the detector.

In Grid implementation, non-self-Grids cover the total non-self-region. If only a fixed percentage of the anti-password region is desired to be covered, some detectors are marked for removal. In order to do that, detectors with minimum run length are removed from detector list at first and the newly achieved coverage is calculated. This process will go on until the expected coverage is achieved. This whole process can be performed with a single scan of the whole Grid and remove the detectors with the given run length calculated at first. In the Grid-based implementation, it is possible to increase or decrease the detectors' coverage with only a single scan through the Grid. It also indicates another advantage over the hypersphere-based implementations [10], where the detectors are required to be generated from the beginning to get the required coverage value.

■■ **Saving the Detector Space**

Detectors generated from the Grid space implementation are stored by their run length and starting index. In the implementation presented in this work, the detectors are saved in row-wise format and, in each row, the detector is stored with the starting column index and run length. The benefit of using row-wise format is to find a particular point in the space, the row of that point can easily be determined and the search space is reduced to that row only to check whether it falls in a detector Grid or self-point Grid. This approach also does the compression of the space to store the whole detector region and it can be easily reconstructed from the saved information of the detector space.

■■ *Visual Representation of a Simulation of G-NAS*

As discussed above, initially self-points are mapped in the Grid space. The simulation involves a Grid with 32×32 dimension and 500 self-points. It uses 256-bit SHA-256 encryption algorithm to get self-points. Then, the detectors are created to cover the non-self-space. After the creation of detectors, some detectors are removed from the detector list. This is done so that detectors do not cover the entire non-self-region. Some regions are left uncovered to induce ambiguity for attackers so that they cannot find out the self-region easily. ◘ Figure 3.28 shows the generated self-points (dark gray), detectors (black), and candidate detector for removal from detector list (light gray—all detectors with run lengths 1 and 2 in this example) in the Grid. Only the surviving detectors (black blocks) will be saved for the NAS. To increase the obfuscation level, smaller detectors (light gray blocks) can be omitted for saving.

Mapping of Self-points to Grid

Three different approaches of Grid-based implementation have been implemented and tested. They are

— Mod-based approach
— Two-layer approach

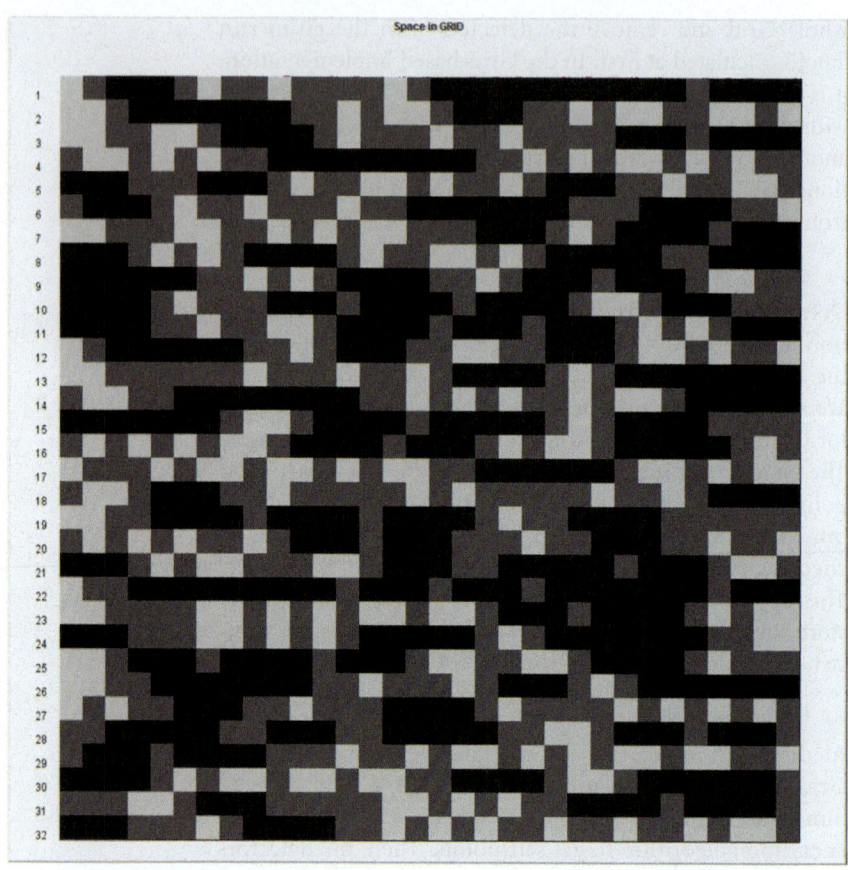

Fig. 3.28 Self-points (*dark-gray*), detectors (*black*), and candidate detectors marked for removal (*light gray*) of run length up to 2 in the 32 × 32 Grid. Here 390 grids are dark gray grids, 412 grids are black grids, and 222 grids are light-gray grids with run lengths 1 and 2

XOR-based approach

All Grids are two-dimensional with dimensions of $2^i \times 2^i$, where the value of i is dependent on the number of user information (self-points). In each of the cases, the hashed self-point string is divided into two equal halves. But the data in each half is processed differently in these implementations.

Mod-Based G-NAS

In this method, each part of the divided hashed string of the self-point is considered. Last i bits are taken from the binary representation of each of the parts. This value denoted by these i-bits actually provides the modulus if the string was divided by 2^i. Then each of the parted string is converted into corresponding values. This self-point is

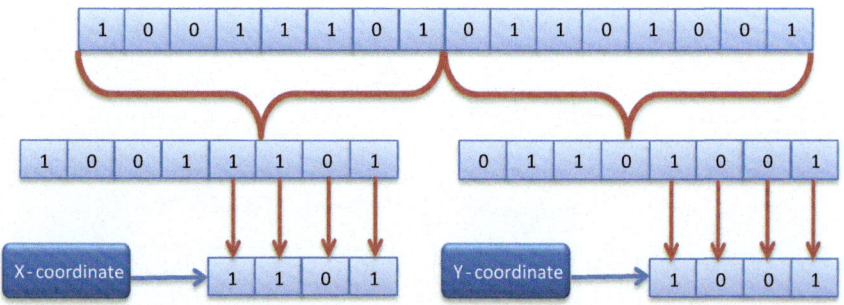

Fig. 3.29 Mod-based approach of mapping points in the grid

Input:
S: set of self-points in binary vector format
d: *Integer* – dimension of the self-point vector
N: *Integer* – dimension of the Grid
Output:
A two-dimensional array representing Grid space
1. $grid \leftarrow \Phi$
2. $len_{half} = d/2;$
3. $for\ i \leftarrow 1\ to\ N;$
4. $for\ j \leftarrow 1\ to\ N;$
5. $grid[i][j] \leftarrow 0;$
6. Repeat for every S_i in S
7. $X \leftarrow S_i[1:len_{half}];$
8. $Y \leftarrow S_i[len_{half}+1:d];$
9. $row \leftarrow$ convertToInteger(X);
10. $col \leftarrow$ convertToInteger(Y);
11. $grid[row][col] \leftarrow grid[row][col] + 1;$
12. end Repeat
13. return $grid;$

Fig. 3.30 Pseudocode for detector generation algorithm in negative detection using mod-based G-NAS. The convert-to-integer function takes a string input consisting only numbers and produces the integer value of the string

mapped in the Grid using those computed values. Other bits are simply discarded from any computation. Here the projection scheme is fixed for every possible self-point unless the Grid dimension is changed.

An example of the mod-based approach of the Grid-based implementation is illustrated in ◘ Fig. 3.29. For example, it is assumed that the hashing function produces a binary string of length 16. The string is divided into two parts. Suppose the dimension of the Grid is $16 \times 16 (= 2^4 \times 2^4)$. So the last four bits from each part are considered. The value of these four bits resembles the remainder if the part is divided by $16 (= 2^4)$. In this example, the first part produces $(1101)_2$, i.e., 13 and the second part

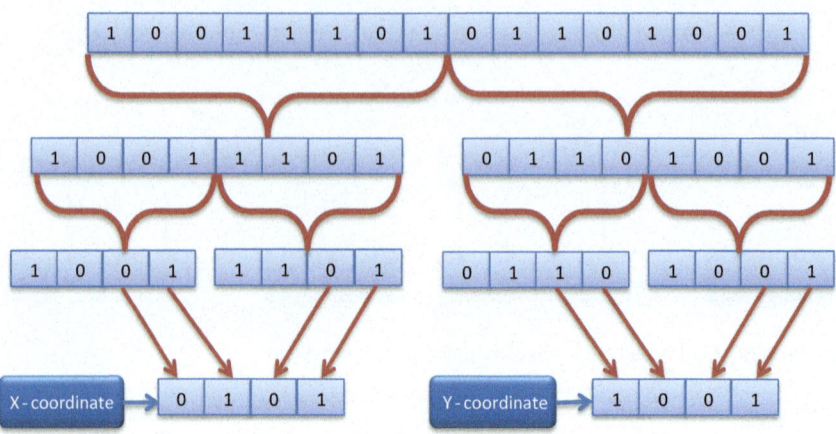

Fig. 3.31 Two-layer-based approach of mapping points in the grid

produces $(1001)_2$, i.e., 9, so, the string will map to the cell (13, 9) in the Grid. The pseudocode of the Mod-based projection algorithm is shown in **Fig. 3.30**.

Two-Layer-Based G-NAS

Each part of the divided hashed string is again divided into two equal sub-parts. Last $i/2$ bits are taken from the binary representation of each of the sub-parts. Both of the $i/2$ bit length binary sub-strings are concatenated to form an i bit length binary string. This binary string is converted into a corresponding value. The same procedure is done with the other part of the hashed string. Both the values are then used to map the self-point in the Grid. All other bits are discarded. Again, the projection scheme is fixed for every possible self-point unless the Grid dimension is changed.

An example of the two-layer-based approach of the Grid-based implementation is illustrated in **Fig. 3.31**. For example, it is assumed that the hashing function produces a binary string length of 16. The string is divided into two parts. Each part is again divided into two equal sub-parts. Suppose the dimension of the Grid is $16 \times 16 (= 2^4 \times 2^4)$. So the last two bits from each subpart are taken into consideration. The same procedure is repeated in the other part of the string. In this example the first part produces $(0101)_2$, i.e., 5 and the second part produces $(1001)_2$, i.e., 9. So the string will map to the cell (5, 9) in the Grid. The pseudocode of the two-layer-based projection algorithm is shown in **Fig. 3.32**.

```
Input:
S: set of self-points in binary vector format
d: Integer – dimension of the self-point vector
N: Integer – dimension of the Grid
k: Integer – square root of dimension of the Grid (Grid dimension: 2^k)
Output:
A two-dimensional array representing Grid space
    1.  grid ← Φ
    2.  len_half = d/2;
    3.  for i ← 1 to N;
    4.    for j ← 1 to N;
    5.      grid[i][j] ← 0;
    6.  Repeat for every S_i in S
    7.    X1 ← S_i[1 : len_half/2];
    8.    X2 ← S_i[len_half/2 + 1 : len_half];
    9.    Y1 ← S_i[len_half + 1 : 3 * len_half/2];
    10.   Y2 ← S_i[3 * len_half/2 + 1 : d];
    11.   S1 ← substring(X1, k/2);
    12.   S2 ← substring(X2, k/2);
    13.   S3 ← substring(Y1, k/2);
    14.   S4 ← substring(Y2, k/2);
    15.   row ← convertToInteger(concate(S1, S2));
    16.   col ← convertToInteger(concate(S3, S4));
    17.   grid[row][col] ← grid[row][col] + 1;
    18. end Repeat
```

◻ **Fig. 3.32** Pseudocode for detector generation algorithm in negative detection using two-layer-based G-NAS. The convertToInteger function takes a string input consisting only numbers and produces the integer value of the string. The *concate* function concatenates two strings into one single string

▪▪ *XOR-Based G-NAS*

Initially, i positions are randomly selected from the hashed string. Then, for each half of the hashed string, every j (power of 2) consecutive bits are XOR-ed and replaced by the result. That leaves each of the halved hashed binary string compressed in a ratio of j to 1. From the compressed string, bits from the selected positions are concatenated. That produces a binary string of length i. This binary string is converted into corresponding value. Both values are used to map the self-point mapped in the Grid. All the self-point hashed strings use the same i positions selected at the beginning. More bit information is involved in the mapping of self-points in this projection approach. This projection scheme chooses new random positions every time the scheme is initiated. So this scheme will also produce a different projection for the same password in separate implementations.

An example of the XOR-based approach of the Grid-based implementation is illustrated in ◻ Fig. 3.33. For example, it is assumed that the hashing function produces a binary string length of 32. This example illustrates a

3

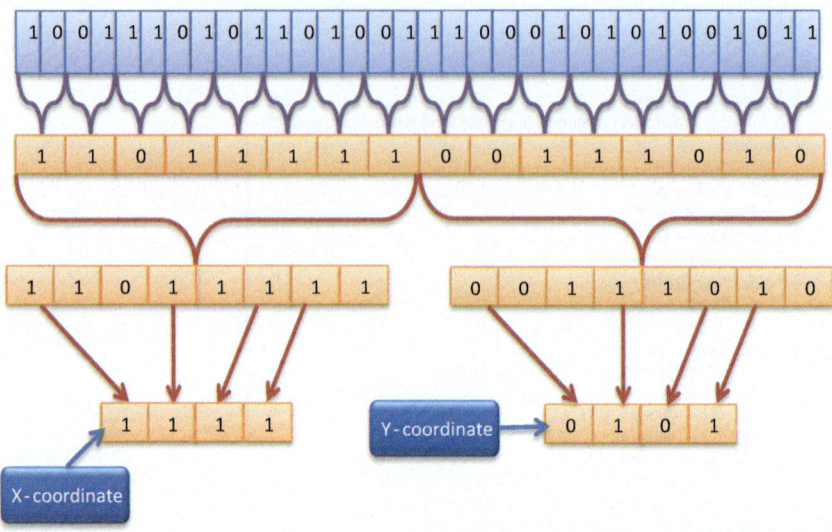

■ **Fig. 3.33** The XOR-based approach of mapping points in the grid

scenario with a compression ratio of 2:1. Every two consecutive bits are XOR-ed to produce the new string. So the new string has a length of 16. The string is divided into two parts. Suppose the dimension of the Grid is $16 \times 16 (= 2^4 \times 2^4)$ and four random bit positions are 1, 4, 6, and 7. So, bits from those positions are taken in order. The same procedure is repeated in the other part of the string. In this example, the first part produces $(1111)_2$ i.e., 15 and the second part produces $(0101)_2$, i.e., 5. So the string will map to the cell (15, 5) in the Grid. The pseudocode of the XOR-based projection algorithm is shown in ■ Fig. 3.34.

Experimental Results for G-NAS

Experiments are carried out using the dataset mentioned for B-NAS and R-NAS to generate the statistical properties of the discussed algorithms. For each simulation, 20,000 passwords and usernames are randomly chosen from the datasets and combined to form the authentication credential (Self-points). SHA-256 is used to hash the authentication credentials so that the spread of the hashed string is even over the space. Along with detector coverage, detection rate and false rate are also calculated to indicate the performance of the implementation. The detection rate is the calculated detector coverage with trials. The false rate is the percentage of invalid credentials not detected by the detectors. Grid dimension is

```
Input:
S: set of self-points in binary vector format
n: Integer – dimension of the self-point vector
r: Integer – compression ratio
N: Integer – dimension of the Grid
k: Integer – square root of dimension of the Grid (Grid dimension: 2^k)
Output:
A two-dimensional array representing Grid space
    1.  grid ← Φ
    2.  len_half = d/ 2;
    3.  for i ← 1 to k
    4.  pos (i) ← random(1, len_half);
    5.  for i ← 1 to N;
    6.  for j ← 1 to N;
    7.  grid[i][j] ← 0;
    8.  Repeat for every S_i in S
    9.  X ← S_i[1 : len_half];
    10. Y ← S_i[len_half + 1 : d];
    11. j ← 1;
    12. for i ← 1 to len_half
    13. X_cmprsd(j) ← XOR(X(i), X(i + 1), ..., X(i + r - ));
    14. Y_cmprsd(j) ← XOR(Y(i), Y(i + 1), ..., Y(i + r - ));
    15. j ← j + 1;
    16. i ← i + r;
    17. for i ← 1 to k
    18. S1(i) ← X_cmprsd (pos(i));
    19. S2(i) ← Y_cmprsd (pos(i));
    20. row ← convertToInteger(S1);
    21. col ← convertToInteger(S2);
    22. grid[row][col] ← grid[row][col] + 1;
    23. end Repeat
    24. return grid;
```

◻ Fig. 3.34 Pseudocode for detector generation algorithm in negative detection using XOR-based G-NAS. The *convertToInteger* function takes a string input consisting only numbers and produces the integer value of the string. The *concate* function concatenates two strings into one single string

varied from 2^6 to 2^{11} to get the detection rates and false rates for each algorithm.

Different sizes of self-point samples were projected into Grids of different dimensions to get the appropriate dimension for a specific size of the self-point list. The dimensions of Grids were varied from $\left(2^5 \times 2^5\right)$ to $\left(2^{11} \times 2^{11}\right)$. Coverage of the self-region is calculated in each simulation and results of these simulations are shown in ◻ Fig. 3.34. These simulations are done with the mod-based approach. Two other approaches also show very similar patterns with the mod-based approach. Another important point is that encryption size does not matter in the self-Grid coverage in Grid implementation for any approach. The reason is, always, some specific bits are used from the whole encrypted bit string for mapping. From ◻ Fig. 3.35, the appropriate Grid dimension

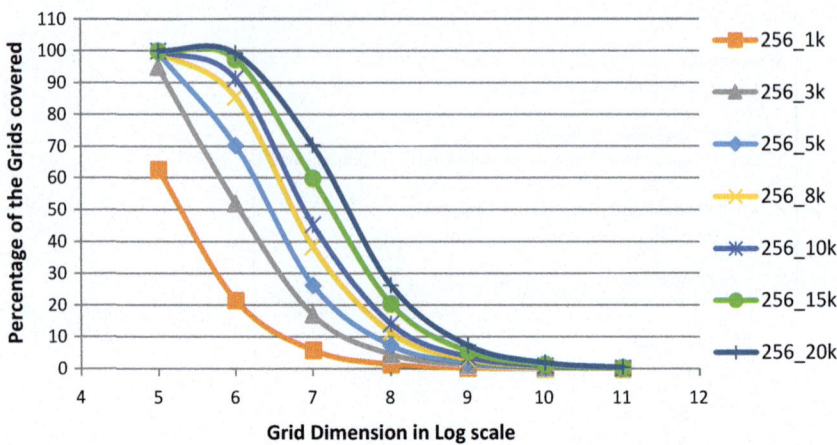

Fig. 3.35 Coverage of self-region in different grid sizes for different sample sizes of self-points for 256-bit encryption

for some specific size of self-points can be found. For example, if the targeted coverage by self-region is below 10%, then for a self-region of 10,000 points, the Grid dimension will be $2^9 \times 2^9$ or 512×512.

For a given number of password points, the detection rate (the percentage of non-password points is detected by the available detectors) increases with the increase of Grid dimension. Also, the false positive rate (the percentage of non-password points falls in the self-Grids) decreases with the Grid dimension.

The trend for detection rate and false positive rate follows the same pattern in the case of removing of some small detectors (e.g., removal of detectors with run length 1 and 2). ◻ Figure 3.36 shows the effect of change of Grid size on detection rate and false positive rate without any removal of detectors and removal after detectors of RL1 and RL2 for a fixed set of 20,000 password points in mod-based projection. From ◻ Fig. 3.36, it can be seen that the detection rate graphs before removal of any detector, after removal of RL1 detectors, and after removal of RL2 detectors almost overlapped over each other. It is also true for the false positive rate graphs. Therefore, from ◻ Fig. 3.36, it can be concluded that after removing the small length detectors the Grid still shows the same pattern for detection rate and false positive rate as before.

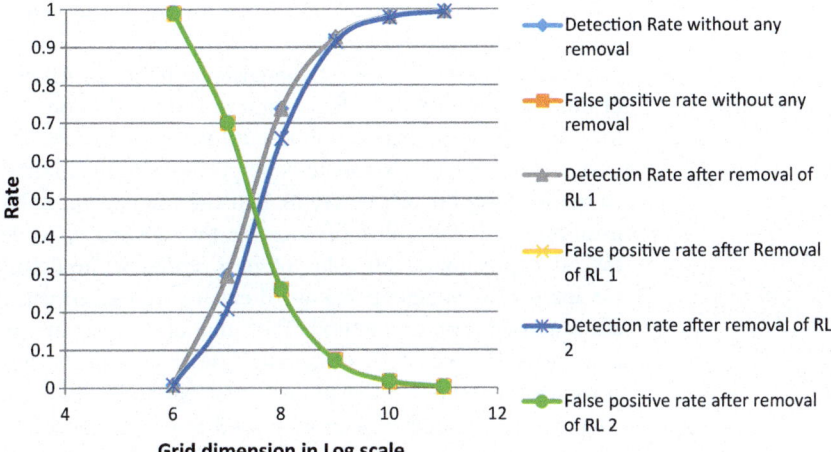

- Detection Rate without any removal
- False positive rate without any removal
- Detection Rate after removal of RL 1
- False positive rate after Removal of RL 1
- Detection rate after removal of RL 2
- False positive rate after removal of RL 2

■ **Fig. 3.36** Effect of grid size on detection rate and false positive rate without any removal of detectors and removal after detectors of RL1 and RL2 for a fixed set of 20,000 password points in mod-based projection

00100101100000010 11000
0100111000000000011001
001100000110100100101
0000110000000000101100
010010100011010001010

■ **Fig. 3.37** Matrix-based representation of the space after authentication data point mapping

Matrix-Based NAS Approach

The Matrix-based approach (GM-NAS) uses a bit matrix of dimension 'k' by 'M' to generate the space and also the detectors. The advantage with this approach is that detector space is built along with the space creation. ■ Figure 3.37 shows the space after the username–password hashes are mapped. The optimum Matrix dimension 'k' and 'M' can be chosen to achieve required level of obfuscation. The complexity of the algorithm to generate the total space is $O(kN)$. As 'k' is very insignificant in comparison to N, the complexity can be said to be $O(N)$, viewed as 'k' layers of one-dimensional Grid. Each access request has to pass all the 'k' layers for authentication. The required storage is dependent on k and M. The value of M is dependent on N (approximately $2 * N$) and value of k is dependent on the required obfuscation level.

Therefore, no more than $2 * k * N$ bits are needed to define the space for N password points.

The Matrix is stored as 'k'-dimensional bit array. All bits of the Matrix are set to 0. Self-points are mapped to one bit for each row of the Matrix and that bit is set to 1. Each self-point will be mapped to exactly k bit positions in the Matrix. At most, $k * N$ bit will be set to 1 in the Matrix. These bit positions with the value of 1 create the self-space, while the other positions with value of zero represent the detector space. The username–password entries are hashed using any standard hashing method. The value of M is set as any i-th power of 2, i.e., 2^i. For mapping, last i-bits of the hash is considered to map that self-point in first row and second last i-bits are considered to map in second row. This process is repeated up to k times for k rows in the Matrix. Selection of bits can be altered to gain different mappings to self-point in the Matrix. When all self-points are mapped to the Matrix, the self-space along with the detector space will be created simultaneously. The whole k by M Matrix is stored containing the detector within it. For authorization, any request is mapped and compared with the created Matrix. If any of the k mappings positions of the access request is found to be 0 in the Matrix, the access request is rejected. The request will only proceed for positive authentication if all the mapped positions in the Matrix are found to be 1.

▪▪ *Details of the Matrix-Based NAS Method*

In this work, a hash function $h: P \rightarrow Z_H = \{0, 1, \ldots, H - 1\}$ is used over a list of N passwords, $p_1, p_2, \ldots, p_N \in P$, to obtain hash values $h_i = h(p_i)$, as integers from 0 to $H - 1$. Typically, H will be a large power of 2 (e.g., $= 2^{128}$). It considers h_i being effectively random and uniformly distributed over Z_H, and constructs a decision function $t: Z_H \rightarrow \{0, 1\}$ with the following properties:

$$(A) \ t(h_i) = 1 \quad \forall i = 1, \ldots, N,$$

$$(B) \ t(h) = 1 \text{ for at most a proportion } p \text{ of all } h \in Z_H.$$

It is assumed that $pH \gg N$, so most hash values $h \in Z_H$ accepted by t do not come from valid passwords. However, p should be significantly less than 1, so that many non-passwords are rejected. In fact, it is expected that $H \gg (2N)^{\left\lceil \log_2 \left(\frac{1}{p}\right) \right\rceil}$. For reasonable values of $p (\geq 1/16)$ and $N (\leq 106)$, it is likely to hold $(H \gg 2^{84})$. The function t

will belong to a class T of such functions with the property that:

(i) For any choice of password $p_1, p_2, \ldots, p_N$ (or equivalently hash values $h_1, h_2, \ldots, h_N$), there exists a $t \in T$ such that (A) and (B) hold.

(ii) The knowledge of $t \in T$ should not allow us to identify any particular password p_i or its hash value h_i. Now,

Input: $h_1, h_2, \ldots, h_N$ hash values of valid passwords.

Output: $k \times M$ bit array A

Set all bits in A to 0.

For $I = 1$ to N

      Set $h = h_i$

      For $j = 1$ to k

            Set $A\,[j, h \bmod M] = 1$

            Set $h = h\,/\,M$

      Next

Next

Now the function t is defined as follows:

Input: $h =$ hash code of password to check

Output: $1 = pass\,/\,0 = fail$

Set $x = h$

For $i = 1$ to k

      If $A\,[i, x \bmod M] = 0$ then $Return\ 0$

      Set $x = x\,/\,M$

Next

Return 1

1. Given $h_1, h_2, \ldots, h_N$, find a $t \in T$ satisfying (A) and (B) in reasonable time.

2. Given $h \in Z_H, t \in T$, calculate $t(h)$ in reasonable time.

Ideally, it is possible to add a small number of additional passwords $p_{N+1}, p_{N+2}, \ldots$ and recalculate a suitable $t \in T$ more quickly than in 1. An implementer may also like to store the choice of $t \in T$ efficiently so that the set T of test functions should not be too large.

Also two integer parameters, M and k depending on N and p are selected. Let A be a $k \times M$ bit array constructed as follows:

This algorithm returns 1, if h is one of h_i. If $M = 2N$ then at most half of the entries in A are set to 1 for each value of the first coordinate i. Thus, if we assume H is divisible by Mk then at most a proportion of $1/2$ of all h will pass the test for each i. The overall proportion of h passing this test is at most 2^{-k}. Hence we choose k so that $2^{-k} \leq p$. Hence $k = \lceil \log_2(1/p) \rceil$, and we wish H to be divisible by $(2N)^k$, which is ensured if N and H are powers of two and

$$H \geq (2N)^{\left\lceil \log_2 \left(\frac{1}{p} \right) \right\rceil}.$$

If we assume that h_i is random (a reasonable assumption for good hash functions), then one can take a slightly smaller M, namely $M = N/\log_e 2 \approx 1.4N$ as typically then a proportion of $e^{-N/M} = 1/2$ entries in A are 0. Indeed a bound on the number of nonzero entries in A is all we need, and can be verified in practice by keeping a count of the number of nonzero entries in A.

In practice, it is appropriate to modify M to be a power of 2 slightly larger than N, making the mod and division operations even simpler. For random h_i we then need to choose k so that $\left(1 - e^{N/M}\right)^k \leq p$. In the deterministic worst case, we need $(N/M)^k \leq p$. In both cases this only incurs a small performance penalty if M is chosen suitably.

The addition of new passwords to the system can be easily accomplished by the first algorithm just for the extra passwords; at least until the acceptance probability exceeds p.

▪▪ *Optimality Criteria*

It needs $kM = 2kN$ bits of memory to implement the above algorithm, and the algorithms run in time $O(k)$ for calculating $t(h)$ and $O(kN)$ for generating t. It is shown that these are all close to optimal.

For memory usage, consider a fixed algorithm t satisfying (A) and (B). Assuming the hash values $h_1, h_2, \ldots, h_N$ are random, the probability that t satisfies (A) and (B) is clearly at most p^N, as each h_i is uniformly and independently chosen at random and has a probability at most p of passing test t. Thus, it is seen that t is an appropriate test for p^N proportion of choices in sets of passwords $\{p_1, p_2, \ldots, p_N\}$; we need at least $1/p^N$ possible values of $t \in T$ to cover all possible choices of passwords. So $|T| \geq 1/p^N$ and the number of bits needed to specify a particular $t \in T$ is at least $\log_2\left(\frac{1}{p^N}\right) = N\log_2\left(\frac{1}{p}\right) \approx kN$.

Calculation of $t(h)$ can be done at best in time $O(1)$. Determining $t \in T$ must result in an output of length at least $\log_2 |T|$ which must be of order at least $\log_2\left(\frac{1}{p^N}\right) \approx kN$. Hence, these are optimal up to constants, except for possibly the time to calculate $t(h)$ where there is a factor k difference between the algorithm and the theoretical lower bound. However, k is generally considered to be *less than* 10; this does not seem to be much of a loss. It is also clear that these algorithms are very practical. For a million passwords say $p = 1/16$, one megabyte of data would be needed. Processing the passwords to determine the array A takes less than 1 second, and testing a presented password takes a few microseconds or so.

■■ *Choosing the Parameters*
Given N, the number of valid passwords, and p, the desired proportion of possible passwords passing the test, the matrix algorithm requires that one chooses parameters M and k so that

$p \approx \left(1 - e^{-N/M}\right)^k$, noting that $1 - e^{-N/M}$ is the approximate probability that an entry in the matrix equals 1. This may be accomplished by taking $k = \left\lceil \log_2\left(\frac{1}{p}\right) \right\rceil$ and $M \approx (N/\ln 2)$, where M is preferably taken to be a power of 2.

To ensure that the expansions used in the construction of the matrix are uniformly distributed, one must also require that Mk divides the number of possible hash values, having both as powers of 2 with a large hash function, an easily satisfied condition for realistic values of p and N. The Grid implementation corresponds to $k = 1$ in the matrix implementation.

■■ *Performance and Optimality*
From the construction of the appropriate array under either method, it is clear that generation and testing require equal time, and that adding valid passwords is equally easy.

Since corresponding implementations satisfy $L = Mk$, storage requirements are also the same and, in fact, of optimal order, as verified by the following argument:

Consider a fixed test T that admits at most a proportion p of possible hash values. Assuming that the hash function is random, the probability that T admits all N valid passwords is at most pN, so that T is a valid test for at most a proportion pN of possible sets of passwords. Thus at least $p - N$ tests are needed to cover all possible sets, and the number of bits required to specify a particular test is at least $\log_2 p^{-N} \approx Nk$.

Comparisons Among Different Approaches of NAS

Each of the different implementations of NAS has its own advantages and disadvantages. ◘ Table 3.6 summarizes the comparison among different methods of NAS. As shown in ◘ Table 3.6, real space and binary space models use probabilistic approach for detector generation and map the same region of the model space for the self-points every time if the parameters to create the self-points remain the same. Detectors are generated as hyperspheres for both models and detectors may overlap over each other. To update the existing detectors and self-points, comparison with all existing self-points and detectors is needed for both real and binary space models. The Grid space model uses a deterministic approach for detector generation and some variant implementation (XOR method) of Grid space model maps different regions for the same set of self-points each time the self-points are generated. Detectors and self-points are generated as non-overlapping Grid cells in the Grid space, and only one comparison with a Grid cell is needed to update an existing Grid space model.

◘ Table 3.6 shows the coverage by detectors and number of detectors generated for each space model implementations for 10,000 self-points with SHA-256 hashing. The real and binary space implementation generates a lot of detectors but achieves detector coverage of about 68%. These two implementations need more space to store the higher number of detectors in comparison with Grid implementation. The Grid implementation generates a relatively small number of detectors and achieves very high detector coverage of about 90%. Binary and real space implementation has very low false positive, while Grid space implementation has a relatively higher false positive rate. Grid space implementation

◻ Table 3.6 Comparison of different NAS approaches

	Binary space	Real space	Grid space
Approach	Non-deterministic	Non-deterministic	Deterministic
Mapping	All info	Curtailed info. (can be varied)	Curtailed info. (can be varied)
Update of password profile	Comparison with all existing self-points and detectors	Comparison with all existing self-point and detectors	Comparison with at most one self-grid or one detectors
Regeneration of password profile	Maps same region for self-points	Maps same region for self-points	Maps different region for self-points (random bit)
Space size	Fixed—huge	Can be varied (decimal places)—huge	Varied (on password profile size)—small
Detector and self-point	Hypercube	Hypersphere	Rectangles
Overlap of detectors	Considered	Considered (50% of radius)	No overlap
Coverage	67.26%	68.17%	95% (with 95% level of significance)
Dimension of Space	256	4	2
No. of detectors	116,382	100,000	514
Space size	2^{256} points	1000^4 points	218 cells
Mapping of points	1:1	$1:2^{216}$	$1:2^{238}$
Self-Region coverage (%) or false positive rate (%)	7.85×10^{-22}	4.93×10^{-6}	1.42

■ **Table 3.7** Time required to create 1 million negative detectors using three different versions **of NAS**

Three implementations of NAS	Time taken (ms)
B-NAS	135.35
R-NAS	53.17
G-NAS	13

always generates the same set of detectors for the same set of self-points, but real and binary space implementation will generate different sets of detectors each time. However, in different runs of G-NAS, the configuration can be changed to set different sets of detectors for the same login information. Binary space implementation maps each different username–password pair to a unique point in the model space, real space implementation maps 2^{216} different username–password points to a single point (for four-dimensional model considering four points after decimal point) and Grid implementation maps 2^{238} different username–password points to a single Grid cell (for a $2^9 \times 2^9$ Grid) in the corresponding space models.

B-NAS and R-NAS approaches require more space and time to generate detectors than G-NAS. G-NAS approach (deterministic solution) can be configured to a specific Grid size that can control the amount of space used by the detectors and self-points. Hence, system administrators can easily tune the parameter (Grid size) that best fits the requirement of any system. Moreover, the detection rates of the G-NAS are 95% with 95% levels of significance that can be found using χ^2-test of the goodness of fit, which proves the effectiveness and validity of the proposed approach.

The time requirement for generating a large number of negative detectors using three different versions of NAS is also tested. The negative detectors are created in the server-side applications and pushed to the first layer of NAS to handle all the incoming password requests. To test the time requirement, DELL PowerEdge R530 Server with 44 cores at 2.2 GHz and 320 GB RAM is used. The results are shown in ■ Table 3.7. According to the table, the required time is quite reasonable for generating 1 million detectors. These generated detectors will be stored on the server side, and hence there is no burden on the client side regarding storage space. In addition, due to the lower requirement of detectors' generation time, system administrators can easily generate a new set of detectors and thereby reduce the chance of compromising the mapping of detector space in the first layer. This also proves its efficiency in terms of computational cost and the proposed approaches can perform better than other existing approaches.

■■ *Reducing Side-Channel Attacks in Authentication Systems*

The proposed NAS approaches have the ability to prevent side-channel attacks as all the access requests are first verified in the first layer of NAS. If the first layer cannot catch an incoming request, it moves the request to the second layer for additional checking. From an intruder perspective, the information contained in the first layer is the commonly available information to be compromised. The first layer of NAS contains only the detectors information and no information regarding the positive password space. In three implementations of NAS, it is evident that the total number of detectors is exponentially larger in comparison with the number of self-points (passwords); however, the created set of detectors is hashed output. Hence, compromising the set of detectors provides the attackers a list of hashed outputs. The strength of the hashed output can be increased by choosing the higher valued hash functions (SHA-256 or SHA-512). As the chance of hash collision is very low, the probability of the valid passwords having the same hash as the compromised detector points is close to zero. Again as the hashing is considered as an irreversible process, it is not possible for the attackers to get the actual string of characters that construct the detectors. Again, if the set of detectors is updated on a regular basis (they will change after every run of B-NAS and R-NAS due to the non-deterministic approach), any extracted information regarding the possible password space may not be applicable after a certain period of time. Hence, with the incorporation of two layers in NAS and putting only the non-self-information in the first layer, the effects of side-channel attacks can be greatly reduced. It is true that the NAS may not prevent all the types of side-channel attacks but with the two-layer-based approach, it makes harder to access and compromise the positive information of the genuine users. Another advantage of this two-layer approach is that the unsuccessful access requests can later be used for forensic analysis to extract any pattern of the attackers to further strengthen the design of NAS.

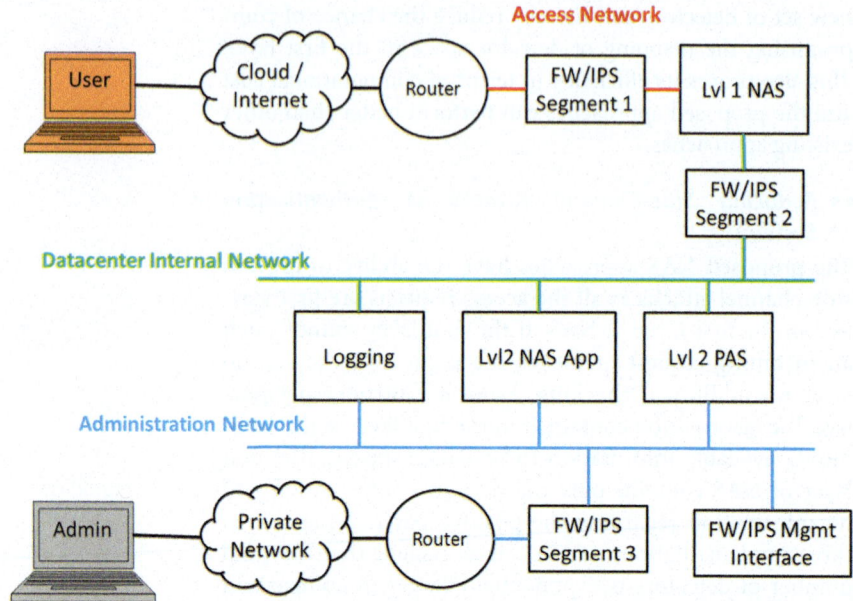

□ Fig. 3.38   Component network layout of NAS prototype

NAS Architectural Design and Prototype System

For prototype purposes, all the implementation has been designed and implemented inside a virtual infrastructure. Every server and access terminal is deployed as a virtual machine with network interfaces. Through these interfaces, each VM is connected to virtual switches that represent the network segments as shown in □ Fig. 3.38.

In the network layout, there are three main network segments, which are an administrative network, datacenter internal network, and access network. These network segments are created using three Firewalls (FW)/Intrusion Prevention Systems (IPS), which are present in all three segments.

The administrative network is basically a private network consisting of an administrator machine, routers, FW/IPS Segment 3, and FW/IPS Management Interface. The administrator machine has access to Lv2 NAS App server to create user accounts as well as to create or update the anti-password database. Any newly created or modified user accounts are stored in Lv2 Positive Authentication Server (PAS). These user accounts are used to generate anti-password data using Lv2 NAS App server and pushed to Lv1NAS Server.

The datacenter internal network has four components which are Logging Server, Lv2 NAS App Server, Lv2 PAS, and FW/IPS Segment 2. Lv2 NAS App Server and Lv2 PAS are mentioned while describing the administrative network, whereas these two components are crucial for the functionality of the system.

As for the access network, it has a direct access to the clients from the outside network such as the Internet. This network has three components which are Lv1 NAS Server, FW/IPS Segment 1, and router. As mentioned above, Lv1 NAS Server receives anti-password data from Lv2 NAS App server, and this data is used to validate the authenticity of the clients at the access network.

The NAS is composed of seven different components. These are as follows:

1. NAS Clients
2. Lvl1 NAS Detector Server
3. Lvl2 NAS Application Server
4. Lvl2 Positive Authentication Server
5. NAS Log Server
6. NAS Scheduler Database
7. NAS Database.

■■ Layer 1 Components

■■ NAS Clients

■■ NAS Web Client

The login page requests username and password for users that want to access the system. After the credentials are submitted, the page then uses them to generate the self-point for that user using the same process that was used by the Level-two server that generates the detectors. The dynamic salt for each user is looked-up in mem cached using the digest that results from hashing the concatenation of user and static salt. After obtaining the dynamic salt, it is concatenated to the input password, and then it is hashed and the result is concatenated to the username and hashed again to obtain the digest. Depending on the type of implementation chosen, the appropriate algorithm to process the digest is used. The resulting self-point is then compared against the set of detectors to look for a match. If a match is found, the login is unsuccessful. If it does not match any detectors, it passes on to the next layer. The same web page executes a call to a stored procedure on the second layer to get the positive authentication of the user.

The login page returns the result of both the negative and positive authentications along with the summary of the detector results and the self-point. In a production implementation, this data should only be used for diagnostic purposes and the user should only see the result of the authentication, either success or failure.

■■ NAS Windows Client

The Windows 8 Client is running the pGina Credential Provider replacement to allow easy access to the login interface. This setup is coded a pGina plugin in C# to contact the Lvl1 NAS Detector Server and the Lvl2 Positive Authentication Server. This allows us to use any authentication mechanism on the server and return if the user is allowed or denied access. Due to the simple socket connection of the client, the Lvll1 NAS Detector Server can be reused from previous proof of concept with only minor code changes.

■■ Lvl1 NAS Detector Server

Lvll1 NAS Detector Server is a PHP server that communicates with Mem cached in order to get the current NAS Detectors. The user account information that is provided to NAS Windows Client is validated against the detectors. If there is a match, user access is not allowed. If there is not a match, the user credentials are then compared to the Lvl2 Positive Authentication Servers (PAS) user account information, where a Microsoft Active Directory server for positive authentication is used.

■■ *Layer 2 Components*
■■ *Lvl2 NAS Application Server*

Here the Layer2 Negative Authentication Application (L2NAS) for the Linux2.4/2.6 platform including CentOS 6.4 is implemented. The implementation of the L2NAA is designed to manage user account, create an anti-password database, push anti-password database to Layer 1 Server, view statistics and logs, and configure input and preference parameters.

■■ *Lvl2 Positive Authentication Server*

In the NAS architecture, Microsoft Active Directory Server fulfills the requirement of a positive authentication source. As part of the NAS implementation, Active Directory LDIF entries were updated to contain the *user Password* attribute.

This attribute was used to contain the user account's hashed password. The hashed password will be used to create the self-points needed by the NAS application. In order to use the *user Password* attribute (i.e., to make it retrievable by the directory search), it is required to set the *dSHeuristic* to false for the *f UserPwdSupport* entry.

■■ NAS Log Server

This module shows the number of anti-password based on different algorithms. Logs are stored in ArcSight Logger. Logs stored in ArcSight Logger can be used for future forensic analysis. ArcSight Logger dashboards have been configured to show the most relevant information from the status messages. Column graphs show users trying to login in Lvl-1NAS and Lvl2PAS. Pie charts show the successful versus failed login attempts in Lvl1 and Lvl2. Going forward, a better coverage of Lvl1 will reduce the failures in Lvl2.

■■ NAS Scheduler Database

The quartz framework is used to schedule and run the jobs in the system, where the number of jobs is fixed in the system. These jobs are configurable based on the valid 'cron' expression.

■■ NAS Database

This module is used to generate and store anti-password database using different algorithms such as B-NAS, R-NAS, and G-NAS.

■■ *Account Creation and Authentication Flow*

The account creation and authentication processes are shown in ■ Fig. 3.38. There are two sections below that give a description of the account creation and authentication flow. To achieve account creation, the administrator must successfully complete the five steps mentioned below to upload the negative database to NAS L1 Server.

When clients request to access resources they go through layers 1 and 2, which is called a client authentication process. Below there is a description of this process given in three steps, and there is also an illustration found in ■ Fig. 3.39.

■■ *Account Creation Flow*

The administrator creates user accounts using steps 1, 2, 3, 4, and 5, and these steps are described below:

1. After the user account is created, static salt is concatenated with Username (ID) and password for self-point creation.

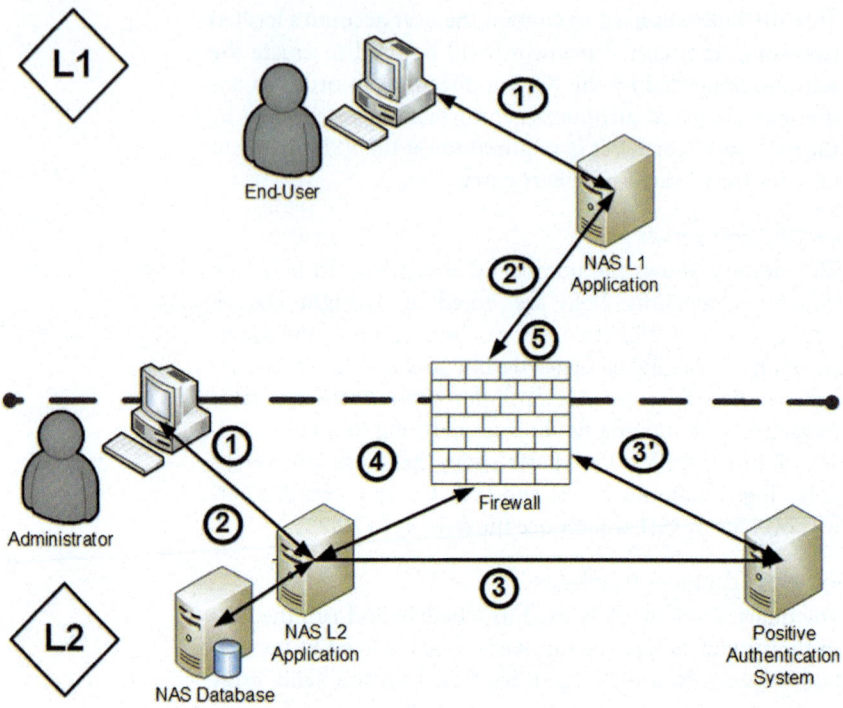

□ Fig. 3.39   Two-step NAS authentication process

2. Hashed ID (i.e., static salt, username, and password) passed to NAS Database.
3. User account information is passed to the Positive Authentication System (PAS).
4. Detectors are created using the Hashed user information from NAS Database.
5. Detector list is passed to NAS L1.

■■ Authentication Process Flow

The clients are authenticated using steps 1′, 2′, and 3′, and these steps are described below:

1′. ID and password are provided to NAS Ll for authentication.
2′. ID and password are concatenated with static salt (i.e., application private key). This information is converted to a point and the list of detectors is checked. If the point does not fall into any detectors, the user credentials are passed through the firewall. Otherwise, this access request is denied.
3′. PAS user credentials are validated, and if valid, an authentication token is passed.

This implementation of NAS prototype illustrates one possible way to introduce it in existing network infrastructure; however, there may be other ways to incorporate such a complementary scheme to make password-based or any positive authentication system more robust.

Chapter Summary

The negative authentication approach can detect and filter out most of the invalid requests, and hence lower the probability of making guessing requests to access the positive authentication data. In the case of compromising of the first layer (negative detector region), an attacker can have access to the file containing anti-password information, which can reveal very little on individual password. Hence, exposing the negative detector upfront reduces the overall password cracking risks. Also, the password file that contains the authentication data is kept in another secure server and is not exposed to the outside of NAS, which assures the security of the password file (positive information). The detailed designs of different versions of NAS and their implementations are illustrated. The experiments with real-world password data and the results are described in the context of current cyberattacks to demonstrate the benefit of using NAS. All these models of NAS reduce the false positive rate of authentication requests (deterministic method performs best among the three) and increase the detection rate of invalid requests. The last section illustrates the deployment of NAS prototype and shows that it can easily be integrated with the current password-based authentication system by providing an additional layer before the positive password database. The results demonstrated that combining the negative approach with the existing authentication system provides significant benefits to overcome certain password-based cyberattacks.

Acknowledgements Some parts of NAS research work are supported by IARPA Seedling program and Cooperative Agreement (Number N66001-12-C-2003) and conducted in collaboration with experts from the Massachusetts Institute of Technology (MIT) in particular, Prof. John Williams, Dr. Abel Sanchez, and Alvaro Madero. However, points of view and opinions on this chapter are those of the author(s) and do not necessarily represent the position or policies of the United States. Authors would like to acknowledge the contributions of all NAS project members and other students who contributed in prior publications.

Review Questions

1. Explain the concept of the Negative Authentication Algorithm (NAS).
2. What are different NAS approaches based on representation?
3. Explain the Binary-valued Negative Authentication System (B-NAS).
4. Explain the Real-valued Negative Authentication System (R-NAS).
5. Explain the Grid-based Negative Authentication System (G-NAS). Describe different implementations of G-NAS.
6. Compare among three approaches of NAS.
7. How to calculate the distance between two points in the four-dimensional positive space for the Real-valued Negative Authentication System (NAS)?
8. If there are three users in a system and all of them are using the same password «ABC», what is the probability for breaking an R-NAS? Calculate the same for a B-NAS.
9. Calculate the confusion parameter for a G-NAS.
10. Present a hypothetical attack scenario to break R-NAS.
11. Illustrate (with diagram) an implementation of NAS.
12. Illustrate (with diagram) the authentication process of NAS.

References

1. Khalil G (2016) Password security thirty-five years late. Url: ▶ https://www.sans.org/reading-room/whitepapers/basics/password-security-thirty-five-years-35592, 2014. Date accessed 01 Sept 2016
2. Perlroth N (2013) Hackers in china attacked the times for last 4 months. NY Times, 30 Jan 2013
3. Verizon Enterprise: 2016 Data Breach Investigations Report. Url: ▶ www.verizonenterprise.com/resources/reports/rp_DBIR_2016_Report_en_xg.pdf
4. Brants T, Franz A (2006) The Google web 1t 5-gram corpus version 1.1. LDC2006T13
5. Bond M (2016) Comments on grid sure authentication. Online at ▶ http://www.cl.cam.ac.uk/mkb23/research/GridsureComments.pdf, 2008. Date accessed 01 Sept 2016
6. Bonneau J (2012) Guessing human-chosen secrets. Ph.D. thesis, University of Cambridge

7. Schechter S, Herley C, Mitzenmacher M (2010) Popularity is everything: a new approach to protecting passwords from statistical guessing attacks. In: Proceedings of the 5th USENIX conference on Hot topics in security, pp 1–8. USENIX Association

8. Forrest S, Perelson AS, Allen L, Cherukuri R (1994) Self-nonself discrimination in a computer. In: IEEE computer society symposium on research in security and privacy, pp 202–202. Institute of Electrical and Electronics Engineers

9. De Castro LN, Timmis J (2002) Artificial immune systems: a new computational intelligence approach. Springer

10. Hofmeyr SA, Forrest S (2000) Architecture for an artificial immune system. Evol Comput 8(4):443–473

11. Dasgupta D, Ji Z, Gonzalez F (2003) Artificial immune system (AIS) research in the last five years. In: The 2003 congress on evolutionary computation, 2003 (CEC'03), vol 1, pp 123–130

12. Dasgupta D, Forrest S (1999) An anomaly detection algorithm inspired by the immune system. In: Artificial immune systems and their applications. Springer, pp 262–277

13. Esponda F, Ackley ES, Helman P, Jia H, Forrest S (2006) Protecting data privacy through hard-to-reverse negative databases. In: Information security, pp 72–84. Springer

14. Dasgupta D, Azeem R (2008) An investigation of negative authentication systems. In: Proceedings of the 3rd international conference on information war-fare and security, pp 117–126

15. Ji Z, Dasgupta D (2009) V-detector: an efficient negative selection algorithm with probably adequate detector coverage. Inf Sci 179(10), 1390–1406. doi: 10.1016/j.ins.2008.12.015. Url: ▶ http://dx.doi.org/10.1016/j.ins.2008.12.015

16. Ji Z (2006) Negative selection algorithms: from the thymus to v-detector. Ph.D. thesis

17. Dasgupta D, Ferebee D, Saha S, Nag AK, Madero A, Sanchez A, William J, Subedi KP (2014) G-NAS: a grid-based approach for negative authentication. In: Symposium on computational intelligence in cyber security (CICS) at IEEE symposium series on computational intelligence (SSCI), IEEE, Orlando, Florida

18. Williams JR, Perkins E, Cook R (2004) A contact algorithm for partitioning N arbitrary sized objects. Eng Comput 21(2/3/4):235–248

19. Dasgupta D, Saha S (2010) Password security through negative filtering. In: 2010 international conference on emerging security technologies (EST). IEEE, pp 83–89

20. Metropolis N, Ulam S (1949) The monte carlo method. J Am Stat Assoc 44(247):335–341

21. Kanerva P (1988) Sparse distributed memory. MIT press

22. Skull Security: Password-skull security. Url: ▶ https://wiki.skullsecurity.org/Passwords

23. Cubrilovic N (2016) Rockyou hack: from bad to worse. Url: ▶ http://techcrunch.com/2009/12/14/rockyou-hack-securitymyspace-facebook-passwords, 2009. Date accessed 01 Sept 2016

24. Butler R (2009) List of the 1000 most common surnames in the US. Url: ▶ http://names.mongabay.com/mostcommonsurnames.htm, 2009. Date accessed 01 Sept 2016

25. Dasgupta D, Saha S (2009) A biologically inspired password authentication system. In: Proceedings of the 5th annual workshop on cyber security and information intelligence research. ACM, p 41

Pseudo-Passwords and Non-textual Approaches

Beyond passwords—graphical authentication

© Springer International Publishing AG 2017

D. Dasgupta et al., *Advances in User Authentication*, Infosys Science Foundation Series,
DOI 10.1007/978-3-319-58808-7_4

This chapter describes various complementary approaches of passwords, namely, Honeywords, Cracking-Resistant Password Vaults using Natural Encoders, Bloom Filter, and non-textual and graphical passwords to protect user identities against any type of credential breaches. At the end, a comparison of various non-textual passwords is provided by highlighting their strength and weaknesses.

As mentioned in the first chapter, the system maintains a profile (F) containing information of legitimate users (u_i) and their passwords p_i as username/password–hash pairs of the form

$$(u_i H(p_i))$$

On some UNIX/Linux systems, the profile F might be labeled as /etc./passwd or /etc./shadow.

Users frequently choose weak passwords and consequently, an adversary who steals a (hashed) password profile of users, often performs brute-force search to find a password, p whose hash value $H(p)$ equals the hash value stored for a given user's password, thus allowing the adversary to impersonate the user. One approach to improve the situation is to make password hashing more complex and time-consuming to break. As the selection of the hash function plays an important role in password security, a brief history of hash functions is given below.

Evolution of Hash Functions

Hashing is the core of password-based authentication systems. Over the years, many one-way hash functions were proposed, in particular, the MD4 (Message Digest, version 4) algorithm was designed by Ron Rivest in 1990 [1]. He also invented several versions of MD for different purposes, and one of this is an improved version, called MD5, and was first applied to password hashing. It can produce 64 bits hash value for a password of any length. MD5 hash was broken in 2000. The National Security Agency (NSA) published SHA (Secure Hash Algorithm), a hash function similar to MD5 and later addressed some weakness to SHA; a new algorithm was released in 1995, called SHA-1, which has been the most popular hash function till today. SHA-1 produces a 160-bit hash value and is supposed to be collision-free which means that it is impossible to find two messages that hash to the same hash value. However, in 2013, Marc Stevens gave a theoretical proof that SHA-1 is not collision-free [2]. In February 2017, three cryptographers showed a practical technique to create an SHA-1 collision and developed an algorithm for finding SHA-1 collisions faster than brute-force [3]. The team highly recommended the technical community to move to safer cryptographic hashes such as SHA-2 (with different block sizes *SHA-256* and *SHA-512*) and *SHA-3* having a different internal structure from the rest of the SHA family.

Honeyword Approach

Similar to Negative Authentication System (NAS) described in ▶ Chap. 3, another complementary approach that was introduced in 2013 as a defense against stolen password profile is called *Honeywords*, an approach developed by Juels and Rivest [4]. In order to hide legitimate account/passwords, several fake user accounts (honeypot or decoy accounts) are set up; an alarm is raised when an adversary uses any of these fake accounts indicating an illegal access attempt. Since there is no such legitimate user, the adversary's attempt can reliably be detected when such an attempt occurs. However, the adversary may be able to differentiate real usernames from fake usernames (accounts), and thus avoid being caught.

The honeyword approach can be viewed as an extension of this basic idea applicable to all users (i.e., including the legitimate user accounts) in order to detect guessing password attacks by inverting a hashed password profile and attempts to login. Particularly, in honeyword approach, multiple possible (pseudo) passwords are stored for each account, but only one of which is the actual password. ◻ Figure 4.1 shows a conceptual diagram of honeywords and password points.

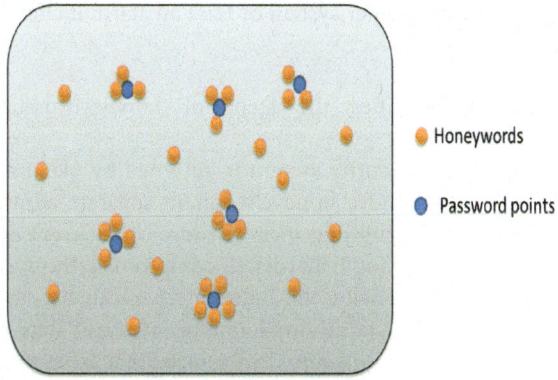

Honeywords

Password points

◻ **Fig. 4.1** Illustration of honeywords where pseudo-passwords (variations of each actual password) are created and stored to confuse the attackers

The attempted use of a honeyword to login sets off an alarm; it puts the adversary at risk of being detected, so honeywords are used to detect illegitimate users. There exist similar ideas in the literature; for example, the term «honeyword» first appeared in the work of Bojinov et al. [5] developing a Kamouflage system. Another closely related work to the honeyword is the «honeyfile» [6], a bogus password file, used as a bait file that is intended for hackers to access, and when the file is opened, an alarm is set off signaling an intrusion. Lastly, a patent application by Rao [7] describes the use of per-account decoy passwords called "failwords" used to trick an adversary into believing that he has logged into successfully when he has not. All these works tried to address some attacks relating to passwords such as stolen files of password hashes and easily guessable passwords.

Honeychecker

An auxiliary server (the «honeychecker») can distinguish the user password from honeywords for the login routine, and can set off an alarm if a honeyword is submitted. The honeywords-based authentication process is shown in a flow diagram (◘ Fig. 4.2).

Honeychecker maintains a single database value $c(i)$ for each user u_i; the values are small integers in the range 1 to k (e.g., $k = 20$). The honeychecker accepts commands of the following two types:

— Set: i, j
  Sets $c(i)$ to have value j.
— Check: i, j
  Checks that $c(i) = j$. May return result of check to requesting computer system or raise an alarm if check fails.

◘ Figure 4.3 describes the steps of honeywords-based approach [4].

A distributed security system is designed by placing the computer system and a honeychecker in separate administrative domains or running their software on different operating systems—making it harder to compromise the system as a whole. Additionally, the honeychecker can be configured to manage only a minimal amount of secret state. The index $c(i)$ for each u_i is stored in a table that is encrypted and authenticated under keys stored by the honeychecker. It can be assumed that the computer system communicates

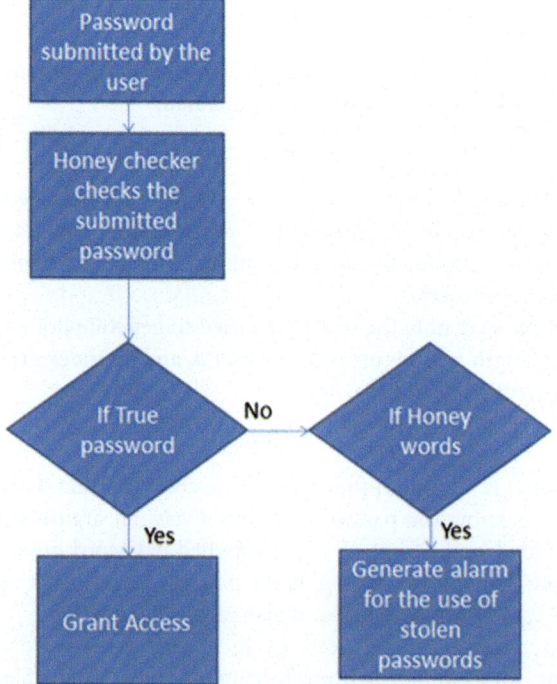

Fig. 4.2   Flow diagram of the honeychecker process

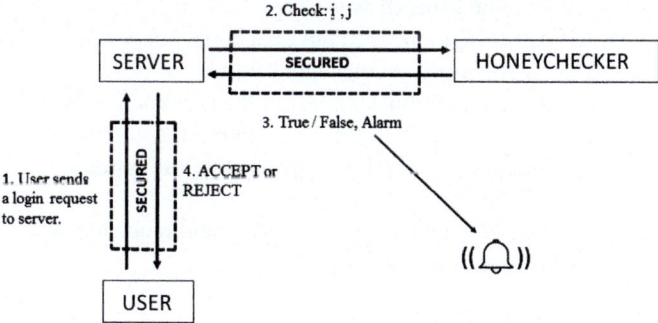

Fig. 4.3   Illustrates the steps of the honeyword approach for authentication

with the honeychecker when a login attempt is made on the computer system, or when a user changes her password. This communication is considered to be over dedicated lines and/or encrypted and authenticated. The honeychecker should have extensive instrumentation to detect the anomalies of various sorts.

This subsection describes how the honeyword approach works in its simplest form.

For each user u_i, a list W_i of distinct words (called «potential passwords» or «sweetwords») is generated:

$$W_i = (w_{i,1}, w_{i,2}, w_{i,3}, \ldots, w_{i,k})$$

Here, k is a small integer, such as $k = 20$, where k is a fixed system-wide parameter chosen in a per-user manner—for example, $k_i = 200$ for the system administrator(s) and from $k = 2$ for guest users.

Let **Gen**(k) denote the procedure used to generate both a list W_i of length k of sweetwords for user u_i and an index $c(i)$ of the correct password p_i within W_i:

$$(W_i, c(i)) = \textbf{Gen}(k)$$

Here, **Gen**$()$ is a random function but needs to interact with the user to know the password. Alternatively, an argument can be added in the form of a user-supplied password, p_i, so that **Gen**$(k; p_i)$ ensures that p_i is the password in W_i; that is $p_i = w_{i,c(i)}$. Use of **Salt** per user is also recommended.

If an adversary enters one of the user's honeywords, obtaining, for example, by brute-forcing the password profile (F), then an appropriate action is taken (which is determined by the policy) as follows:

- setting off an alarm or notifying a system administrator,
- letting login proceed as usual,
- letting the login proceed but on a honeypot system,
- tracing the source of the login carefully,
- turning on additional logging of the user's activities,
- shutting down that user's account until the user establishes a new password (e.g., by meeting the system administrator),
- shutting down the computer system and requiring all the users to establish new passwords.

If a relatively small set of honeywords are randomly selected from a larger class of possible sweetwords, it can limit the impact on guessing attacks against chaffing-by-tweaking. For example, we might use take-a-tail with a three-digit tail ($t = 3$), yielding $|T(p_i)| = 1000$. Setting $k = 20$, then, means randomly picking $k - 1 = 19$ honeywords from $T(p_i)$. Knowing the correct password p_i only gives an adversary (or malicious user) a chance of $(k - 1)/1000 \approx 0.02$ of hitting a honeyword in this case, greatly reducing the ability to trigger an alarm. The vast majority ($\approx 98\%$) of user attempts will be passwords in $T(p_i)$, but not in W_i. Further details on honeywords-based works including the study of different attack models to defeat honeyword approach are available elsewhere [4].

Use of Password Vaults

This section describes another frequently used complementary approach known as password vaults or wallets or managers based on secure encoding system [8]. Password vaults are increasingly popular applications that store multiple passwords in an encrypted form under a single master password that the user only needs to memorize. The justification of using a password vault is that it can greatly reduce on a user the burden of memorization [9] of many passwords for different accounts such as user's website passwords, along with the associated domains, in a small database that can be encrypted with a single master password.

Vaults enable users to select stronger passwords for each website (since they need not have to memorize these) or use cryptographically strong, randomly chosen keys as passwords. Many modern vault applications also back up encrypted vaults with an online storage service, ensuring that as long as users remember their master passwords, they can always recover their stored credentials. These vault management services provide attackers with a rich target for compromise and also introduce a single point of failure. The popular password vault service «LastPass», for example, reported a suspected breach in 2011 [10]. Security analysis of several popular web-based password managers revealed critical vulnerabilities, including bugs which may allow a malicious website to exfiltrate the encrypted password vault from the browser [11]. An attacker that captures users' encrypted vaults can mount offline brute-force attacks against them, attempting decryption using (what the attacker believes to be) the most probable master password. So the password vault is as vulnerable as the regular passwords, with a higher risk of losing all user passwords if a vault is compromised.

Available evidence (c.f., [12]) suggests that most master passwords for vaults selected by users are weak in the sense that even a modestly resourced attacker can easily crack them in a matter of minutes or hours. This state of affairs prompts the following core question addressed by later work [5]: Can password vaults be encrypted under low-entropy passwords, yet resist offline attacks?

Kamouflage: According to this approach, the master password for a vault may include a large number of decoy/honey vaults and decoy master passwords that make the offline attack *appear* to be cracked. The attacker would still have to try each of these master passwords in online attacks to find the correct vault. This approach is very similar to Honeywords, while the application is not at the individual password level.

Kamouflage stores $N - 1$ decoy vaults encrypted under decoy master passwords that are generated by parsing the true master password and using its structure to help select the decoys.

The drawback of using Kamouflage is that it actually degrades overall security relative to traditional vaults. Because the decoy vaults are based on the structure of the true master password, so cracking ANY of the decoy vaults reveals details about the structure of the true master key, which can then be cracked much faster and with higher confidence. The authors state that Kamouflage actually degrades overall security relative to traditional password-based encryption. For example, if the master password has the format password!12345, the template is 8 letters, 1 symbol, 5 digits. ALL decoy master passwords will have this same format. Therefore, cracking ANY of the decoy passwords exfiltrates the format of the master password, reducing the search space significantly; in fact it becomes easier to crack ALL master passwords (decoys and the real one) than cracking a conventional password-based encryption (PBE) ciphertext. Accordingly, an attacker needs to do significantly less computational work (on average about 40% less for number of decoys, $N = 1,000$) and make a few online queries. In practice, an attack against Kamouflage would therefore require less total computational resource than one against traditional PBE.

Another issue is that the decoy master passwords are chosen uniformly with respect to the master password template. Real user-selected passwords are distributed differently. The cracked passwords can be ranked by popularity, and the real master password is likely to be assigned a high rank.

▪▪ Breaking Kamouflage:

The authors of a study [5] crafted an attack that employs a simple password likelihood-based model that is used to generate some trial master passwords in decreasing order of probability until one decrypts the vault successfully (may be a decoy or real). At this point, the template for the master password is discovered. The cracked master password is then used in an online attack attempt. If successful the algorithm halts, else it resumes the offline attack. Tests were conducted using $N = 10^3$ and $N = 10^4$ where N is the number of vaults (1 real and $N - 1$ decoys). The model was trained using the RY-tr set and tested with all remaining sets.

The authors use the RockYou (32.58 million passwords partitioned into training (RY-tr: 90%) and test (RY-ts: 10%) sets), Myspace, and Yahoo leaked datasets. The language model is trained using only the RY-tr set, but ALL sets are used for adversarial training and testing. The RY-ts set is used to model test scenarios in which the adversary and defender have the same knowledge. In addition, MySpace and Yahoo sets are used to model additional challenging scenarios in which the adversary has more precise knowledge of password distribution than the defender.

Their experiments on breaking a Kamouflage vault (for $N = 10^3$) require on average only 44% of the computational cost of breaking traditional PBE and incur only 11 online queries if master passwords are sampled from the Myspace dataset. For $N = 10^4$, only 24% of the computational cost of breaking PBE is required. Therefore, with *increasing N*, computational cost *decreases* because the chance of decrypting one of the decoys, and therefore finding the pattern, increases. However, the number of online queries increases with N, requiring an expected 108 for Myspace set or 219 for the Yahoo set, both of which are an order of magnitude less than the expected number of queries for brute-forcing ($N/2 = 5000$ for $N = 10^4$).

■■ Natural Language Encoders (NLE)

Another work introduces a secure encoding scheme, called a natural language encoder (NLE) [8]. An NLE permits the construction of password vaults, which when decrypted with the wrong master password produce plausible looking decoy passwords. They showed how to build NLEs using existing tools from natural language processing, such as n-gram models and probabilistic context-free grammars (PCFG), and evaluate their ability to generate plausible decoys. Finally, they presented, implemented, and evaluated a full, NLE-based cracking-resistant vault system called «NoCrack».

■■ **Details on NLE Approach**

This system stores an explicit ciphertext and has the property that decryption with the wrong key generates decoy vaults on the fly without leaking information like Kamouflage. Based on previous studies (cryptographic camouflage and honey encryption), but unlike these schemes, which draw plaintexts from simple (uniform) distributions, the authors introduce natural language encoders (NLEs) that can encode natural language texts (lists of human-selected passwords in this case) such that decoding a uniformly selected bit of string results in a fresh sample of natural language text. Instead of creating NLEs from scratch, they convert models such as n-gram Markov or probabilistic context-free grammars (PCFG) which have both been used to build password crackers from statistical NLP to NLEs.

Instead of explicitly storing these decoy vaults, they construct a ciphertext, which when decrypted appears to have been sampled from the distribution of plaintext vaults across the entire user population.

Readers interested in further details of prior works can find on cryptographic camouflage system by Hoover and Kausik [13] and a generalization called honey encryption (HE) [14] for the restricted domain of passwords [15–17].

NLEs from the n-Gram Model Concept

n-gram models are used widely in natural language domains [17–19]. For natural language encoders, an n-gram is a sequence of characters contained in a longer string. For instance, the 4-grams of the word «password12» are {pass, assw, sswo, swor, word, ord1, rd12} [8]. In building models, it is better to add two special characters to any string: ˆ to the beginning and $ to the end. Accordingly, the 4-gram set mentioned in this example also includes ˆpas and d12$.

An n-gram model can be represented as a Markov chain of order $n - 1$. The probability of a string is estimated by the following formula:

$\Pr[w_1 w_2 \ldots w_j] \approx \prod_{i=1}^{j} \Pr[w_i | w_{i-(n-1)} \ldots w_{i-1}]$, where the individual probabilities in the given formula are calculated for a given model empirically via some text database. In case of RockYou password leak, we assume $c(\cdot)$ as the number of occurrences of a substring in the leak. The empirical probability of the previous term is then calculated using the following formula [8]:

$$\Pr\left[w_i | w_{i-(n-1)} \ldots w_{i-1}\right] = \frac{c\left(w_1 w_2 \ldots w_j\right)}{\sum_x c\left(w_1 w_2 \ldots w_j x\right)}$$

for any string $w_1 w_2 \ldots w_j$ of any length i. Let $F_{w_{i-(n-1)} \ldots w_{i-1}}$ represent the cumulative distribution function (CDF) associated with the probability distribution for each scenario. The Markov chain associated with such an n-gram model can be shown as a directed graph with nodes labeled by n-grams. In that directed graph, an edge from node $w_{i-(n-1)} \ldots w_{i-1}$ to $w_{i-(n-2)} \ldots w_i$ is labeled with w_i and $F_{w_{i-(n-1)} \ldots w_{i-1}}(w_i)$. To sample from the model, the process starts at node ^ and samples from $[0, 1)$. Then it requires to find the first edge whose CDF value is larger than the considered sample, follow it to move to the next node, and repeat the process. The process completes at a node that represents stop symbol. The sequence of w_i values mentioned on the edges is the final string.

In this current context, a Markov chain may not have edges and nodes sufficient to represent all possible strings. For encoding purpose, it is required to extend the Markov chain such that each node has an edge labeled with each character. The probabilities for these edges are set to be as small as possible, and re-normalize the other edge weights appropriately. To apply this same concept for a new node, set all its output edges with equal probability.

A distribution-transforming encoder (DTE) can be built by encoding strings as their path in a Markov chain. To encode a string $p = w_1 w_2 \ldots w_j$ it is needed to process each $w_i w_i$ in turn by choosing their values randomly from $[0, 1)$. It results in picking the edge labeled w_{i+1}. In order to decode, it requires the input as the random choices in a walk. Both encoding and decoding are fast, namely $O(n)$ for a password of length n.

NLEs from the PCFG Model Concept

A Probabilistic Context-Free Grammar (PCFG) is a five-tuple grammar, $G = \left(N, \sum, R, S, p\right)$ where N represents a set of nonterminals, $\sum$ represents a set of terminals, R represents the set of relations $N \rightarrow \{N \cup \sum\}$, $S \in N$ is the start symbol, and p is a function of the form $R \rightarrow [0, 1]$, which represents the probability of applying a given rule. In case of NLE, for any nonterminal, $Y \in N$, it holds that $\sum_{\alpha \in N \cup \sum} p(Y \rightarrow \alpha) = 1$. A PCFG generally represents the distribution of strings in a language in a correct form. A PCFG associates a probability to derive a member in a given language defined by underlying CFG.

Encoding or Decoding of the Sub-grammar (SG)

Having a canonical representation of the trained PCFG (mentioned previously), a sub-grammar can be specified by encoding for every nonterminal (except catch-all rule, T) the number of corresponding rules (size of the rules) used by that sub-grammar followed by a list of such rules. Each rule in that list is encoded in similar fashion as a derivation rule for a specific password.

To encode the size of a rule set the following steps are adopted. Using a set of leaked password vaults, the sub-grammar, PCFG is generated for all the vaults of size ≥ 2. A histogram of the number of the nonterminal rules used by each sub-grammar is then created for all nonterminals of PCFG (except T). This histogram provides an empirical distribution on the number of rules used for every non-terminal, which is later used as the distribution for sizes. This distribution helps in decoding a random string to a sub-grammar. The distribution-transforming encoder (DTE) encodes this distribution using the inverse transform sampling mechanism as mentioned in [13].

It is to be noted that SG (for sub-grammar) encodes an input in a way that identifies structure across various passwords, making SG a good candidate to encode password vaults. SG later pads out all encodings of input to a constant length with random bits. This is important to prevent any possible leak of the size of sub-grammar.

Evaluating the Encoders

Earlier subsections have explained how to construct functional NLEs that model real-world password selections by human users of password vaults.

To evaluate the quality of these NLEs and resilience of both the n-gram Markov chain model and the PCFG model, machine learning techniques are applied. The objective of the adversary is to identify the *true* plaintext when decrypting the vault. Suppose $q =$ the number of guessing/decrypt attempts. If the true master password is among these guesses, the result will be $q-1$ random samples from the NLE, as well as the true plaintext, thus q plaintext candidates in total. The adversary is assumed to order these q plaintexts from highest to lowest likelihood of being the true vault. The adversary's best strategy is to then make online authentication attempts using one password from each decrypted vault in order. The position of the true vault indicates the number of online attempts that must be made. The authors use machine learning to order the list instead of password likelihood.

Evaluating Single Passwords

Security goal for single passwords is to build decoys that are indistinguishable from real passwords. This is evaluated in two ways. First, the accuracy of a binary classifier in assigning the correct label («true» or «decoy») is measured. Second, the ability of such a classifier to assign a high rank to a true password in an ordered list of plaintexts is evaluated. They use supervised learning with two training sets: passwords labeled «true» and passwords labeled «decoy». They report that random forest classifiers with 20 estimators worked best. The following features were used:

(1) *Bag of characters*: Captures the frequency distribution of characters in a password. They are represented as a vector of integers of size equal to the number of printable characters. The length of the password was appended to the vector.

(2) *n-gram probability*: Two 4-gram models were trained separately on each of the two classes of passwords. For a given password, the probability yielded by each of the two models was used as a feature.

The authors evaluate 5 NLEs as sources of decoys (WEIR-UNIF, WEIR, UNIF, NG, and PCFG), all trained using the RY-tr set (mentioned in an earlier section), as well as sampling directly from RY-tr. WEIR and WEIR-UNIF are grammars proposed by Weir, et al. [19], whereas PCFG and NG represent the models described above. In addition, the RY-ts, Myspace, and Yahoo sets were used as sources for true passwords. For each experiment, given a true/decoy source pair, t passwords were sampled from the true set uniformly without replacement to obtain a derived set of true passwords. Then a set of t decoy passwords is generated using the appropriate NLE or sampling from the decoy set RY-tr. They then do tenfold cross-validation of the random forest classifier with 90/10% training/test splits.

NLEs	Myspace		Yahoo		RY-ts	
	α	$\bar{r}$	α	$\bar{r}$	α	$\bar{r}$
WEIR-UNIF	66	24	72	13	82	5
WEIR	63	35	54	36	60	25
UNIF	86	2	97	<2	97	<2
NG	70	22	68	41	61	41
PCFG	70	26	58	39	57	39
RY-tr	70	22	64	50	50	50

Fig. 4.4 For different decoy/true password source pairs, percentage classification accuracy (α) and percentage average rank ($\bar{r}$) of a real password in a list of $q = 1000$ decoy passwords for ML adversary. Lower (α) and higher ($\bar{r}$) signify good decoys [8]

α the average accuracy of the classifier on testing data

$\bar{r}$ estimate of the number of online authentication attempts required for a brute-force attacker that uses the classifier to identify the true password

An effective classifier has high α and low $\bar{r}$ (■ Fig. 4.4).

NoCrack Password Vaults

The authors present a prototype password vault system called NoCrack that they claim resists offline brute-force attacks better than other schemes and addresses other challenges that arise in deployment of a honey vault system: concealing website names associated with vault passwords, incorporating computer-generated passwords in addition to user-selected ones, and authenticating users to the service while ensuring compromise does not permit offline brute-force attacks.

They avoid password-based authentication for the fear that people will choose the same password as the master vault password. Instead, they enroll specific devices using an email address, to which a temporary token is sent. The temporary token is then entered and a permanent token is generated. Multiple devices can be enrolled, and the process can be repeated if the token or device is lost. The plaintext vault is never stored in the clear on the client's persistent storage, but the encrypted vault is cached locally. The vault is synced at the beginning of a browser session. The NoCrack system may store decoy passwords for websites and domains to which the user does not have an account. Therefore it is the user's responsibility to determine which entries in the vault are valid and which are not.

The performance of the prototype implementation, where s is the size of the vault (■ Fig. 4.5).

Operation	$s = 2$	200	2,000	20,000
Recover password	6.34 ms	6.41 ms	6.42 ms	6.50 ms
Add password	0.13 s	0.68 s	1.11 s	9.25 s
Vault size on disk	4.71 KB	164.00 KB	1.55 MB	15.26 MB

◘ **Fig. 4.5** Running times (median over 100 trials) of operations for different vault sizes, $s = s_1 + s_2$. The *final row* is size of encrypted vaults on disk [8]

The author's note is that the current implementation is naïve and many performance improvements can be made. Time for password recovery is very fast, with a negligible difference as s increases. However, time taken to add passwords increases significantly as s increases because all entries in the vault are decrypted, decoded, re-encoded, and re-encrypted each time a password is added. They conclude that, while NoCrack does incur some time and space overhead, the absolute performance is more than sufficient for the envisioned scenarios.

Bloom Filter

It is named after the inventor, Burton Howard Bloom in 1970 [20]; it is a space-efficient probabilistic data structure that is used to determine whether an element belongs to a set or not. In this approach, a query returns either «inside set (may be wrong)» or «definitely not in set», i.e., while false positives are possible, false negatives are not. Also, elements can be added to the set, but not removed (though this can be addressed with a counting filter). It can answer membership queries in $O(1)$ time. Bloom filters have found wide-ranging applications, for example, this approach was used to address the weakness of password-based authentication [21]. It works as follows: a Bloom filter represents a set $S = f\{x_1, x_2, \ldots, x_n\}$ of n elements is described by an array of m Boolean values, with all array elements initially set to 0. Then use k independent, uniform hash functions $h_1, h_2, \ldots, h_k$ generating integer values such that these hash functions map each item in the universe to a random number uniform within the range $\{1, \ldots, m\}$, i.e., each hash function randomly maps to one of the m array positions. For each element $x \in S$, the bits $h_i(x)$ are set to 1 for $1 \leq i \leq k$. Any new element is passed through all k hash functions and thus mapped to k different positions setting each bit to 1. A location can be set to 1 multiple times,

4

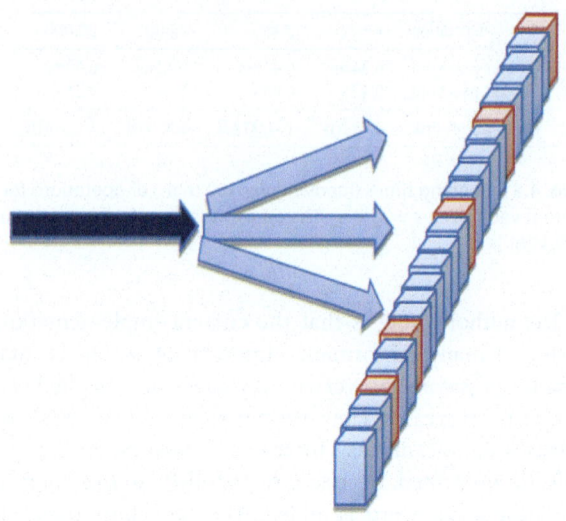

■ **Fig. 4.6** Concept of a Bloom filter. The access request (*Deep Blue*) is authenticated by *n* hash functions

but only the first change has an effect. ■ Figure 4.6 shows a conceptual diagram of Bloom Filter.

Let us assume we have a hash table of m bits, and independent, uniform k hash functions. From [21, 22], with n words we have the proportion of bits left unset (μ), then $\mu = \left(1 - \frac{k}{m}\right)^{n}$.

A word will be falsely shown as present in the dictionary if and only if it hashes to a set of bits that are all set. The expected proportion, P of words in the input space will be mistakenly shown as in the dictionary is $P = (1-\mu)^{k}$.

Using these equations, Eugene Spafford [21] has shown how to derive appropriate values to choose for a filter and hash functions. For example, suppose we pick $n = 250{,}000$ words for the dictionary, and we wish to have a 0.5% chance ($P = 0.005$, i.e., 1 out of 200) of false positives on any arbitrary text string chosen from the entire input alphabet. If six uniform hash functions are chosen, then 62,800,000 bits of storage will be needed and achieve $\mu = 0.586$, i.e., a file of 350 k bytes.

Constructing Bloom Filters

Consider a set $A = \{a_1, a_2, \ldots, a_n\}$ of n elements. Bloom filters describe membership information of A using a bit vector V of length m. For this, k hash functions, $h_1, h_2, \ldots, h_k$ with $h_i : X \rightarrow \{1 \ldots m\}$, are used as described below:

The following procedure builds an m bits Bloom filter, corresponding to a set A and using $h_1, h_2, \ldots, h_k$ hash functions:

```
Procedure BloomFilter(set A, hash_func-
tions, integer m)

            returns filter

    filter = allocate m bits initialized to 0

    foreach a_i in A:

      foreach hash function h_j:

        filter[h_j(a_i)] = 1

    end foreach

    end foreach

return filter
```

Therefore, if a_i is a member of a set A, in the resulting Bloom filter V all bits obtained corresponding to the hashed values of a_i are set to 1. Testing for membership of an element *elm* is equivalent to testing that all corresponding bits of V are set:

```
Procedure MembershipTest (elm, filter, hash_
functions)

            returns yes/no

    foreach hash function h_j:

      if filter[h_j(elm)] != 1 return No

    end foreach

    return Yes
```

Nice features: filters can be built incrementally: as new elements are added to a set the corresponding positions are computed through the hash functions and bits are set in the filter. Moreover, the filter expressing the reunion of two sets is simply computed bitwise OR applied over the two corresponding Bloom filters.

The request is fed through the hashes to get the k positions for authentication. If any of the bits at these positions are 0, the request is denied. There are several variations of Bloom filter. Counting filters provide a way to implement a delete operation on a Bloom filter without recreating the filter again [23]. In a counting filter, the array positions (buckets) are extended from being a single bit to being an n-bit counter. Regular Bloom filters can be considered as counting filters with a bucket size of one bit [23].

Stable Bloom filter is a variant of basic Bloom filters for streaming data [23]. State information is continuously evicted to make room for more recent elements. There exist possibilities of false negatives as all the information is not stored. In Scalable Bloom filter, the structure of the filter is dynamically adapted to the number of elements assuring minimum false positive rate. Counting Bloom filters can be used to approximate the number of differences between two sets [24].

A different data structure used is based on «d-left» hashing that is functionally equivalent but uses approximately half as much space as counting Bloom filters [25]. The keys could be reinserted in a new hash table of double size if the designed capacity is exceeded. In Lattice-based Bloom filter, a compact approximator associates to each key to an element of a lattice. Furthermore, a Bloomier filter generalizes a Bloom filter to compactly store a function with a static support.

Comparison of Complementary Approaches: NAS (G-NAS, GM-NAS), Honeywords, and Bloom Filter

In ▶ Chap. 3, Grid and Matrix-based NAS models are designed to deny illegal access requests and shield the positive authentication information. While Bloom filter model acts as a filter for the positive authentication mechanism, the «Honeyword» system does not give any extra security layer for the authentication system.

Grid model uses a two-dimensional grid for mapping self-points. Each self-point is mapped to only one cell, and access requests need to pass one check in this layer. Matrix model uses a k by M matrix for mapping, and self-points are mapped to k positions in that matrix. So access requests are needed to pass k checks. In Bloom filter model, self-points are mapped by n hash functions to M-length string. So each request goes through n checks. «Honeyword» model does not have any preliminary checks for the authentication requests and there are some extra false credentials information.

Detector space is segregated in Grid model and saved separately from self-points. In case of Matrix-based model, whole data space is saved and self-points are not separated from that space. Bloom filter does not have any idea of self-points or detectors. In Honeyword model, information about the honeypots is saved separately. Matrix model and Bloom filter model need relatively small memory or space size to achieve similar detection rate as the Grid model.

Matrix model also utilizes more information from the hashed strings than the Grid and Bloom filter model. As Matrix and Grid models use only one hash function while the Bloom filter needs n hash functions, Matrix and Grid models need less time for computation also.

While authentication, Grid model needs $O(N)$ time for authentication using its detectors. Matrix model needs $O(1)$ time for detection and Bloom filter need, $O(1)$ time for filtering. False positive rate for authentication purpose is relatively lower in Matrix and Bloom filter model than Grid model. A strong point of Matrix and Grid model is that there is no chance of false negative. This is also true for the basic model of Bloom filter. For Honeyword model, there is a probability that an alarm may be raised by a valid user by mistake. Above discussions are summarized in ◨ Table 4.1.

Non-textual Passwords

The most common and widely used form of authenticating individuals is to use *user-id* and passwords. However, this knowledge-based factor poses several drawbacks due to poor selection of passwords by their users. Also, various password breach incidents prove that it is mostly the target for the cyber criminals to compromise sensitive information. If a password is hard for attackers to guess, it is also hard for legitimate users to remember. Security researchers have developed various authentication approaches that use pictures as an alternate for textual passwords. This approach is mostly known as graphical passwords or non-textual passwords. Basically, a graphical password serves as an authentication mechanism that prompts users to select from a set of images in a particular order in a graphical user interface (GUI). This type of authentication approach can be used in web-based login application, ATMs, and mobile devices where users can interact with the GUI easily to authenticate themselves. Graphical passwords [26–28] are classified into two categories—recognition-based and recall-based approaches. Some of the most common ways of the above two categories are described in a later section.

□ Table 4.1   Qualitative comparison among password complementary methods

Characteristics	Grid-based NAS model	Matrix-based NAS model	Bloom filter	Honeywords
Approach	Implements NA algorithm	Implements NA algorithm	Filter for Positive authentication	Extra false credentials
Space	Two-dimension grid	k by M matrix	M-length string	n times false credentials
Mapping	One cell	k-bit positions	n-bit positions for n hashes	Not applicable
Checks per access request	One check	k checks	n checks for n hash functions	Alarm raised if «honey-word» is used
Detector space	Separate from the self-points	Not separated from self-points	No detector space available	No detector space but «honeywords» are marked
Efficiency	Relatively large grid needed	Relatively small matrix needed	Relatively long string is needed	Not applicable
Information utilization	Low	High	Low	Not applicable
Hash function used	One	One	N hash functions	Depends on the implementation
Complexity of authentication	$O(N)$	$O(1)$	$O(1)$	Not applicable
False positive rate	Relatively higher	Relatively lower	Relatively lower	Not applicable
False negative rate	No false negative	No false negative	No false negative	Chance of false negative

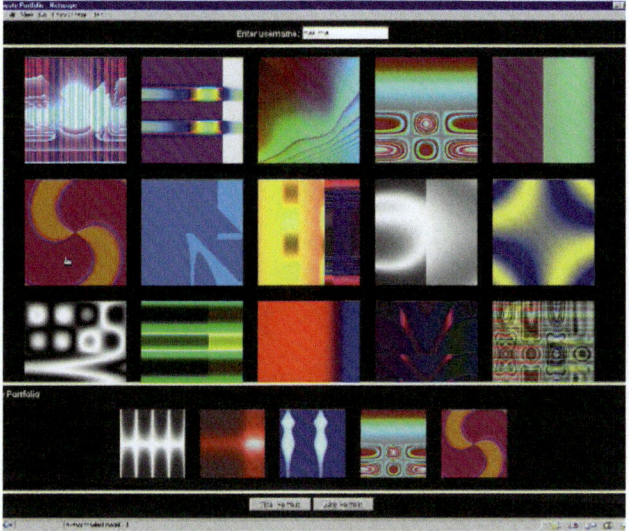

▣ Fig. 4.7 Random images used by Dhamija and Perrig Scheme [34]

Recognition-Based Techniques

In this technique, a user is presented with a list of images, and the user recognizes and identifies those images that he selected during the registration phase [29–32]. If the right images are identified, the user is authenticated successfully. One research [33] shows that 90% of users can remember their passwords after one or two months. Some of the recognition-based techniques are described below.

Dhamija and Perrig [34] developed a graphical authentication scheme based on hash visualization technique [35]. In the proposed system, a user is prompted to select a certain number of images from a random set of pictures during registration phase as shown in ▣ Fig. 4.7. During authentication phase, the user needs to identify the same set of images to verify his/her identity. Using this approach, 90% of the participants succeeded to identify themselves. One of the drawbacks of this system is, it takes more time for users to select a set of pictures from the database and the average login time is longer than the traditional password-based approach. Akula and Devisetty's algorithm [36] is also similar to the previous approach. They use an SHA-1 hash function to make the authentication process secure and take less memory. But due to recent hash collisions, SHA-1 is no longer a recommended hashing function to be used in the authentication process.

4

Fig. 4.8 A shoulder-surfing resistant graphical password scheme [37]

Sobrado and Birget [37] developed a graphical password technique to overcome the most common shoulder-surfing problem. The proposed system will display some pass-objects (selected by users before) to the users. Users need to identify those pass-objects and click inside the convex hull which is bounded by them as shown in ◘ Fig. 4.8. The proposed system can use 1000 objects to make them indistinguishable from one another, and it eventually increases the search space large enough to prevent other possible attacks. The major drawback of this proposed approach is that the time required to login is slow when compared to traditional password-based methods.

Man et al. [38] proposed another graphical password-based approach to prevent shoulder-surfing attacks. In this approach, a user selects a set of pictures as pass-objects. Each of them has several variants, and each variant comes with a unique code. During authentication phase, the user is prompted to type a string of unique codes corresponding to the pass-object variants shown in the GUI screen and those variants' relative positions. The drawback of this approach is that it requires users to memorize the alphanumeric code for every variant of a pass-object.

«Passface» is a type of graphical password-based authentication, where human faces are used as passwords

◘ **Fig. 4.9**   A typical example of Passface-based authentication technique [60]

(◘ Fig. 4.9). During registration phase, a user is asked to choose four human face images as his/her password. During authentication, the user is provided with nine images where one of them is chosen previously by the user and others are not. This process is repeated multiple times to verify all the selected images. The idea behind this approach is that people can recall human face easily rather than any random set of pictures. A study conducted by Valentine [39, 40] showed that users could remember the Passfaces for a longer period of time in comparison with textual passwords. Davis et al. [41] studied the graphical passwords created using the Pass-face technique and reported that this technique has some obvious patterns among these passwords. For instance, the majority of the users choose the list of faces from the same race, due to which attackers can easily predict them.

4

◘ Fig. 4.10   An example of a graphical password

Jansen et al. [42–44] proposed a graphical password-based approach for handheld devices. During registration phase, a user chooses a theme consisting of thumbnail photos and then selects a list of images as a password as mentioned in ◘ Fig. 4.10. During authentication phase, the user must enter the registered images in the correct order to identify accurately. One drawback of this technique is that a total number of possible thumbnail images are 30 which makes the password space very small. Every thumbnail image is assigned a numerical value, and the sequence of user selection will generate a numerical password.

Takada and Koike [45] mentioned about a similar graphical password technique for mobile devices. According to this approach, it allows users to select their favorite image for authentication. Users first register their favorite images (pass-images) with the authentication server. During authentication phase, a user has to go through several verification phases. At each phase, the user selects a pass-image among several decoy-images or picks nothing if no pass-image is present on the screen. The process will authorize a user only if all verification steps are successful. Allowing users to register their own images helps them to remember their pass-images quickly. A study done by Davis [41] showed that when users are given the option to choose pictures by themselves, the choices are sometimes predictable if attackers are familiar with users to some extent.

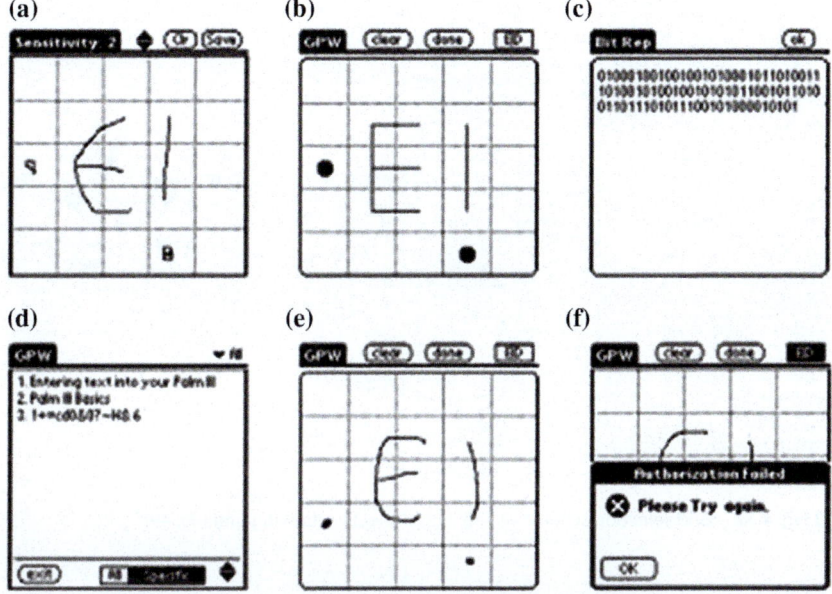

◘ Fig. 4.11   Draw-a-secret (DAS) Scheme. **a** User inputs desired secret, **b** internal representation, **c** raw bit string, **d** interface to database, **e** re-entry of (incorrect) secret, **f** authorization failed [46]

Recall-Based Techniques

In this technique, a user is prompted to reproduce the pattern that he/she created or selected earlier during the registration phase. In this section, two types of graphical password technique will be illustrated, namely reproducing a drawing and repeating a selection.

▪▪ Reproducing a Drawing

Jermyn, et al. [46] first proposed a scheme called «Draw-a-secret (DAS),» which allows a user to draw a simple picture on a two-dimensional square grid (◘ Fig. 4.11). The coordinates of the grids occupied by the image are saved in the order of the drawing. During authentication phase, the user is prompted to redraw the picture. If his drawing touches the same grids in the same sequence, then the user is authenticated successfully. The drawing grid can be of different dimensions to make sure that the full password space of DAS is larger than that of the full-text password space.

4

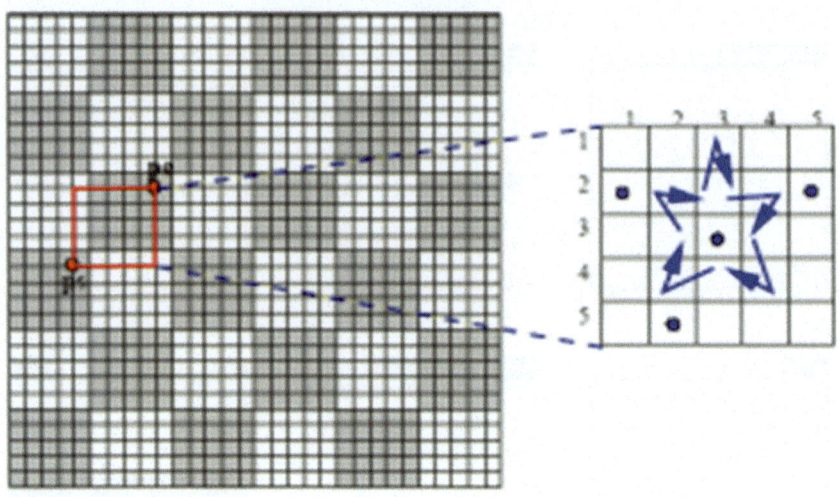

Fig. 4.12   Grid selection scheme where a user selects a drawing grid [48, 49]

Thrope and van Oorschot [47, 48] analyzed the effect of brute-force attack using graphical dictionaries against DAS. According to the study, it is found that DAS passwords with length 8 or larger on a grid size of 5 × 5 can be less susceptible to dictionary attack than text-based passwords. Their analysis also showed that the space of mirror symmetric graphical password is significantly smaller than that of Full DAS. The length of password and stroke-count are one of the important features of DAS scheme to define how complex the password is. The size of DAS password space reduces significantly with fewer strokes for a given password length. Thrope and van Oorschot [47] also proposed a «Grid Selection» strategy where users can zoom in any rectangular region of their interest to enter their graphical password (❏ Fig. 4.12). This approach will significantly increase the DAS password space. Goldberg et al. [49] conducted a user study using a technique called «Passdoodle». This technique is one form of graphical password comprised of handwritten designs or text, drawn with a stylus onto a touch-sensitive screen. That study concluded that users could remember doodle images in full extent as accurately as alphanumeric passwords.

Nali and Thorpe [50] proposed a further study using DAS scheme. In their study, they asked 16 participants to draw a DAS password (six doodles and six logos in 6 × 6 grids). The purpose of the survey was to find any predictable characteristics in the graphical passwords. These drawings

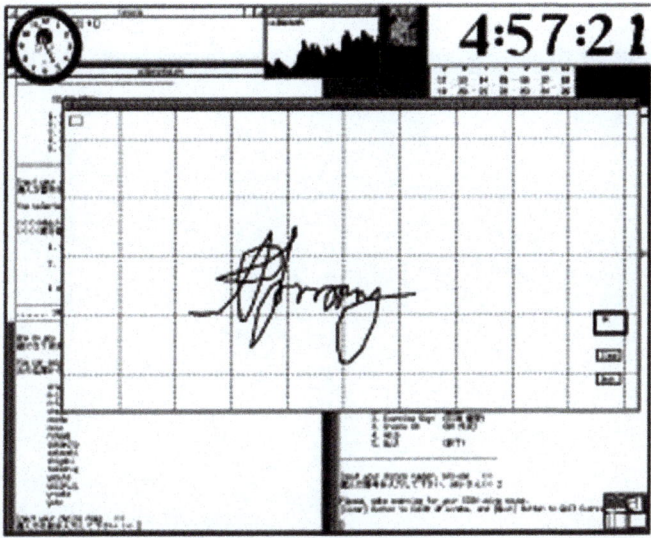

☐ **Fig. 4.13** A signature is drawn by a mouse device [51]

were visually inspected for symmetry and total count of pen strokes. They found that participants inclined to draw symmetric images with a few pen strokes (1–3) and to put their drawing approximately in the center of the grid. This user study only asked the users to draw a memorable password but did not perform any recall-test on whether the passwords were actually memorable or not.

Syukri et al. [51] mentioned a system where a user draws his signature using a mouse as part of authentication (☐ Fig. 4.13). One advantage of this approach is there is no need to memorize the signature and signatures are hard to counterfeit. However, people generally use the mouse as a pointing device, and it requires a right amount of training for people to use the mouse to draw on the screen. In recent technology, the stylus is using this idea to allow users to authenticate themselves in any smart screen devices.

▪▪ Repeating a Sequence of Actions

Blonder [52] implemented a graphical password scheme by having the user clicks on several locations on an image to create a password. During authentication phase, the user should click on the approximate areas of those preselected locations. The benefit of this approach is that the image can assist users to recall their passwords easily compared with other graphical password approaches. Passlogix [53] has also developed similar type of graphical password system.

4

Fig. 4.14   A recall-based technique implemented by Passlogix [53]

In the verification phase, a user must click on several items in the image in the right sequence. Some invisible boundaries are defined for each of these items to detect all the clicked items by a user. It was reported that Microsoft had also developed a similar graphical password technique [54].

Wiedenbeck et al. [55–57] designed PassPoint system that extended the previous approach with allowing no boundaries for items as well using any random image. Because of that, a user can click any place on the image to create a password. A degree of tolerance is calculated for each selected pixel. During authentication phase, the user must click within the predefined tolerance of their chosen pixels in the correct sequence (Fig. 4.14). The idea of this approach is based on the discretization method proposed by Birget et al. [58]. The password space is very large because of the flexibility of choosing any random image, and any image can contain hundreds to thousands of memorable points.

Some studies are conducted by Wiedenbeck [56, 57] regarding the usage and effectiveness of graphical password-based systems. One study showed that graphical password took fewer attempts for a user compared with alphanumerical passwords. However, users authenticating through graphical passwords faced more difficulties in learning the password, and they took more time to input their passwords in comparison with the set of users doing authentication through alphanumeric passwords.

Security Aspects and Possible Attacks on Graphical Passwords

This section discusses some standard threats of password-based systems and how graphical-based approach can handle those attacks. Some of the possible techniques for breaking graphical passwords are mentioned in the following section, and a tabular comparison with text-based passwords has been provided (◘ Table 4.2) [29] (◘ Fig. 4.15).

▪▪ Brute-Force Search

In order to protect against brute-force attack is to have a large enough password space. Some graphical password-based approaches provide password space similar to or larger than that of alphanumeric-based password systems [46–49, 51, 58]. In general, recognition-based approaches have smaller password space than that of recall-based approaches. To launch a brute-force attack against graphical passwords is usually harder than text-based passwords. The attacker program needs to replicate the exact mouse motion as the human input. Hence, a graphical password is less vulnerable to brute-force attacks.

▪▪ Dictionary-Based Attacks

In general, graphical password-based system uses mouse input instead of a keyboard. Hence, it is tough to simulate any dictionary-based attacks on graphical passwords. Some recall-based techniques [45, 51] can be susceptible to dictionary-based attacks, but fully automated dictionary-based attack is not possible for graphical password-based authentication.

▪▪ Password Guessing

Graphical passwords are often predictable for password guessing same as text-based passwords. For instance, studies on the Passface technique showed that people often chose a weak and a predictable graphical password [41],

◻ Table 4.2 Comparison among different Non-password based techniques regarding usability and security

Techniques	Usability		Security issues	
	Authentication process	Memorability	Password space	Possible attack methods
Text-based password	Type in password can be very fast	Depends on the password Long and random passwords are hard to remember	$94 \wedge K$ (there are 94 printable characters excluding SPACE, N is the length of the password). The actual password space is usually much smaller	Dictionary attack, brute-force search, guess, spyware, shoulder-surfing, etc.
Non text-based password				
Perrig and Song [35]	Pick several pictures out of many choices. Takes longer to create than text password	A limited user study showed that more people remembered pictures than text-based passwords	$N!/K! (N-K)!$ (N is the total number of pictures; K is the number of pictures in the graphical password)	Brute-force search, guess, shoulder-surfing
Sobrado and Birget [37]	Type in the code of preregistered picture objects; can be very fast	Users have to memorize both picture objects and their codes. More difficult than text-based password	Same as the text-based password	Brute-force searches, spyware
Passface [60]	Recognize and pick the preregistered pictures; takes longer than a text-based password	Faces are easier to remember, but the choices are still predictable	$N \wedge K$ (K is the number of rounds of authentication, N is the total number of pictures at each round)	Dictionary attack, brute-force search, guess, shoulder-surfing
Jansen et al. [42–44]	User registers a sequence of images; slower than text-based password	Pictures are organized according to different themes to help users remember	$N \wedge K$ (N is the total number of pictures, K is the number of pictures in the graphical password. N is small due to the size limit of mobile devices)	Brute-force search, guess, shoulder-surfing

(Continued)

Table 4.2 (Continued)

Techniques	Usability		Security issues	
	Authentication process	Memorability	Password space	Possible attack methods
Takada and Koike [45]	Recognize and click on the pre-registered images; slower than a text-based password. Slower than a text-based password	Users can use their favorite images; easy to remember than system-assigned pictures	$(N + 1)^{\wedge}K$ (K is the number of rounds of authentication, N is the total number of pictures at each round)	Brute-force search, guess, shoulder-surfing
Jermyn, et al. [46], Thorpe and van Oorschot [47, 48]	Users draw something on a 2D grid	Depends on what users draw on the grid. User studies showed the drawing sequence was difficult to remember	Password space is larger than a text-based password. But the size of DAS password space decreases significantly with fewer strokes for a fixed password length	Dictionary attack, shoulder-surfing
Syukri et al. [51]	Draw signatures using a mouse. Need a reliable signature recognition program	Very easy to remember, but hard to recognize	Infinite password space	Guess, dictionary attack, shoulder-surfing
Goldberg et al. [49]	Draw something with a stylus onto a touch-sensitive screen	Depends on what users draw	Infinite password space	Guess, dictionary attack, shoulder-surfing
Blonder [52], Passlogix [53, 54], Wiedenbeck, et al. [55–57]	Click on several preregistered locations of a picture in the right sequence	Can be hard to remember	$N^{\wedge}K$ (N is the number of pixels or smallest units of a picture, K is the number of locations to be clicked on)	Guess, brute-force search, shoulder-surfing

4

Fig. 4.15   An image used in the passpoint System [55]

which can be vulnerable to guessing attacks. Nali and Thorpe's study [50] mentioned the same type of predictability among the graphical passwords created with the DAS scheme [46]. A good research effort is needed to properly determine the nature of graphical passwords used by a real set of users.

■■ Spyware

Spyware attack can be done using keylogger software in text-based password system. Mouse tracking spyware can be a good alternate to attack against graphical passwords. However, knowing only the mouse motion does not reveal enough information to break graphical passwords. The attackers need to know window position, size, and timing information to launch a successful attack against the graphical password-based system.

■■ Shoulder-Surfing

As users provide login information, an attacker can extract knowledge about their provided credentials by direct observation or external recording devices like digital cameras. Few recognition-based techniques are designed to prevent shoulder-surfing attacks [38, 59]. However, recall-based techniques are prone to shoulder-surfing attacks. Phishing and Pharming can also be possible in graphical password-based system.

▪▪ Social Engineering

In comparison with a text-based password, graphical passwords are less vulnerable for social engineering. For instance, it is extremely hard to provide graphical passwords over the phone. Setting up a phishing website to capture graphical passwords requires more time and resource for attackers, and the chance of completely replicating the service and database is very hard. A comparison of various graphical-based approaches regarding usability and security perspective is mentioned in ▪ Table 4.2 from Suo et al. [29].

Chapter Summary

This chapter describes some complementary approaches, namely Honeywords, Bloom filter, non-textual, and graphical passwords. All these complementary approaches have the potential to make the authentication process more secure than using ordinary password-based systems. The Honeyword approach creates pseudo-passwords in the close vicinity (negative space) of the actual password to lure imposters to use these fake passwords similar to NAS discussed earlier (in ► Chap. 3). However, in the case of NAS approaches, these negative and positive spaces are typically stored on different layers of defense. While the term honeyword was recently introduced, similar concepts with different names (camouflage system, honeyfile, failwords, etc.) existed in the literature for more than a decade.

Password vaults (also called managers, wallets) are increasingly becoming popular to store multiple passwords encrypted under a single master password, and the justification of using a password vault is that a user does not need to remember many passwords for different accounts such as user's website passwords, along with the associated domains, in a small database that can be encrypted under a single master password. Since the master password can also be compromised, similar concepts such as honeywords, honey encryption, Kamouflage, and natural language encoders are also proposed.

Bloom filter approach fools illegitimate users by applying different hash operations on actual passwords. It was also used to identify weak and reusable passwords, not allowing to use these poorly chosen reusable passwords; it is possible to prevent intruder login to one account (by breaking a weak password) and then gain access to other accounts and increase amounts of privilege.

Instead of using text-based passwords, non-textual approaches such as sketches, drawing patterns, graphics and images are in use for user verification and validation. Graphical passwords are also being increasingly adopted in mobile devices (such as smart phones, tablets) where touch screen technologies are used for input.

Review Questions

1. Explain the concept of the «*Honeyword*» approach.
2. Describe an algorithm for «*Honeyword*» generation.
3. Illustrate Honey checker systems with diagram.
4. Explain the *n*-gram model related to non-password system (Resistant Password Vaults using Natural Language Encoders).
5. What is Bloom Filter? Explain its working principles.
6. Give a qualitative comparison of NAS, Honeywords, and Bloom Filter.
7. Describe different graphical password-based approaches.
8. Mention some possible attacks on graphical passwords.
9. Make a comparative table of graphical passwords in terms of usability and security issues.
10. Provide a qualitative comparison among different non-password-based authentication methods.

References

1. Rivest RL (1990) The MD4 message digest algorithm. In: Proceeding proceedings of the 10th annual international cryptology conference on advances in cryptology (CRYPTO'90), 11–15 Aug 1990, pp 303–311
2. Stevens M (2013) New collision attacks on SHA-1 based on optimal joint local-collision analysis. A chapter in advances in cryptology (EUROCRYPTO), Volume 7881 of the series lecture notes in computer science, pp 245–261
3. Google security blog (2017) Announcing the first SHA-1 collision, 23 Feb 2017. ► https://security.googleblog.com/2017/02/announcing-first-sha1-collision.html
4. Juels A, Rivest RL (2013) Honeywords: making password-cracking detectable. In: Proceedings of the 2013 ACM SIGSAC conference on computer & communications security, pp 145–160, ACM, 2013

5. Bojinov H, Bursztein E, Boyen X, Boneh D (2010) Kamouflage: loss-resistant password management. In: European symposium on research in computer security, 2010, pp 286–302

6. Yuill J, Zappe M, Denning D, Feer F (2004) Honeyfiles: deceptive files for intrusion detection. In: Proceedings from the fifth annual IEEE SMC information assurance workshop, June 2004

7. Rao S (2006) Data and system security with failwords. U.S. Patent Application US2006/0161786A1, U.S.PatentOffice. ▶ http://www.google.com/patents/US20060161786. 20 July 2006

8. Chatterjee R, Bonneau J, Juels A, Ristenpart T (2015) Cracking-resistant password vaults using natural language encoders. In: IEEE symposium on security and privacy, IEEE, 2015, pp 481–498

9. Bonneau J, Herley C, Van Oorschot PC, Stajano F (2012) The quest to replace passwords: a framework for comparative evaluation of web authentication schemes, In: IEEE symposium on security and privacy, May 2012

10. Whitney L (2011) LastPass CEO reveals details on security breach, CNet, May 2011

11. Li Z, He W, Akhawe D, Song D (2014) The emperor's new password manager: security analysis of web-based password managers. In: 23rd USENIX security symposium (USENIX security), 2014

12. Bonneau J (2012) The science of guessing: analyzing an anonymized corpus of 70 million passwords. In: IEEE symposium on security and privacy, 2012, pp 538–552

13. Hoover D, Kausik B (1999) Software smart cards via cryptographic camouflage, In: IEEE symposium on security and privacy, IEEE 1999, pp 208–215

14. Juels A, Ristenpart T (2014) Honey Encryption: beyond the brute-force barrier. In: Advances in Cryptology—EUROCRYPT, Springer, pp 523–540

15. Kelley P, Komanduri S, Mazurek M, Shay R, Vidas T, Bauer L, Christin N, Cranor L, Lopez J (2012) Guess again (and again and again): measuring password strength by simulating password-cracking algorithms, In: IEEE symposium on security and privacy (SP), 2012, pp 523–537

16. Ma J, Yang W, Luo M, Li N (2014) A study of probabilistic password models, In: Proceedings of the 2014 IEEE symposium on security and privacy, IEEE computer society, 2014, pp 689–704

17. Castelluccia C, Durmuth M, Perito D (2012) Adaptive password-strength meters from markov models. In: NDSS 2012

18. Devillers MM (2010) Analyzing password strength. Radboud University Nijmegen, Tech. Rep, 2010

19. Weir M, Aggarwal S, De Medeiros B, Glodek B (2009) Password cracking using probabilistic context-free grammars. In: IEEE symposium on security and privacy (SP), 2009, pp 162–175

20. Bloom B (1970) Space/time tradeoffs in hash coding with allowable errors. Commun ACM 13(7):422–426

21. Spafford Eugene H (1992) Opus: preventing weak password choices. Comput Secur 11(3):273–278

22. Mullin JK (1983) A second look at bloom filters. Commun ACM 26(8):570–571

23. Deng F, Rafiei D (2006) Approximately detecting duplicates for streaming data using stable bloom filters. In: ACM SIGMOD international conference on management of data (SIGMOD '06), New York, NY, USA: ACM, 2006, pp 25–36

24. Agarwal S, Trachtenberg A (2006) Approximating the number of differences between remote sets. In: Information theory workshop, Punta del Este, Uruguay, 2006, pp 217–221

25. Bonomi F, Mitzenmacher M, Panigrahy R, Singh S, Varghese G (2006) An improved construction for counting bloom filters. In: Azar Y, Erlebach T (eds) A book chapter in Algorithms ESA 2006 Springer Berlin Heidelberg, 2006, pp 684–695

26. Biddle R, Chiasson S, Van Oorschot PC (2012) Graphical passwords: learning from the first twelve years. ACM Comput Surv (CSUR) 44(4), 2012

27. Jermyn I, Mayer AJ, Monrose F, Reiter MK, Rubin AD (1999) The design and analysis of graphical passwords. In: Usenix Security, 1999, pp 1–14

28. Werner S, Hauck C, Masingale M (2016) Password entry times for recognition-based graphical passwords. In: Proceedings of the human factors and ergonomics society annual meeting, 60(1), 2016

29. Suo X, Zhu Y, Owen GS (2005) Graphical passwords: a survey. In: 21st annual computer security applications conference, IEEE, 2005, pp 10–pp

30. Sreelatha M, Shashi M, Anirudh M, Ahamer MS, Kumar VM (2011) Authentication schemes for session passwords using color and images. Int J Netw Secur Appl 3(3):111–119

31. Towhidi F, Masrom M (2009) A survey on recognition based graphical user authentication algorithms, arXiv preprint arXiv: 0912. 0942 (2009)

32. Dunphy P, Heiner AP, Asokan N (2010) A closer look at recognition-based graphical passwords on mobile devices. In: Proceedings of the sixth symposium on usable privacy and security, ACM, 2010, pp 3

33. Lin PL, Weng LT, Huang PW (2008) Graphical passwords using images with random tracks of geometric shapes. 2008 Congr Images Sig Process, 2008

34. Dhamija R, Perrig A (2000) Deja Vu: a user study using images for authentication, In: Proceedings of 9th USENIX security symposium, 2000

35. Perrig A, Song D (1999) Hash visualization: a new technique to improve real-world security. In: International workshop on cryptographic techniques and E-commerce, 1999, pp 131–138

36. Akula S, Devisetty V (2004) Image based registration and authentication system. In: Proceedings of midwest instruction and computing symposium, 2004

37. Sobrado L Birget JC (2002) Graphical passwords, The rutgers scholar, An electronic Dashin for undergraduate research, 4, 2002

38. Man S, Hong D, Mathews M (2003) A shoulder-surfing resistant graphical password scheme. In: Proceedings of international conference on security and management, Las Vegas, NV, 2003

39. Valentine T (1998) An evaluation of the passface personal authentication system, technical report. Goldsmiths College, University of London 1998

40. Valentine T (1999) Memory for passfaces after a long delay, technical report. Goldsmiths College, University of London 1999

41. Davis D, Monrose F, Reiter MK (2004) On user choice in graphical password schemes. In: Proceedings of the 13th usenix security symposium. San Diego, CA, 2004

42. Jansen W (2004) Authenticating mobile device users through image selection, in data security, 2004

43. Jansen W, Gavrila S, Korolev V, Ayers R, Swanstrom R (2003) Picture password: a visual login technique for mobile devices, National Institute of Standards and Technology Interagency Report NISTIR 7030, 2003

44. Jansen WA (2003) Authenticating users on handheld devices. In: Proceedings of Canadian information technology security symposium, 2003

45. Takada T, Koike H, Awase-E: Image-based authentication for mobile phones using user's favorite images. In: Human-computer interaction with mobile devices and services, Springer-Verlag GmbH, 2003, 2795:347–351

46. Jermynv I, Mayer A, Monrose F, Reiter MK,. Rubin AD (1999) The design and analysis of graphical passwords. In: Proceedings of the 8th USENIX security symposium, 1999

47. Thorpe J, van Oorschot PC (2004) Graphical dictionaries and the memorable space of graphical passwords. In: Proceedings of the 13th USENIX security symposium. San Deigo, USA: USENIX, 2004

48. Thorpe J, van Oorschot PC (2004) Towards secure design choices for implementing graphical passwords. In: Proceedings of the 20th annual computer security applications conference. Tucson, Arizona, 2004

49. Goldberg J, Hagman J, Sazawal V (2002) Doodling our way to better authentication. In: Presented at proceedings of human factors in computing systems (CHI), Minneapolis, Minnesota, USA, 2002

50. Nali D, Thorpe J (2004) Analyzing user choice in graphical passwords, technical report. School of Information Technology and Engineering, University of Ottawa, Canada May 27 2004

51. Syukri AF, Okamoto E, Mambo M (1998) A user identification system using signature written with mouse. In: Third Australasian conference on information security and privacy (ACISP): Springer-Verlag Lecture Notes in Computer Science (1438), 1998, pp 403–441

52. Blonder GE (1996) Graphical passwords, in Lucent Technologies, Inc., Murray Hill, NJ, U. S. Patent, Ed. United States, 1996

53. PassLogix. ▶ http://www.oracle.com/us/corporate/Acquisitions/passlogix/passlogix-general-presentation-189464.pdf Date Accessed: September 01, 2016

54. Paulson LD (2002) Taking a graphical approach to the password. Computer 35:19

55. Wiedenbeck S, Waters J, Birget JC, Brodskiy A, Memon N (2005) Authentication using graphical passwords: basic results, In: Human-Computer Interaction International. Las Vegas, NV

56. Wiedenbeck S, Waters J, Birget JC, Brodskiy A, Memon N (2005) Authentication using graphical passwords: effects of tolerance and image choice. In: Proceedings of the 2005 symposium on usable privacy and security, ACM, 2005, pp 1–12

57. Wiedenbeck S, Waters J, Birget JC, Brodskiy A, Memon N (2005) PassPoints: design and longitudinal evaluation of a graphical password system. Int J Hum Comput Stud 63(1):102–127

58. Birget JC, Hong D, Memon N. (2003) Robust discretization, with an application to graphical passwords, Cryptology ePrint archive 2003

59. Hong D, Man S, Hawes B, Mathews M, A password scheme strongly resistant to spyware. In: Proceedings of international conference on security and management. Las Vergas, NV, 2004

60. Passfaces: two factor authentication for the enterprise. Available at "▶ www.realuser.com," last accessed in March 2017

Multi-Factor Authentication

More secure approach towards authenticating individuals

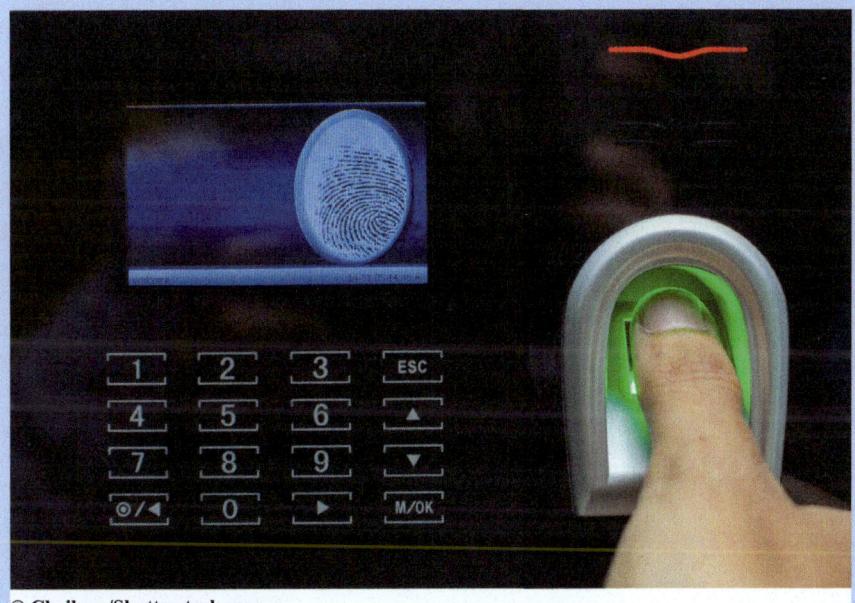

© Chaikom/Shutterstock.com

© Springer International Publishing AG 2017

D. Dasgupta et al., *Advances in User Authentication*, Infosys Science Foundation Series,

DOI 10.1007/978-3-319-58808-7_5

Introduction

Multifactor authentication (MFA) is a secure process of authentication which requires more than one authentication technique chosen from independent categories of credentials. Like single factor, multifactor is increasingly used to verify the users' identities in accessing the cyber system and information. MFA combines two or more types of authentication to provide better and secure way of authenticating users. As mentioned in ▶ Chap. 1, the most common four types of authentication factors are:

- What the user knows—usually the cognitive information of the users (example: passwords)
- What the user has—usually the items that a user possesses (example: smart cards)
- What the user is—a user's physiological and biometric traits (example: face, fingerprint, and voice)
- Where the user is—a user's location information (example: GPS, IP address)

The goal of MFA is to validate legitimate users to secure their sensitive information by providing a layered defense and at the same time, make it harder for unauthorized individuals to gain access. The benefits of adopting MFA are to provide a resilient way of authenticating users. Due to compromise or data breach, any factor is not usable at any point of time; the system still can provide authentication service using the non-compromised authentication factors. From the attackers' perspective, they have to overcome multiple barriers to break into a target system.

◘ Figure 5.1 shows a process of MFA to access a secure facility, where a legitimate user can gain access through providing multiple factors at the entrance of the secure location. On the other hand, an intruder having the password only gets the password matched at the entrance; but with the verification of other factors, the person cannot enter the secure facility (shown in ◘ Fig. 5.2). These two figures illustrate the usage of MFA to provide a secure access to a facility.

The current trend of moving data servers to cloud-based platforms, strong service is now moved from «nice to have» to an «absolute necessary» technology, and MFA is considered as one of the most cost-effective mechanisms [1, 2].

One of the benefits of MFA is narrated in the following example. Bill works for a well-known bank which uses passwords and fingerprint verification to initiate and accept bank transfers. He recently transferred money from a customer's

Fig. 5.1 A legitimate user is authenticating using multiple authentication factors to access a secure area. © aurielaki/Shutterstock.com

Fig. 5.2 An intruder can easily be caught using multifactor-based authentication system. © aurielaki/Shutterstock.com

account into his own off-shore account using those credentials. The bank was able to trace the transfer back to him and confront him. He denied the accusation by saying that his password was hacked, but he had no explanation for his fingerprint verification. Hence, MFA supports non-repudiation which provides a robust way to authenticate users.

Issues of Single-Factor Authentication (SFA)

SFA is the approach to secure a system through only one category of credentials. The well-known approach of SFA is the password-based authentication. Some drawbacks of SFA (specifically password-based authentication) are noted as below:

1. Users need to remember the authentication factor to access different services. If the same password is used for multiple accounts, there is an inherent risk of compromisation of all these applications due to breach of the password.
2. In the case of using different passwords fall on the users and also password fatigue is observed where only one password is used for authentication by some users.
3. If this factor is compromised or breached, then the user cannot access the service until the system is repaired. This incurs a significant delay in accessing necessary information at the time of need.
4. If a single factor is compromised without the knowledge of the actual user, its effect can be devastating.

Hence, moving towards MFA approach is an urgent need in today's cyber system.

Two-Factor-Based Authentication

Two-factor-based authentication is a way of authenticating the legitimate users' through two different authentication factors; mostly from separate categories of credentials. For example, a person can be verified using his/her own password and a hardware token. ▪ Figure 5.3 shows an example of authentication using two-factors. In all ATM booths, two authentication factors are needed. First one is the ATM card (need to swipe the card) and then PIN code needs to be entered. The combination of two factors makes it difficult for intruders to access the bank account. They need to possess

☐ **Fig. 5.3** Two-factor authentication with password and token generated in an external hardware. © Lim Yong Hian/Shutterstock.com

the physical card (*what you have* factor) and know the PIN code (what you know factor).

After a series of incidents and events of password hacking, LinkedIn, Twitter, Evernote, etc., adopted two-factor authentication and expected decrease in incidents like online identity theft and online financial fraud events. In particular, the soft token is introduced as a second factor in two-factor authentication. The soft token is sent through a smartphone application, SMS, or automated phone call. Several web services provide two-factor-based authentication. Some of them are Celestix's HOTPin, Microsoft's PhoneFactor, RSA's Authentication Manager, Google Authenticator, Apple's iOS, iTunes, etc. ☐ Figure 5.4 shows a scenario when a user has to provide a soft token that comes through SMS to his mobile phone.

In addition to the soft tokens, sometimes challenge-response questions are asked as the second factor to verify users' identities. In some cases, image response confirmation of previously saved image is also considered as the second factor.

Two factor Used in Mobile Authentication

All the smartphones in use today (for example: Apple iOS, Google Android, Microsoft Windows, Blackberry OS 10) have different apps that use two-factor-based authentication to identify the owner of that device. Some mobile devices can verify the fingerprint, the front facing camera to identify the user's face, and sometimes the microphone to verify the voice of the user. Along with that the passwords, pins, patterns are also used to protect the devices from fraudulent

usage. Some phone apps need GPS information to track the location of the user, and that can be considered as another authentication factor to identify legitimate users. Some apps provide one-time password tokens that are sent to the phone through SMS, which verify the user identity with the provided token and the possession of the device at the same time.

For instance, Google Authenticator is a two-factor authentication application used in the smartphone and available in Google play store and Apple App Store. To access different websites or web services, a user needs to type his/her username and password at first. Then the user needs to type a six to eight digit one-time password (OTP) which was delivered to his device in response to his login attempt. In general, the OTP is valid for 30–60 s and one needs to attempt for another OTP if the timeout occurs. This OTP provides a second factor toward two-factor authentication.

Although two-factor is apparently more secure that single factor (using a password), it is always a good practice to use a stronger form of passwords whenever required. Using two-factor authentication does not imply that a user can use a weak and easily guessable password as another factor.

Different Authentication Factors Considered in MFA

An authentication factor is defined as a category of a credential used to identify the legitimate users to access a service or system. In MFA, each authentication factor increases the level of assurance in different categories of users [1]. There are basically four to five factors, which are described in ▶ Chap. 1:

Knowledge factor—information that a user should provide at the time of authentication. Examples include passwords, PIN, answers of secret questions.

Possession factor—object that a user should have in his/her possession at the time of authentication. Examples include security token, driving license, one-time password (OTP), etc.

Inherence factor—biological trait a user has to provide at the time of authentication. Examples include face, fingerprint, voice, etc.

Location factor—current location of a user at the time of authentication. As most of the users have a smartphone with GPS built-in, it can serve as the fourth factor of authentication of a user. Most mobile applications use GPS locations to provide a customized response to the users.

Time factor—present time when authentication is taking place can be utilized as a separate factor to authenticate the user in real time. Verification of employees' identification in different office hours can prevent many kinds of serious data breaches. The time factor can easily prevent online banking fraud events to a great extent.

Two or more combinations of the factors mentioned above (i.e. MFA) can provide a robust and secure authentication to the end users.

Necessity to Move Toward MFA

Multifactor authentication is considered as the best practice nowadays to protect users' information from fraudulent access. The following points highlight the need of MFA in the current industries [3]:

1. Millions of records are exposed by hackers every day. Losses range anywhere from $16 billion dollars to 12.7 million US consumers in 2014 [4].
2. Identity theft is the fastest growing crime in recent years. In 2014, out of 2.5 million complaints, more than 70% of them were related to frauds and identity theft [5].
3. People are using smartphones to access their emails, financial transactions, and social communications. Hence, the incorporation of smartphone-based authentication

mechanism is the utmost need to facilitate the users to access their mission-critical online services.

4. In recent online cyberattacks, both enterprises and individuals are affected by the loss of their confidential information and go through financial loss. The only way to protect individuals from identity theft is to increase the way they authenticate in online systems.

5. Cyber criminals not only steal the confidential information, but also delete and modify the other infrastructure of the programs and services. They inject malicious code into the system and make the system stalemate for a longer period.

6. With the recent increment of usage of computing systems, phishing, pharming, keyloggers, and other approaches are also substantially scaled to steal sensitive passwords from the users.

7. Dynamic authentication techniques provide a continuous way of protecting users' identity and prevent major security breaches. This type of authentication is only possible with multiple layers of verifying users' credentials.

According to SafeNet Survey [6], the percentage of organizations embracing the MFA to identify the users will increase by 87% in 2016. Similarly, businesses that have less than 10% of their workforce using MFA are expected to decline more than one-third by 2016.

Different Characteristics of a Good MFA

A good MFA should follow the following characteristics.

Easy to Use

Any MFA should be simple to set up and easy to use. It should be easily integrated with any existing system, i.e., it can be set up with just a few simple clicks.

Scalable

The MFA should support all the major operating systems, file systems, devices, media, and various authentication modalities. It will rather reward if it supports cloud platform.

Resilient

The MFA should provide strong authentication using the highest industry standards. If any of the authentication modality is compromised, the system should be capable of authenticating users with other available non-compromised authentication modalities.

Reliable

The MFA must be reliable in authenticating legitimate users. The authentication decisions of MFA must provide correct authentication decision in all operating conditions.

Single Sign-On

In modern days most of the MFA products support Single Sign-On (SSO). SSO is an authentication process that allows a user to use the same user-id and password credentials to log on to multiple applications within an organization. This process authenticates a legitimate user for all the applications for which he/she has the appropriate right to access the services. It thus eliminates the need for further attempt when the user switches applications in an ongoing session.

Single sign-on reduces the chance of human error which is a primary component of system failure. Social Sign-on is the most common form of Single Sign-on found in daily life, which uses a centralized account and allows you to use it to log in an application. Standard forms of this are Facebook, Google, and Twitter. More and more services are currently allowing users to sign in with their social media accounts.

According to a statistic in Meditech's system, the following information shows how SSO is beneficial in terms of saving time [7]:

- 2 s minimum saved per login
- 70,000 average logins per week
- 140,000 minimum seconds saved per week
- 2333 minimum minutes saved per week (roughly 39 h)
- 2022 minimum hours saved per year (roughly 84 days).

Advantages of Using SSO

The reasons for the companies to adopt SSO are highlighted below [8]:

Reduce Help Desk Maintenance Costs: SSO saves the end users to remember a large list of passwords. As most of the IT help desk call are related to resetting the password (according to Gartner estimate [9], 20–50% of all help desks calls are related to password resets), incorporation of SSO reduces the number of calls. According to Forrester estimate [9], the average help desk labor cost for a password reset is about $70, and it takes around 40 min. This seems more time consuming and more expensive in terms of money. Again this price will increase with the total number of the users for the service. As the users are accessing 5–20 applications per day and if every application needs separate passwords to access, the condition of resetting password is worse. Hence, the inclusion of SSO helps the organizations to reduce the user burden to remember passwords and also economically benefit to reduce the overall cost of help desk maintenance.

Improve the level of customer satisfaction: SSO is designed to excel user satisfaction during the process of authentication to make it simple and faster than typical methods. A user-friendly login process is mandatory to the users with SSO as they use their credentials to login to multiple web services at the same time. Recently, most of the sites offering SSO are social networking websites.

Boost the productivity of the organization: In an enterprise, a user needs to put the passwords correct in first three/five tries (mostly that is the frequency of a number of tries allowed) to access the services. In the case of failure, he/she needs to call the help desk to reset the password, and until that is done, he is on hold to access the service. SSO brings the productivity of the organization as the users are required to remember only one password which significantly reduces the time for login and also a chance of failed login.

Improve compliance and security proficiencies: As SSO allows the users to remember a single password to access multiple web services, it is easier to recommend security policies to use strong passwords and store the password in a secure way.

Assist Business-to-business (B2B) collaboration: With the use of SSO, businesses can centralize the authentication management and make the users to login to the central system once and can access all the available applications with the same credentials. This will allow the users to perform their duties in a quicker way and can increase the productivity of the organization.

SSO Categories

According to Peterson [10], there are five categories of SSO:

1. Password Synchronization (same sign-on):
   This approach needs an individual to login each application that he needs to access but every time he enters the same password regardless of the application.
2. Password Relay (aka enterprise SSO):
   It is a flexible design where the password is stored securely across different systems, and it populates automatically at the time of user's login.
3. True SSO:
   In this approach, a single login and credential are used to login additional systems and applications. Microsoft's active directory (AD) is the example of true SSO. Microsoft has now optimized its environment for online applications including Office 365 and SharePoint.
4. Federation:
   In this approach, an authentication request coming from outside of the organization is trusted in order to access application or data. It uses standards such as SAML or Windows Federated Authentication to create a relationship of confidence between two parties or systems.
5. Web SSO:
   This is a similar type of single login with the access ability on the web regardless of the hosting environment.

Implementing SSO

Peterson [10] has made four recommendations for the organizations to implement SSO:

1. First, it is needed to sort out which systems and/or applications require the immediate access services and then choose the right SSO approach for those services.
2. It is needed to leverage the existing technologies of the company and use them if possible.
3. If a company goes for federation-type SSO, always consider the big picture which can be needed shortly to choose the appropriate SSO method.
4. To remember the fact that SSO is a process of authenticating users with ease but it is not an ultimate solution. This approach may be required to change as new technologies are coming.

Another important consideration regarding the implementation of SSO is to decide which department within an organization will take responsibility for the underlying technology.

Challenges with SSO

Although SSO comes with a set of advantages to the end users, it also embraces some challenges throughout the process.

One of the biggest problems with SSO is that data that identifies a particular user is not always the same throughout the enterprise [11]. For instance, user *Robert Marshall* might be identified as *Robert Marshall* in one application, *R. Marshall* in the second one and *Marshall, Robert* in the third one. It is possible to identify a person based on his last name, job title, and job description. But there is a chance that more than one person with the same name and title exists in an organization.

A global profile needs to be created for every employee that incorporates additional attributes of the users. These attributes are the location of the office, the supervisor name, authorizations to access certain type of data, and unique ID for the organization.

One major consideration for SSO is that if the single point of entry is somehow compromised, it will compromise all the personal and commercial data contained in other connected systems as well. Hence, it is a crucial requirement to secure the login process as secure as possible for SSO. Enabling multiple factors during logon to verify credentials can increase the security of SSO and prevent compromise of identity.

Fast Identity Online (FIDO)

FIDO (Fast ID Online) is a group of technology-driven security specifications that address the shortcoming of the current method of authentication and provide a robust way for strong authentication. The FIDO Alliance [12, 13] was formed in 2012 with PayPal, Lenovo, Nok Nok Labs, Validity Sensors, Infineon, and Agnitio as the founding companies. The Alliance was publicly launched in February 2013.

FIDO specifications support multifactor authentication (MFA) and public key cryptography. A great benefit of FIDO-compliant authentication is that users do not require complex passwords, remember complex strong password rules, and or go through the traditional password recovery procedures. FIDO stores personally identifying information (PII) such as biometric data of the users locally on the user's device to protect it. FIDO's local storage of biometrics and other personal identification is intended to ease user concerns about personal data stored on an external server in the cloud.

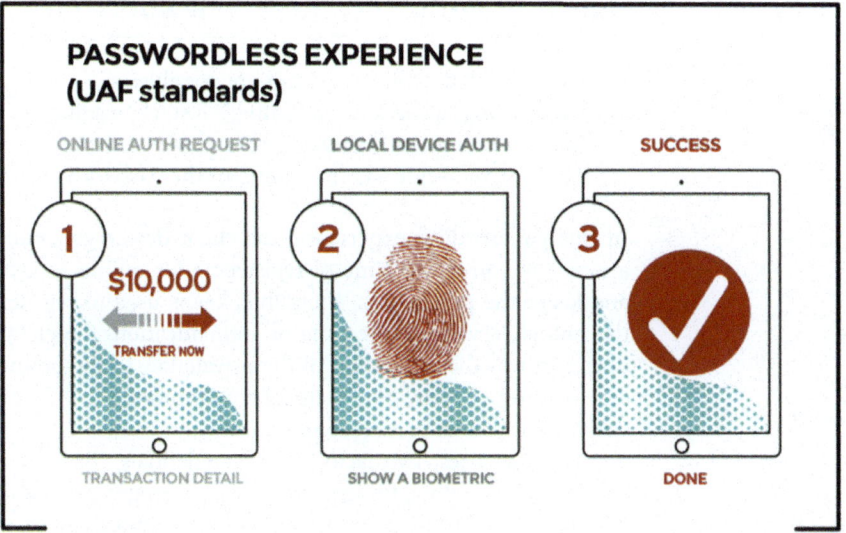

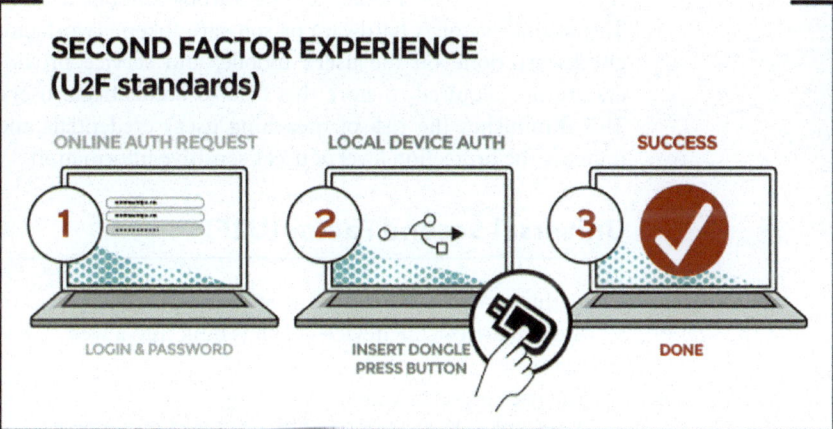

Fig. 5.5 The user experience with FIDO framework [14]

The technical specifications of FIDO highlight a common interface for user authentication on the client device using biometrics, PINs, and two-factor methods. These factors support data privacy and stronger authentication for online services. FIDO supports the Universal Authentication Framework (UAF) protocol (aka passwordless experience) and the Universal Second Factor (U2F) protocol (aka second-factor experience) (See Fig. 5.5). FIDO protocols are based on public key cryptography and are strongly resistant to phishing attacks.

Universal Authentication Framework (UAF)

In this protocol, the following steps are possible:
- A user brings a client device with UAF stack installed
- The user presents a PIN code or biometric
- Website can decide whether it retains the password.

In this passwordless experience, the client device generates a new key pair at the time of registering for online service and keeps the private key. The public key is used to register the online service. At the time of authentication, the client device proves the possession of the private key to the online service through providing a fingerprint, entering a PIN, or speaking as part of voice recognition.

As UAF uses two-factor-based credentials, biometric factors include retina, fingerprint, or voice recognition as the first factor. Due to the incorporation of latest smartphones, it is possible to provide biometric input through mobile devices. The second factor is hardware or software token. Two-factor checks are done on the user's mobile, and service providers are not required to store this data to authenticate users. This diminishes the risk of breaching user's credentials and increase the protection level of user's sensitive information.

Universal Second Factor (U2F)

Using this protocol, the following steps are possible:
- A user carries U2F device which is built into a web browser.
- The user presents U2F device.
- Accessed websites can simplify password if required.

In this second-factor experience, it requires a strong second factor for authentication. For instance, near-field communication (NFC) or USB security tokens. During registration and authentication, a user uses the second factor by pressing a button on a USB device or tapping over NFC. Users can use their U2F devices across different online services that support FIDO framework.

How FIDO Works

FIDO framework combines different vendors under the same platform to use the standardized techniques to provide a stronger authentication. Details regarding FIDO operating principles [14] are briefly mentioned below.

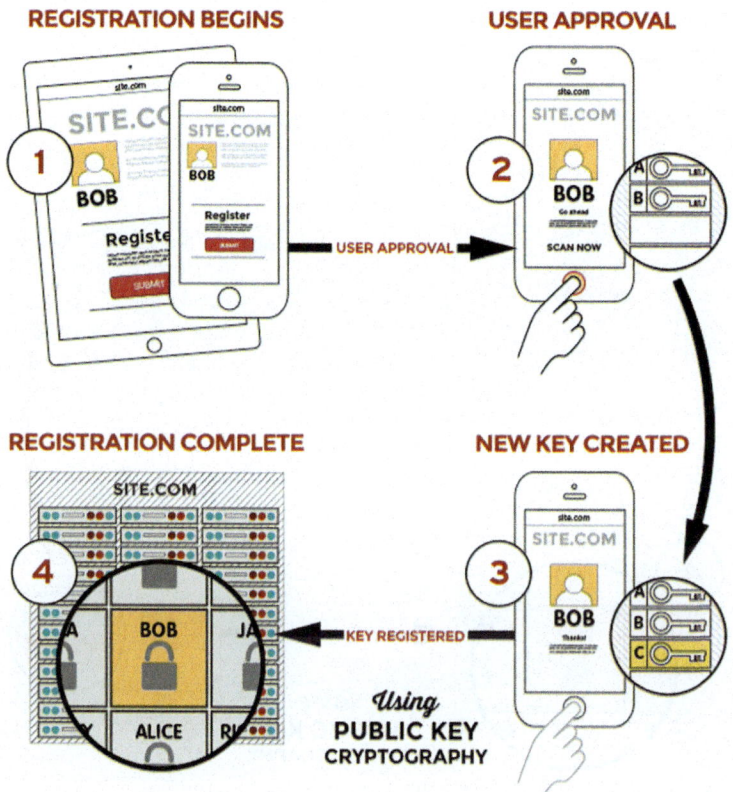

Fig. 5.6 FIDO registration process as mentioned in [14]

FIDO protocols use standard public key cryptography techniques to provide more reliable authentication. FIDO protocols are designed in a way to protect user privacy. They do not share information with other online services and also track a user across other services. On top of that, biometric data do not go out of the user's device.

FIDO Registration

During the registration phase (See ◻ Fig. 5.6), a user can choose any available FIDO authenticator to match with the online service he/she is using. He then unlocks the FIDO authenticator using the fingerprint reader, the button on second-factor device or PIN. The device then creates a new public/private key pair which is unique for that device, online service, and user's active account. The public key is sent to the online service showing the associativity with the user's account. The private key and other authentication-related information do not leave the local device.

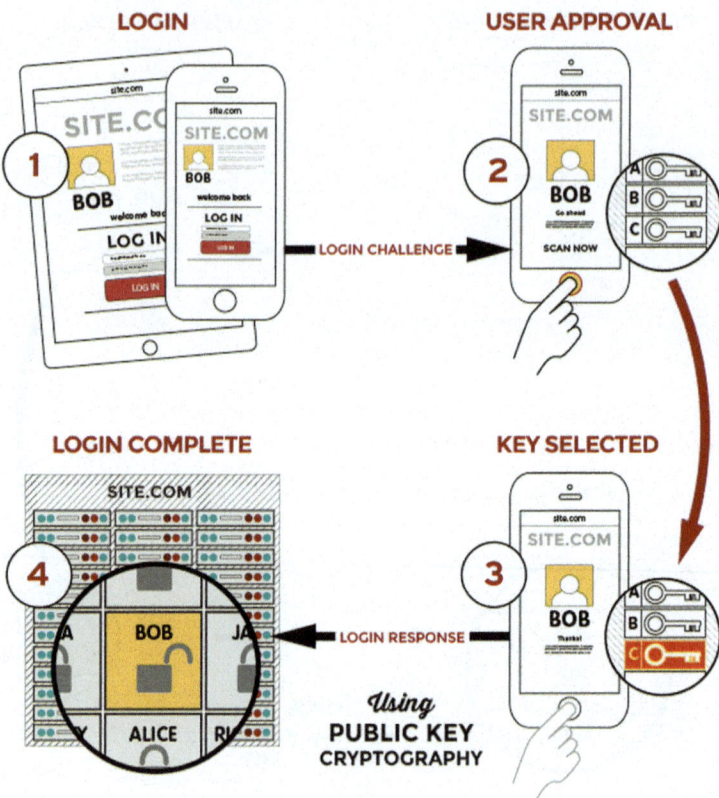

◻ Fig. 5.7   FIDO login process as referred to in [14]

FIDO Login

During the login phase (See ◻ Fig. 5.7), the user is challenged with online service challenges with the previously registered device. The user then unlocks the FIDO authenticator using the same approach done in the registration process. The device then selects the correct key and signs the service's challenge. Client's device sends the signed challenge back to the online service. The service then verifies and allows the user to login into the system.

MFA Authentication Products

As passwords are becoming more insecure and breach over recent years, multifactor tools are now widely used in many organizations and by individuals. They are used for identity management, and now it is common in providing a secure layer for SSO products. The most common MFA products in the market are discussed along with their key features and pricing information.

CA Strong Authentication

CA Strong Authentication (formerly CA AuthMinder) is a multifactor authentication system which helps users to manage a wide range of authentication approaches—passwords and knowledge-based authentication to two-factor-based options (software or hardware tokens). It also supports out-of-band authentication methods such as SMS, email as one-time passwords (OTP).

CA Strong Authentication [15, 16] also supports two-factor authentication to VPNs that facilitate the protection of accessing and doing online transactions from PC, laptops, or mobile phones. This product comes with free mobile applications to use it in smartphones and software development kits (SDKs) for the developers to quickly integrate available CA technologies in their mobile applications.

The CA Strong Authentication server supports a flexible administration console along with unique public key infrastructure (PKI) and OTP software-based credentials. It is integrated with CA Risk Authentication for adaptive, context-aware authentication and CA Single Sign-On to offer a robust solution to the organizations to avoid inappropriate access. Different components of CA Strong Authentication are shown in ◘ Fig. 5.8.

CA supports other products that are similar to Strong Authentication. They are CA Risk Authentication, CA Single Sign-On, and CA Privilege Identity Manager.

This product is licensed on a per user basis, not per credential. Hence, it is considered as one of the low-cost solutions on the market. The organization can issue unlimited credentials to support an unlimited number of users and devices.

Key Features of CA Strong Authentication

The following key features of CA Strong authentication are mentioned here [17].

- It supports passwords and knowledge-based authentication (KBA) as well as two-factor software and hardware tokens.
- It comes with one-time passwords (OTP) as out-of-band authentication and delivers it to the users via voice, text, or email.

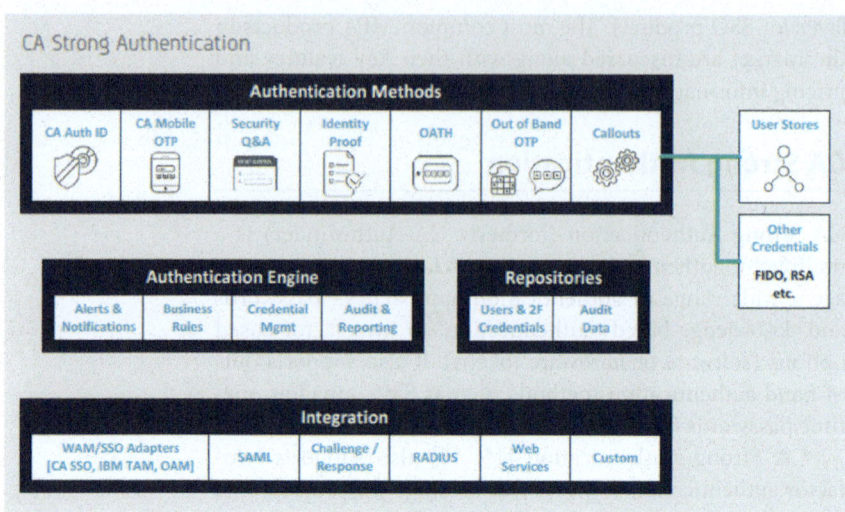

■ **Fig. 5.8** CA Strong Authentication Methods, Authentication Engine, and Integration components [17]

- There is no risk of compromise of passwords as passwords are never stored.
- It is possible to use a workstation, smartphones, or tablets as a second-factor token.
- It supports a good variety of integration options, for instance, SAML, API, and RADIUS.
- It can be used as cloud service, MSP-hosted service, or on-premise solution.

Key Benefits of CA Strong Authentication

- CA Strong Authentication reduces the risk of inappropriate access, any data breaches, and data compromise events.
- This product enhances the level of security without changing the experience of users' login.
- This product reduces the chance of password compromise or steal.
- This product is scalable according to the requirements of the organizations.

Supported Environments

The supported environments in terms of devices, browsers, single sign-on, and different standards are shown in ■ Table 5.1.

■ **Table 5.1**   Supported environments for CA Authentication with examples

Environments	Examples
Devices	Android, iOS, PC, Win8 Mobile
Browsers	Apple Safari, Google Chrome, Microsoft® Internet Explorer®, Mozilla Firefox
Single sign-on	CA single sign-on, IBM® Tivoli® Access Manager, Oracle Access Manager
Standards and protocols	EMV-CAP/DPA, HOTP/TOTP, HTTP/HTTPS, RADIUS, SAML, SOAP.

■ **Table 5.2**   Different features of Okta Verify MFA product

Token types	Voice calls, SMS, mobile app, hardware tokens
Server types	Cloud-based service
MOBILE Phone support	Apple iOS, Android
Authentication methods	Radius, Active Directory, SAML
Pricing	$4/month per user with free thirty-day trial

Okta Verify

Okta Verify provided by Okta is a multifactor authentication product as well as the single sign-on product. The organizations who need both MFA and SSO consider Okta Verify for their users. It is a cloud-based security service.

Some key features of Okta Verify are given in ■ Table 5.2 [18].

Features of Okta Verify

Okta Verify supports Just-In-Time (JIT) [19] provisioning that allows users to create automatic user accounts with one of the following ways:
— Active Directory (AD) Delegated Authentication
— Desktop Single Sign-On
— Inbound SAML

Okta Verify allows the options for the customers to import all their AD accounts and set up correctly so that Okta Verify can authenticate them (customers) with their AD logins and creates their accounts on the fly. This feature is helpful for the organizations that are introducing SSO for a large number of employees at a given time.

This MFA product supports secure authentication features for more than 3000 web and SaaS applications, and it can easily integrate with Radius and AD logins. This product is capable of supporting two AD connectors to the same directory store as a way of allowing redundancy. This helps Okta Verify, as a cloud service, if there is a scenario of a link going down or there is a requirement of having a backup.

«*Okta Mobile*» is a mobile application that allows users to access all their required applications on a mobile device. On the other hand «*Okta Verify*» is a mobile application that allows users to have a second additional factor for doing authentication. Okta offers multifactor authentication and provides a flexible solution to the users. It comes with traditional password and second authentication factor can be selected from SMS, six-digit soft token, or security questions.

Okta Verify is available in iTunes and Google play store from where the users can download and install them. For Windows Mobile users, it is available in Windows Store. After installing, a user needs to login to his Okta Organization on a computer. Then setup procedure is required which varies based on company requirements about their MFA strategies.

Okta Verify comes with a graphical report that allows the users to view last month's app usage and other online activities. It includes how many users signed into the system and which users failed to log in the system over that period.

Okta Modern Factors

The supported factors [18] of Okta Verify application are included here:
- Okta Verify One-Time Passcode
- Okta Verify with Push
- SMS
- Yubikey
- Google Authenticator
- RSA SecureID
- Symantec VIP
- Duo

Okta Verify supports adaptive authentication that uses the factors mentioned above for MFA, and the decisions of the factors depend on the following policies:

- Location
- Group membership
- IP address

Vasco IDENTIKEY Authentication Server 3.8

IDENTIKEY Authentication Server (IAS) is a comprehensive, centralized, and flexible authentication platform that supports MFA software tools and DIGIPASS tokens to provide additional security. This authentication platform supports organizations of all sizes with a state-of-the-art solution for strong authentication.

Product Features

The most significant features of IDENTIKEY Authentication Server (IAS) [20, 21] are listed here:

- DIGIPASS two-factor authentication
- E-signature option to validate data for transaction
- Have the support for EMV-CAP and Hardware Security Module (HSM)
- Have the support of RADIUS and Microsoft IIS web server-based clients
- Have the support of Office365
- Have the support of Internet-hosted applications using SOAP
- Integration of Active Directory and ODBC database
- Virtual DIGIPASS (OTP delivered through SMS or email)

Product Benefits

Some of the potential benefits of IDENTIKEY Authentication Server (IAS) [20, 21] are mentioned here:

- It is easy to implement strong user authentication
- It is robust and scalable as well as expandable with the number of users and applications
- It is designed to meet the requirements of any organization of any size
- It is flexible that supports custom integration
- It is easy to integrate with the existing organizational infrastructure

◻ Table 5.3   IAS Platform Editions [22]

Included authentication clients	Standard edition	Gold edition	Enterprise edition	Banking edition
RADIUS authentication (+WAP)	✓	✓	✓	✓
Webfilters (OWA, CWI, StoreFront, RDWA, Generic)		✓	✓	✓
Desktop windows logon		✓	✓	
SOAP authentication			✓	✓
SOAP e-Signature			Option	✓
EMV-CAP				✓
HSM interface			Option	Option
SBR, ADFS3.0 modules		✓	✓	

- It supports high availability through server replication and load balancing
- It has significantly low cost of ownership
- It is available as appliance or virtual appliance platforms
- It can easily be extended to support Single Sign-on Module

Product Specifications

Product specification for IAS [22] is shown in ◻ Table 5.3.

Supported Environments

The supported environments for IAS [20] are shown in ◻ Table 5.4.

Dell Defender

Dell Defender 8.1 is an MFA product by Dell that supports additional security features along with traditional username–password-based system. It includes biometrics and smartphone-based authentication features. This product utilizes Active Directory (AD) for administration and identity management. The incorporation of AD enhances security and

■ **Table 5.4**   Supported environment for IDENTIKEY authentication server [20]

Operating system (Windows version)	• Windows Server 2008 (64-bit) with SP2, 2008R2 with SP1 (64-bit) • Windows Server 2012 (64-bit), 2012R2 (64-bit) • SBS2008 SP2 + (64-bit), SBS2011 (64-bit) • Windows Server 2012 Essentials (64-bit), 2012R2 Essentials (64-bit)
Operating system (Linux version)	• Suse Linux Enterprise server 11SP3 (64-bit) • Ubuntu Server 12.04.03 LTS, 14.04.02 LTS (64-bit) • Redhat Enterprise Linux version 6.6 (64-bit) • CentOS 6.6 (64-bit)
Supported windows client	• Windows 7 (64 bits) • Windows 8 (64 bits) • Windows 8.1 (64 bits) – Can be used for very small deployments as well as for testing and demonstration purposes
Virtual images	• VMWare ESXi Server version 5.5 • Citrix Xenserver 6.2 • Microsoft Hyper-V (WS2008R2, WS2012, WS2012R2)
Supported webservers	• Apache Tomcat 8.0.21 • IBM WebSphere Applications Server 8.5.5 • Should include Java: JRE8, JSP2, JS2.4
Supported web browsers	• Chrome 41, Firefox ESR31, Internet Explorer 9, 10, 11
Data store (DBMS)	• Oracle 11gR2 (64-bit, Linux, Windows) • Microsoft SQL Server 2008SP3, 2008R2SP2, 2012SP1, 2014 (Windows) • PostgreSQL 9.2.7 (Linux, Windows)
Data store (active directory)	• Windows Server 2008 AD, 2008R2 AD • Windows Server 2012 AD, 2012R2 AD
LDAP back end authentication	• Windows Server 2008 AD, 2008R2 AD • Windows Server 2012 AD, 2012R2 AD • Novell e-Directory 8.8 SP8 • IBM TAM Directory Server 6.3
HSM	• SafeNet ProtectServer Gold, Orange, Express • Thales nShield Connect, Solo (on selected platforms)

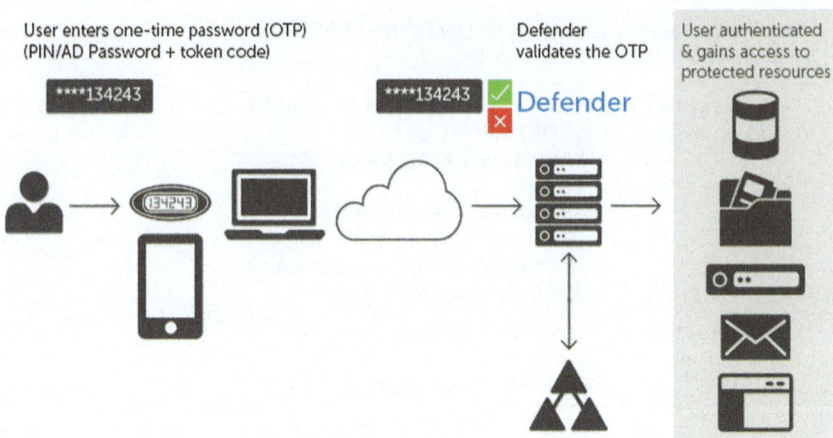

User enters one-time password (OTP)
(PIN/AD Password + token code)

****134243

Defender
validates the OTP

****134243 ✓ Defender
            ✗

User authenticated
& gains access to
protected resources

◻ Fig. 5.9   Dell defender incorporates with the existing username and password-based system and provides two-factor authentication [23]

scalability of the system along with saving money because current employees of an organization can efficiently manage the Defender software.

Moreover, Dell Defender [23, 24] enables the users—

1. request and register the software and hardware tokens easily,
2. reduce the total cost and time for using two-factor-based authentication (shown in ◻ Fig. 5.9),
3. support any OATH-compliant hardware token,
4. allow organizations to increase security and compliance measures in a flexible and cost-effective way.

Benefits of Dell Defender

— It enhances security through strong authentication for a given system or application.
— It leverages the features of Active Directory which provide scalability, security, and compliance.
— It saves time and money by using user token and convenient renewals of user tokens with the expiration of the battery life.
— It comes with the flexibility of standards through supporting any OATH-compliant hardware token.
— It also provides a comprehensive audit trail with the feature of compliance and forensics.

Features of Dell Defender

The following features of Dell Defender makes it a good candidate for two-factor-based MFA [24]:

▪▪ Active Directory-centric

Dell Defender can incorporate with the existing corporate directory of active directory and use the scalability, security, and compliance of Active Directory (AD) for two-factor authentication. Another feature added with AD is user token assignment.

▪▪ Token of self-registration

Dell Defender makes an option for the users to request a hardware or software token. These tokens are based upon policy defined by the system administrators. The request token can easily apply to the users' accounts by a secure way. Hence, this token-based approach reduces the burden of administrative system as well as all sorts of costs related to assigning conventional manual token assignment.

▪▪ Web-based administration

Dell Defender comes with the feature of administering tokens, deployment of tokens, and real-time log viewing using web-based management portal. It helps the Defender administrators, help desk administrators, and end users by providing this useful feature.

▪▪ Token flexibility

Dell Defender supports any OATH-compliant hardware token from any preferred token vendor. It also offers a good number of software tokens from different mobile platforms.

▪▪ Help desk troubleshooter

It has the tremendous flexibility for the administrators to troubleshoot, diagnose, and resolve any authentication-related issues using a web browser. It also supports a viewable list of recent authentication attempts and routes, possible reasons for failures, and possible resolution steps for the administrators. Moreover, the administrator can view the user accounts in details along with the assigned tokens to the users, reset or unlock users' account if required.

▪▪ ZeroIMPACT migration

Dell defender has the option to gradually migrate to Defender from a legacy authentication solution with Zero-IMPACT. This approach allows administrators to migrate users to Dell Defender as their legacy tokens expire.

▪▪ Secure webmail access

Dell Defender has the option to enable secure web-based access to the corporate email system from any web browser.

■■ **Pluggable Authentication Module (PAM)**

Dell Defender has the feature that integrates PAM module with it. It helps the users with different services in the Linux systems.

■■ **Encryption**

Dell Defender can secure communications with the help of management DES with Defender Security Server. It also supports AES or triple DES encryption.

Dell Defender Hardware Tokens

Dell Defender 8.1 supports any OATH-compliant token, and it distributes the following types of tokens [23]:
- Vasco DIGIPASS GO 6 and Vasco DIGIPASS GO 7
- Yubico YubiKey
- Verisign VIP Credentials

Dell Defender Software Tokens

Dell Defender 8.1 supports the following software tokens from different mobile platforms [23]:
- Defender Soft Token for BlackBerry
- Defender Soft Token for iPhone
- Defender Soft Token for Android
- Defender Soft Token for Windows Mobile
- Defender Soft Token for Java
- Defender Soft Token for SMS
- Defender Soft Token for Email
- Defender Soft Token for Windows Desktop
- Defender GrIDsure Web-based token
- Defender Soft Token for Palm

Symantec Validation and ID Protection Service (VIP)

Symantec Validation and ID Protection Service [25–27] is a leading cloud-based strong authentication service that allows the organizations to access their networks and application in a secure manner and prevent all types of malicious unauthorized access. This MFA solution comes with two-factor and risk-based token-less authentication features. It follows the open standards (SAML, OATH) and can be easily integrated with the existing application of the organizations. VIP has recently supported biometrics, which brings an option of no requirement of any passwords.

Features of Symantec VIP

The following features [26] illustrate the utility of the Symantec VIP:

Cloud-based infrastructure—VIP provides secure, reliable, and scalable service to support authentication with no requirement of on-premise server hardware.

Transparent risk-based authentication—VIP comes with the option of device and behavior profiling to prevent risky login attempts without any disturbance of legitimate user's login experience.

Multiple two-factor credential options—VIP deploys one-time password (OTP) credentials that support a good number of hardware, software, or mobile form factors.

Different mobile and desktop device credentials—VIP supports different OS-based systems, namely, Android, iOS, Blackberry, Windows, and J2ME devices.

Support of out-of-band authentication factors—VIP integrates some out-of-band options, such as SMS, voice phone calls, or email.

Provision of self-service credentials—VIP facilitates strong authentication to the customers and makes sure that they do not require IT help desk or administrator intervention.

Support of enterprise infrastructure—VIP integrates the current enterprise VPNs, webmail, SSO applications, and other corporate directories.

Integration of web-based application—VIP comes with the web services API to provide strong authentication in different programming languages.

Benefits of Symantec VIP

Some benefits of incorporating Symantec VIP are mentioned below [26]:

▪▪ Strong Protection

Symantec VIP helps business organizations prevent any unauthorized access to sensitive systems and applications. This prevention is done through intelligent authentication device, behavior profiling, or hardware- or software-based credentials.

▪▪ Reduced costs and complexity

The Symantec VIP cloud-based system allows the enterprises to deploy strong authentication rapidly and quickly without the need of upfront capital expenses.

▪▪ Flexibility to fulfill diverse needs

Symantec VIP balances the usage of strong authentication in terms of cost, convenience, and security. It comes with an excellent variety of authentication options for different users and use cases. The options include one-time password (OTP) credentials as well as risk-based analysis to automatically authenticate known users from their normal login behavior or a combination of the two approaches.

▪▪ Scalable and reliable

The Symantec VIP cloud-based infrastructure supports scalability and reliability to assist different organizations to support millions of users easily and in a cost-effective way. It has the flexibility in its design to support large user bases without the deployment of any additional authentication hardware.

▪▪ Future-proof

The Symantec VIP cloud-based solution comes with new capabilities and integrates with Symantec™ Global Intelligence Network, Symantec™ Endpoint Protection, Norton™, and Intel® Identity Protection Technology (IPT)-enabled computers. Hence, it gives a new edge to the enterprises to stay ahead of new emerging threats.

▪▪ Self-service customer portals

It supports increased customer confidence through addressing security concerns in online service adaptation. It also reduces fraud costs through preventing unauthorized transactions, and thereby diminishes the occurrence of fraud events.

Supported Credential options

Symantec VIP supports the following credential options [26]:
- OTP security token
- Desktop client
- Mobile phone credentials
- OTP security card
- Voice-enabled passcode
- SMS OTP
- OTP flash drive
- Intel IPT

Symantec VIP Access for Mobile

Symantec VIP Access for Mobile [27] is a platform which combines secure access with the convenience of «Bring your own token» on any personal devices. It is a great feature to be noted because the users do not need to purchase any hardware tokens or manage the tokens as inventory. Some of the key benefits of this platform are as follows:

Easy to deploy—a user can download the over-the-air software and run without any hassle or activation codes or IT involvement.

Extensive support—it supports over 850 devices of all the major OS supports.

Globally available—it is a cloud-based service, and hence the up-to-date capabilities are available to mitigate latest security threats.

Cost-effective solution—it is cost-effective as no purchase or distribution cost is required for mobile credentials.

Readily adoptable—it is a good candidate for online transactions to provide convenient, easy to use options.

Supports BYOD—it comes with all sorts of managed and unmanaged devices, the same painless user experience.

RSA SECURID

RSA SecurID solution provides two-factor authentication to about 25,000 organizations and 55 million users [28]. This solution provides an extension of security with the incorporation of BYOD, cloud, mobile, VPN, and web portals as shown in ◘ Fig. 5.10. This solution is composed of three primary components—authenticator, platform, and agents.

In a given two-factor authentication session, a user enters his username and PIN along with current token from his or her RSA SecurID device. The agent then transmits the entered information to the management software that allows the access upon validating the information. The user is granted appropriate access level based on his/her authorization level and this information is saved in RSA Authentication Manager software.

Authenticator

There are three types of authenticators available for RSA SecurID. They are hardware tokens, software tokens, and token-less authentication [29, 30].

RSA SecurID hardware tokens protect application through two-factor authentication, hard disk encryption, email, and transaction signing capabilities using this token. The most common example of hardware tokens is RSA SecurID 700 and RSA SecurID 800. RSA SecurID 800 offers

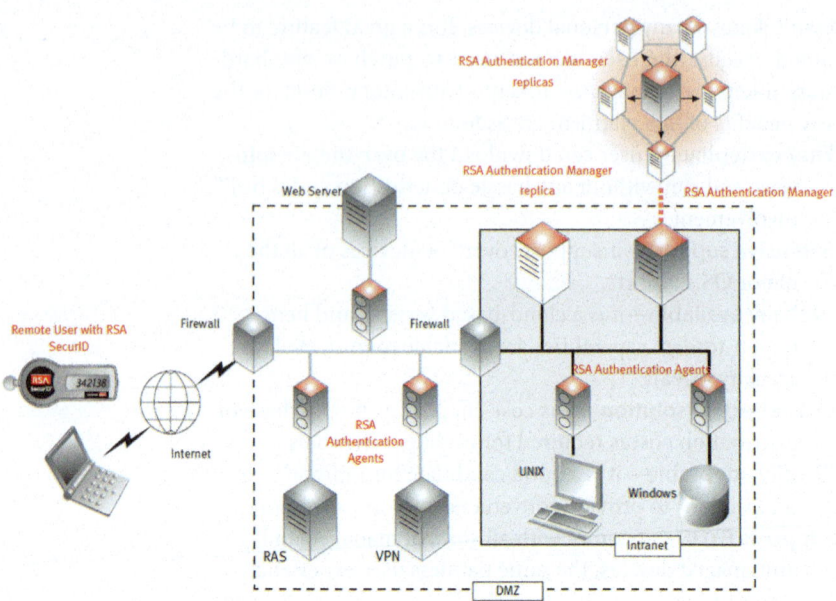

□ **Fig. 5.10**   RSA SecurID different components and how they are connected [58]

one-time password functionality and can be used to store Microsoft Windows password credentials and digital certificates.

RSA SecurID software tokens [30] can make strong authentication to be effective in doing business. It is very convenient to deploy them on mobile devices and transform them into an intelligent security token. It is convenient to the end users to reduce the number of devices that they need to manage to gain secure access to corporate network. This type of token can be reissued if the user leaves the organization. Again, software tokens support a good number of widely deployed mobile platforms to be used.

RSA SecurID Risk-Based Authentication [31] is considered as a token-less, MFA solution which calculates the risk score at the time of providing password credentials. It uses username–passwords, some unique device identification, behavioral profiling of the user to calculate the risk score.

Platform

RSA Authentication Manager is the well-known platform for RSA SecurID to manage RSA SecurID environment which includes authentication methods, applications, users, etc. It supports the following features [32]:

- Centralized Management—provides streamlined management with the browser-based administration console.
- Authentication Choice—supports a good number of authentication methods, namely, hardware tokens, software tokens, and on-demand SMS authentication.
- Scalability—supports high availability replication infrastructure for an organization.
- Interoperability—has partnered with 400 industry-leading solutions including VPN, firewall, and web applications.
- Platform flexibility—supports widely used operating system platforms including VMWare and Microsoft virtual environments.

Agents

RSA SecurID Authentication Agents are the software that does the interception of remote access and local requests from different users and user-groups and sends them to the RSA authentication manager server to do the authentication. This agent saves time and integration costs by deploying different tested and RSA-secured proven solutions. The agent software is embedded into different remote access servers, VPNs, and various wireless devices. More importantly, there is no cost associated with integrating RSA agents and partner agents.

SafeNet Authentication Service

SafeNet Authentication Service (SAS) [33–35] is an MFA solution that is fully automated and highly secure authentication-as-a-service. The goal for SafeNet Authentication is to bring two-factor authentication universally available. It comes with flexible token options to meet the need of the different organizations as well as the total cost of operation.

◘ Figure 5.11 shows how the SAS is connected with cloud applications, on-Premises applications, and administrators.

Features of SafeNet Authentication Service

The main benefits of SafeNet Authentication Service [34] are given below:
Migrate fast—SAS solution comes with a free migration agent that helps the organizations to move quickly with the existing technology to SAS.

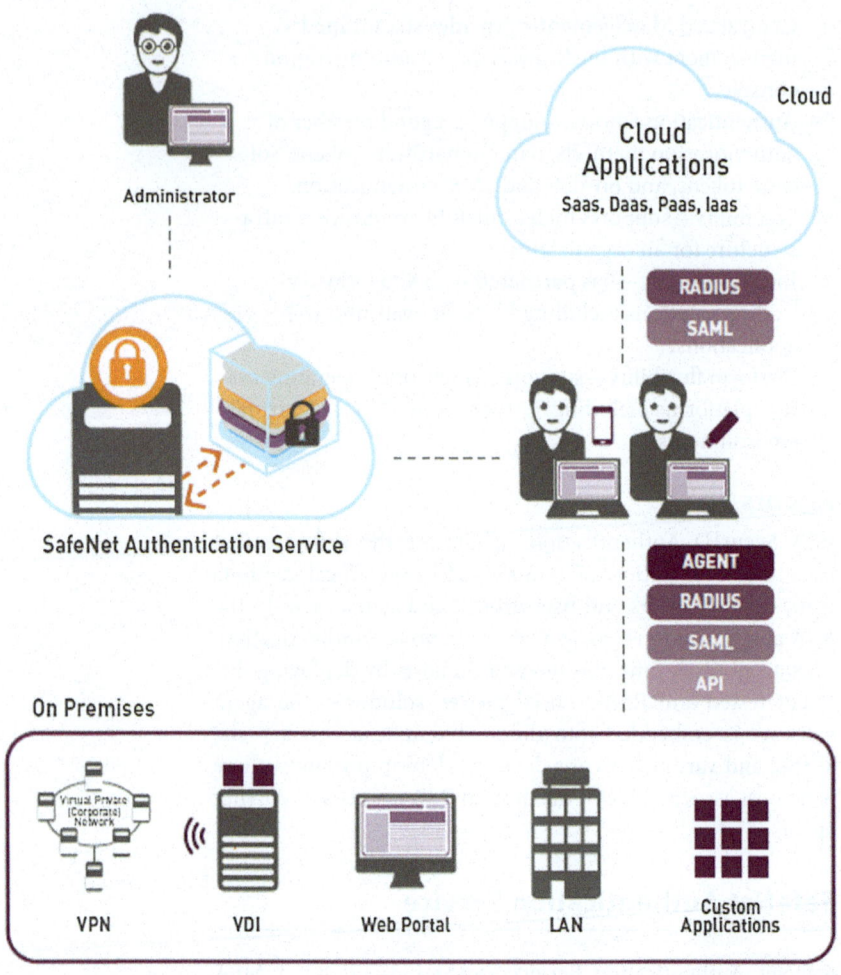

■ **Fig. 5.11** SafeNet authentication service is connected with cloud applications, on-premises, and administrators [34]

Deploy quickly—SAS has the flexibility to allow 20 k users in 20 min.

Automate easily—SAS supports the addition of the users from different user directories, for example, LDAP, Oracle, SQL, and others.

Supporting more options—SAS provides hardware, software, mobile, SMS, and grid-based tokens.

Grow infinitely—SAS can scale an unlimited number of users.

Comply thoroughly—SAS can easily schedule automated compliance, audit, and accounting reports in different formats. These reports follow security standards including SOX, PCI, and HIPAA.

Protect everything—SAS provides flexibility to work across technologies, both on-site and in the cloud. Cloud and web applications are protected with the usage of SAML.

Benefits of SafeNet Authentication Service

Low total cost of operation—it has per user pricing model and no hidden fees involved. The cloud platform reduces the help desk expenses as well as lowers the management time by around 90%. It also supports large-scale automation, user provisioning, and user self-enrollment to ease the tasks of the users.

Quick Migration to cloud—it supports a smooth transition from the different third-party authentication server. It also maintains lower operational cost and reduces administration and management overheads.

Peace of mind—it brings robustness, availability, and protection of the cloud environment and hence brings peace of mind to the enterprises.

SafeNet Authentication Service adopts a price-per-user model for all the authentication products. It reduces the administrative costs through facilitating automated self-provisioning. In case a user loses his token, the system can easily re-enroll a new one in a short time. Also, it offers the service free for 30 days.

According to the recent solution brief [35], this MFA solution provides up to 60% savings in the total cost of operation and 99.999% service availability, and up to 90% reduction in administrative overhead costs.

SecureAuth IdP 8.0

SecureAuth IdP [36] is an identity provider and a unique approach in providing secure user access to the organization. It maintains the secure user access of the resources and data in the cloud, web, using mobile devices or through VPN. This product supports two-factor authentication, adaptive authentication, single sign-on, user self-service, authentication API, and so on. The details are described below.

Two-factor Authentication

SecureAuth IdP supports 20+ authentication methods and provides flexible and scalable two-factor authentication. The list of authentication methods [37, 38] is shown in ▪ Table 5.5.

■ **Table 5.5** 20+ Authentication Methods of SecureAuth IdP 8.0 [37]

SMS OTP	Browser OATH token
Telephony OTP	Windows desktop OATH token
Email OTP	Third-party OATH token
Static PIN	Device fingerprinting
KBA/KBQ	Yubikey (USB)
X.509 native	Password
X.509 Java	Social IDs
CAC	Federated IDs
PIV card	Kerberos
Push notification	Help desk
Mobile OATH token	

The flexible option of two-factor authentication can be used in mobile or desktop device and can be easily configured without requiring any third-party tools.

Adaptive Authentication

On top of the two-factor functionality, SecureAuth IdP comes with adaptive authentication features to increase the security profile of the users. It provides dynamic and risk-based authentication as part of adaptive authentication. With the help of live threat intelligence from Norse DarkMatter™ platform [38], IdP does multifactor IP reputation and risk analysis, which allows the administrators to apply some risk thresholds to the process of user authentication. It can check the IP address for the white and black list as well as ongoing device fingerprint analysis to determine legitimate authentication attempts in real time and adopt appropriate actions accordingly.

Single Sign-on

SecureAuth IdP also supports one password and the unlimited access rule. It provides Single Sign-on feature [39] without any third-party programs or thick clients. This solution supports SSO for web applications, cloud, and mobile apps without additional coding requirements. A user is at first authenticated with the multifactor system provided by SecureAuth IdP. Then it creates an SSO token for the relying party, which can be based on the web, network, cloud, or mobile. Moreover, this solution can authorize SSO into other applications to obtain seamless access.

SecureAuth IdP integrates with the popular SaaS apps [39] including Salesforce, Google Apps, WebEx, SuccessFactors, Workday, etc. In the case of web apps, it includes .NET, J2EE, SharePoint, WebSphere, and much more. More mobile applications [40], iOS, android, windows, and blackberry-based OS are supported. SecureAuth Idp can be easily integrated with Office 365 [41] and provides SSO supports to the application. It also integrates with Active Directory for internal users of Office 365.

User Self-service

SecureAuth IdP provides the option to the users to reset their own passwords and unlock their accounts with well-known two-factor authentication process within seconds. This feature offers the users more flexibility in managing their accounts and reduces the help desk calls for resetting the passwords, which eventually saves cost.

Authentication API

SecureAuth IdP's Authentication API allows the enterprises to embed IdP access control functionality into a custom, in-house built applications to deliver strong and secure authentication. This solution supports the feature to enable risk analysis and two-factor authentication methods and allows the users to unique logins and interfaces that are native to their applications. The API currently supports the following two-factor authentication methods [42]: Phone OTP, OATH OTP, SMS OTP, Hard Token, Email OTP, KBA Verification, Push OTP, and Help Desk (to send the second factor).

Microsoft Azure

Microsoft Azure provides multifactor authentication [43] to verify the users' identity through using factors such as something you know, something you have, and something you are. It helps in securing access to the data and applications and at the same time, it meets user demands by simplifying the login process. This solution provides strong authentication with a good number of verification options—phone call, text message or mobile app verification code, and third-party OATH tokens. Microsoft Azure supports the following benefits to provide secure authentication:

Easy to use—Azure MFA product is easy to set up and use. From users' perspective, the setup can be done with few simple clicks, and it provides additional protection to the data and resources. It also protects the most common applications like Office 365, Salesforce DropBox, and other SaaS applications.

Scalable—Azure MFA product utilizes the cloud service, and it integrates with on-premises Active Directory and different custom applications. This secure access feature can be extended to the mission-critical resources as well. It is also used with VPN, Microsoft IIS, RADIUS, and LDAP.

Always protected—Azure MFA solution comes with real-time fraud monitoring and alerts which help the users and administrators to quickly discover any irregularities of the login attempts and address accordingly.

Reliable—Azure MFA guarantees 99.9% availability for its service in the cloud, mobile applications.

Azure MFA is available as a stand-alone service, and it comes with the following features [44]:

- Protect users of Azure Active Directory, Windows Server Active Directory, LDAP directory, and other users
- Protect applications including cloud and on-premises applications
- Provide security for non-browser clients using application passwords who do not support web-based authentication
- Support advanced features like custom voice greetings, configurable caller ID, trusted IP addresses, etc.
- Support advanced security features like fraud alert, one-time bypass, and block users
- Generate usage reports
- Provide on-premises protection using SDK

Pricing of Azure MFA

The pricing of Azure MFA [44] comes with per user basis or per authentication basis, when it is used as stand-alone service. Per user billing model, the price is $1.40 per month (unlimited authentications), whereas per authentication model, the price is $1.40 per 10 authentications. Azure technical support comes with $29/month, and its billing and subscription management support do not read any further cost.

Azure MFA solution comes in different versions [45]. Azure AD free and Azure AD Basic support MFA are for Azure administrators. Azure AD Premium adds MFA support for users. With the incorporation of multiple layers of protection, it is hard for the attackers to compromise and access the accounts of the users.

Azure AD supports three options for SSO [45]:

- Azure AD Single Sign-on
- Password Single Sign-on
- Existing Single Sign-on

Swivel Secure

Swivel Secure [46, 47] is a two-factor authentication product which has been used in different cutting-edge IT security companies over 30 countries. It comes with the approach that provides a choice of authentication solutions which can be tailored to the requirements of the companies. It comes with different types of authentication methods. Some of them are listed below.

Risk-based Authentication

Swivel comes with a policy engine that makes a decision based on a number of factors to calculate the significant risk factors. These factors are [48]:

- Who the user is
- What service the user tries to access
- What IP address the user is accessing
- The time of accessing the service
- The device used by the users
- The time when the user is last authenticated

This policy engine also supports Single Sign-on feature using the existing browser session. Moreover, this engine is extendable and new policies can be added at any given time. It supports cloud services (SAML), web applications (via Swivel. NET plugin).

SMS-based authentication

The Swivel authentication platform supports the feature to send one-time codes (OTCs) to the users via SMS [49]. The OTC is generally delivered to the users using other channels and security of those channels is maintained. There are several forms of these SMS messages. The first one is advanced SMS that is sent immediately after an authentication attempt, and it is only valid for the next authentication attempt. SMS on demand provides OTC at the time of authentication. SMS inbound will be delivered to a known SMS number provided by the user. This option provides maximum protections against phishing, key-logging, and man-in-the-middle attacks.

Mobile app-based authentication

Swivel MFA product supports two-factor authentication using a mobile device to access remote environments including VPNs, websites, cloud, and virtual desktops [50]. It is available in all the major app stores, and a user can authenticate with one touch of phone's screen. Once the user has provided the username and passwords into the web login screen, a push notification from Swivel is sent to him/her. The user needs to select yes to do the authentication or no if he is not interested in login at that time.

Token-based authentication

Swivel Secure also comes with token-based authentication [51]. The supported tokens are OATH TOTP (time-based) or HOTP (event-based). To do authentication, user has to press the button on the token and an OTC is then displayed to the user. The user enters this OTC on the login page to complete the authentication process. The user's PIN can be combined with the OTC if required.

Customers who require banking-level security and authorization of financial transactions, the Swivel platform also supports the OATH Challenge-Response Algorithm (OCRA). In this scenario, a user enters a specified challenge into an OCRA token (for instance, destination account number) and then enters the required response into the authorization dialog on the transaction screen.

Swivel Secure also supports PINpad [52] and Telephony [53] as part of authentication factors.

DUO Security

DUO security [54, 55] provides two-factor authentication as the second layer of security to VPN, cloud-based apps and on-premise apps. It offers powerful Web APIs that can be integrated with any website for login or transaction verification. A good number of education companies embrace DUO as part of their authentication process. It comes with a good list of features that are listed below.

Universal Phone Support

Duo Security supports all types of devices with and without cell service. It supports smartphone, feature phone, or landline that can act as a second factor for authentication. Users are allowed to choose how to authenticate at each login.

It also requires no additional device for users to carry or administrators to manage.

Duo Mobile Push

The most convenient feature of DUO Security is mobile Push option. It can approve or deny the login request using one-tap, and all the transaction details are also pushed to the phone for approval.

Duo Mobile Passcodes

Duo Security also supports one-time password that are generated in the mobile application. For this type, it supports all major mobile platforms including iPhone, Android, Blackberry, and Windows Mobile, Symbian, Palm WebOS, and Java Mobile. One of the good features of the mobile passcode is to allow for operation without cell service or WiFi connection.

SMS Passcodes

DUO also supports one-time passcodes sent in batch via text message. It also works with any mobile phone and no software is required to install to use this feature. These passcodes can be stored as a text message, and they can be used when there is no cellular coverage.

Phone Call

DUO also comes with voice-based callback feature. One of the major advantages of this feature is it works with mobile or landline phone, and it requires no software to install and no memory or typing burden from the user. This feature also comes with primary and secondary phone numbers for the user to provide a diversity of the authentication approaches.

Web Service Integration

DUO has the option to integrate with web SDK using REST API. Using this API, it is possible to integrate easily with any kind of web application. This feature can protect individual transactions and all types of login sessions. Strong SSL encryption support ensures mutual authentication and privacy of all authentication transactions.

In addition, it provides centralized reporting, real-time fraud alerts, and bypass codes to all individual users of DUO security.

Comparison of Various MFA Products

Multifactor-based authentication is a burning need to ensure individuals' identities in the cyber world. The MFA products described in this chapter provide different varieties of customer solutions to meet the current industry demands. A comparison among various MFA products in terms of their features and benefits is listed in the following ◘ Table 5.6.

■■ NIST-Privileged Accounts Guideline:
NIST guideline for the authentication of privileged accounts [56] emphasized the requirement of higher level of assurance (LOA) by checking user's identity via personal identity verification (PIV) in order to reduce unauthorized access to high-value assets. Accordingly, PIV Cards are required to use by the system, network, security and database administrators, and other information technology (IT) personnel with administrative privileges. The PIV can provide multifactor authentication through indirect verification (Relying Party) employing Kerberos network authentication protocol (to protect the confidentiality and integrity of all communications related to privileged user authentication and privileged sessions).

In some cases such as use of mobile devices where biometrics and other direct credential cannot be presented, the derived PIV credential is recommended [57]. Best security practice along with PIV-enabled systems can provide high level of authentication assurance where the degree of assurance (LOA-4 or LOA-3) may depend on the criticality of the privileged access to the system.

Chapter Summary

This chapter describes several security issues of frequently used single-factor authentication systems (password, PIN, etc.), which could be considered as the motivations for using multifactor authentication. Different characteristics of a good MFA system such as simplicity, scalability, resilience, reliability, and SSO are also described in details. FIDO framework for MFA using UAF and U2F protocols are explained. These protocols implement public key cryptography and are resistant to phishing attacks. UAF uses two-factor-based credentials, like PIN, biometric, different hardware, and software tokens, whereas in U2F, a user uses

Table 5.6 Comparison of various MFA products

MFA product	Available platforms	VPN/Active directory support	Supported authentication factors	Supported protocols	Costs
CA strong authentication [15–17]	Mobile and cloud platform	VPN and active directory	Password, KBA, OTP, hardware tokens	RADIUS, SAML, SOAP	100 users cost $2800, no free trials
Okta verify [18, 19, 59]	Mobile and cloud platform	Active directory	Voice calls, SMS, mobile app, hardware tokens	Radius, SAML	$4 per user month with free 30-day trial
Vasco INDENTIKEY authentication server 3.8 [20–22]	Desktop, web browsers	Active Directory	DigiPass Two-factor, E-signature, Virtual DigiPass (OTP)	Radius, SOAP,	For 100 hardware tokens $6573 with 45 days trial
Dell defender 8.1 [23, 24]	Desktop, Smartphone, cloud platform	Active Directory	Hardware and software token	OATH-compliant	100 software tokens cost $35,000. 90 day free trial for one 25-user license
Symantec validation and ID protection service (VIP) [25–27]	Desktop, Mobile and Cloud Platform	Enterprise VPN	OTP security token, OTP security card, SMS OTP, Voice-enabled passcode	SAML, OAUTH	$15 per hardware tokens, 60 day free trials
RSA SECURID [28–32, 58, 60]	Cloud, mobile and web portals	VPN	Hardware token, software token, token-less authentication		For 100 tokens $11,050 and a free 90-day trial for 25 users

(continued)

5

□ Table 5.6   (continued)

MFA product	Available platforms	VPN/Active directory support	Supported authentication factors	Supported protocols	Costs
SafeNet Authentication Service [33–35]	Cloud and web applications	LDAP	Hardware, software, mobile SMS, and grid-based tokens	SAML	Subscription-based, 30-day free trial
SecureAuth IdP 8.0 [36–42]	Cloud and Mobile applications	VPN, Active Directory	20+ authentication methods, adaptive authentication		$19.50 per user year with a 30-day free trial
Microsoft Azure [43–45]	Cloud and mobile applications	VPN, IIS, LDAP	Phone call, text message, mobile app verification code, password	OATH, RADIUS	Per user basis or per authentication basis
Swivel Secure [46–53]	Cloud and web applications	VPN	SMS, TOTP, HOTP, PINpad, telephony	SAML, OATH	On-premise and cloud-based subscription
DUO Security [54, 55]	Cloud and web applications	VPN	OTP, Phone call, Push notifications	REST	Subscription-based

the second factor by pressing a button on a USB device or tapping over NFC. Toward the end, ten well-known and extensively used MFA products, like Microsoft Azure, Dell defender, CA Strong authentication, etc., have also been illustrated to show the current trend of MFA; this also includes a summary comparison. Lastly, NIST guideline on accessing privileged accounts to provide high level of authentication assurance is described.

Review Questions

Descriptive Questions

Question 1:
    Define Multifactor authentication. Which factors are generally taken for MFA?
Question 2:
    What is the issue of Single factor authentication? How can these issues be resolved through MFA?
Question 3:
    What is FIDO? Briefly describe the key features of FIDO.
Question 4:
    Describe the key features of UAF and U2F of FIDO framework. How does FIDO registration work?
Question 5:
    What is single sign-on? Discuss three advantages of SSO.
Question 6:
    What are the major disadvantages of using SSO? Name three different categories of SSO.
Question 7:
    What are the four implementation steps in SSO? Describe the challenges incurred for enabling SSO.
Question 8:
    Discuss any three MFA products which support SSO.
Question 9:
    Discuss any three MFA products which support Mobile devices.
Question 10:
    Compare the listed MFA products based on their features.

Multiple Choice Questions

Question 1:
    What are the benefits of SSO? (Select all that apply)

A. Fewer credentials that a user has to remember.
B. The amount of time that it takes to log into different services.
C. It does not require as much effort to think of different passwords.
D. Single sign-on allows each service to have its own layer of protection.

Question 2:
What is it called when a user uses a service like Facebook to log into their account on a different website?
A. Single Sign-on
B. Multi Sign-on
C. Social Sign-on
D. Super Sign-on

Question 3:
Bob is about to meet Alice in a coffee shop. They will meet during rush hours. Bob wants to check his bank account on his mobile using the available free WIFI connection. The email service provider supports three types of authentication, namely login password, voice recognition, and SMS message as OTP. Which option would be the best for him to choose in this situation?
A. Password
B. Voice recognition
C. SMS with OTP
D. Security Questions

Question 4:
Amanda pricks her finger and steps away from her station for a few minutes to get something to eat because her blood sugar level was low. While she was gone, her station logged her out, and now she has to go through the process of logging back. Her company uses different biometric systems for authenticating their employees. As it was late, most of the employees left and the cleaning crew has already turned off some of the lights, so the lighting around her is not good. What is the best method of authentication for her to use?
A. Voice Recognition
B. Facial Recognition
C. Fingerprint Recognition
D. Weight Recognition

Question 5:
A construction company is looking to add multiple-factor authentication to one of their construction sites that has hundreds of workers. What would be the best combination of authentication for them to implement?
A. Voice Recognition + Username and Password
B. Swipe Card + Facial Recognition
C. Facial Recognition + One-Time Password Generator
D. Photo ID + A Name

Question 6:
Jason is the type of person who does not like to give out his personal information and is overly suspicious of other people. What would be the best authentication type for Jason?
A. Knowledge-based Authentication
B. Token-based Authentication
C. Biometric Authentication

Question 7:
A hotel uses keys for locking and unlocking their doors. They now want to upgrade their system to make their visitors' life easier and to make their hotel more sophisticated. Which option from below would be the best one for them to pick?
A. One-time token generator
B. Swipe card
C. Fingerprint Scanning
D. Password System

Question 8:
Jack is operating his laptop in the airport terminal and is connected to wireless internet. The lighting conditions are poor in that part of the terminal. Which authentication factor is the best choice for him to verify his identity?
A. Face
B. Voice
C. SMS
D. Fingerprint

Question 9:
Anna is operating her cell phone in a noisy environment and connected to the internet using her phone's data plan. She is trying to access her financial information. Which two factors

are a better choice for her at the given settings? (Select all that apply)

A. Facial
B. **SMS**
C. Voice
D. **Password**
E. Keystroke

Question 10:

A user operates his desktop and is using wired internet connection. The workstation for that user is noisy, and the lighting condition is poor at the time. Which three-factors are better options to choose considering the surrounding conditions?

A. Face
B. **CAPTCHA**
C. Voice
D. **Fingerprint**
E. **SMS**

References

1. Multi-factor Authentication (2016) Accessed date: 01 Dec 2016. ► http://searchsecurity.techtarget.com/definition/multifactor-authentication-MFA
2. Multifactor authentication examples and business case scenarios (2016) Accessed date: 01 Dec 2016. URL: ► http://searchsecurity.techtarget.com/feature/The-fundamentals-of-MFA-The-business-case-for-multifactor-authentication
3. Anderson T (2014) Why multi-factor authentication is a security best practice. Access date: 01 Dec 2016. URL: ► http://www.scmagazineuk.com/why-multi-factor-authentication-is-a-security-best-practice/article/373462/
4. Pascual A, Miller S (2015) 2015 Identity fraud: protecting vulnerable populations. Accessed Date: 01 Dec 2016. URL: ► https://www.javelinstrategy.com/brochure/347
5. Identity Theft and Cybercrime (2015) Access date: 01 Dec 2016. URL: ► http://www.iii.org/fact-statistic/identity-theft-and-cybercrime
6. SafeNet (2014) 2014 authentication survey. Accessed: 01 Dec 2016. URL: ► http://www.safenet-inc.com/resources/data-protection/2014-authentication-survey-executive-summary/
7. Laurello J (2013) Q&A: challenges, benefits of implementing single sign-on in hospitals. Accessed: 01 Dec 2016. URL: ► http://searchhealthit.techtarget.com/feature/QA-Challenges-benefits-of-implementing-single-sign-on-in-hospitals
8. Villanueva JC (2014) 5 big business benefits of using sso (Single Sign-on). Access date: 01 Dec 2016. URL: ► http://www.jscape.com/blog/bid/104856/5-Big-Business-Benefits-of-Using-SSO-Single-Sign-On
9. Blattner N (2014) Password self help—password reset for IBM i. Accessed: 01 Dec 2016. URL: ► http://www.ibmsystemsmag.com/pdfs/PasswordSelfHelp—Password-Reset-for-IBM-i/

10. Peterson T (2013) Moving single sign-on (SSO) beyond convenience. Accessed: 01 Dec 2016. URL: file:///C:/Users/Abhijit/Downloads/moving-single-sign-on-beyond-convenience-13757.pdf

11. Lawton S (2015) Secure authentication with single sign-on (SSO) solutions. Accessed: 01 Dec 2016. URL: ▶ http://www.tomsitpro.com/articles/single-sign-on-solutions,2-853.html

12. FIDO (2015) History of FIDO alliance. Accessed: 01 Dec 2016. URL: ▶ https://fidoalliance.org/about/

13. FIDO (2015) Members: bringing together an ecosystem. Accessed: 01 Dec 2016. URL: ▶ https://fidoalliance.org/membership/members/

14. FIDO (2015) Specifications overview. Accessed: 01 Dec 2016. URL: ▶ https://fidoalliance.org/specifications/overview/

15. CA Technologies (2015) CA strong authentication. Accessed: 01 Dec 2016. URL: ▶ http://www.ca.com/us/securecenter/ca-strong-authentication.aspx

16. CA Technologies (2013) Deliver secure, new business services in a multi-channel customer environment. Accessed date: 01 Dec 2016

17. CA Technologies (2015) CA strong authentication data sheet. Accessed date: 01 Dec 2016. URL: ▶ http://www.ca.com/us/~/media/Files/DataSheets/ca-strong-authentication.PDF

18. Okta (2015) Introducing Okta adaptive MFA. Accessed date: 01 Dec 2016. URL: ▶ https://www.okta.com/product/adaptive-mfa/

19. Okta (2015) Enabling just in time provisioning. Accessed date: 01 Dec 2016. URL: ▶ https://support.okta.com/articles/Knowledge_Article/27715118-Enabling-Just-In-Time-Provisioning?fs=RelatedArticle&l=en_US

20. Vasco (2015) IDENTIKEY authentication server 3.8. Accessed date: 01 Dec 2016. URL: ▶ https://www.vasco.com/Images/IDENTIKEY-Authentication-Server-3.8-Datasheet-(II).pdf

21. Vasco (2014) IDENTIKEY authentication server. Accessed date: 01 Dec 2016. URL: ▶ https://www.vasco.com/Images/Identikey_BR201401-v5.pdf

22. Vasco (2015) IDENTIKEY authentication server. Accessed date: 01 Dec 2016. URL: ▶ https://www.vasco.com/products/server_products/identikey/ik_auth/identikey-authentication-server.aspx

23. Dell Defender (2015) Defender: protect your perimeter with two-factor authentication. Accessed date: 01 Dec 2016. URL: ▶ http://software.dell.com/documents/defender-datasheet-29206.pdf

24. Dell Defender (2015) Two-factor authentication made easy. Accessed date: 01 Dec 2016. URL: ▶ http://software.dell.com/products/defender/

25. Symantec VIP (2015) Symantec validation and ID protection service (VIP). Accessed date: 01 Dec 2016. URL: ▶ http://www.symantec.com/vip-authentication-service/

26. Symantec VIP Data Sheet (2015) Symantec™ validation and ID protection service: prevent unauthorized access to sensitive networks and applications. Accessed date: 01 Dec 2016. URL: ▶ http://www.symantec.com/content/en/us/enterprise/fact_sheets/b-validation_and_id_protection_service_DS_21213686.en-us.pdf

27. Symantec VIP for Mobile (2012) Symantec™ VIP access for mobile. Accessed date: 01 Dec 2016. URL: ▶ http://www.symantec.com/content/en/us/enterprise/fact_sheets/b-verisign_identity_protection_access_for_mobile_DS_21172473.en-us.pdf

28. RSA SECURID (2015) RSA authentication products. Accessed date: 01 Dec 2016. URL: ▶ http://www.emc.com/security/rsa-securid/index.htm

29. RSA SECURID (2011) RSA SECURID® AUTHENTICATORS. Accessed date: 01 Dec 2016. URL: ► http://www.emc.com/collateral/software/data-sheet/h9061-rsa-securid.pdf

30. RSA SECURID (2014) RSA SECURID® software tokens. Accessed date: 01 Dec 2016. URL: ► http://www.emc.com/collateral/data-sheet/h13819-ds-rsa-securid-software-tokens.pdf

31. RSA SECURID (2014) RSA SECURID: risk-based authentication. Accessed date: 01 Dec 2016. URL: ► http://www.emc.com/collateral/data-sheet/h13823-ds-rsa-securid-risk-based-authentication.pdf

32. RSA SecurID (2014) RSA SecurID: management console. Accessed date: 01 Dec 2016. URL: ► http://www.emc.com/collateral/data-sheet/h13822-ds-rsa-securid-management-console.pdf

33. SafeNet (2015) SafeNet authentication service fully automated authentication as-a-Service. Accessed date: 01 Dec 2016. URL: ► http://www.safenet-inc.com/multi-factor-authentication/authentication-as-a-service/sas-safenet-authentication-service/

34. SafeNet (2015) SafeNet authentication service: affordable, flexible, cloud-based authentication. Accessed date: 01 Dec 2016. URL: ► http://www.safenet-inc.com/resources/data-protection/safenet-authentication-service-brochure/?langtype=1033

35. SafeNet (2015) Gemalto SafeNet authentication service: a faster, more effective way to manage authentication deployments. Accessed date: 01 Dec 2016. URL: ► http://www.safenet-inc.com/resources/data-protection/safenet-authentication-service-solution-brief/

36. SecureAuth IdP (2015) SecureAuth IdP 8.0. Access date: 01 Dec 2016. URL: ► https://www.secureauth.com/Product.aspx

37. SecureAuth IdP (2015) Two factor authentication: 20+ strong methods. Access date: 01 Dec 2016. URL: ► http://www.esecuritytogo.com/documents/secureauth_2_factor.pdf

38. SecureAuth IdP (2015) SecureAuth IdP user access control that works for you. Access date: 01 Dec 2016. URL: ► http://www-304.ibm.com/partnerworld/gsd/showimage.do?id=40694

39. SecureAuth IdP (2015) SecureAuth IdP single sign-on. Date: 01 Dec 2016. URL: ► https://www.secureauth.com/SecureAuth/media/Resources/SolutionBriefs/SecureAuth-Single-Sign-on.pdf?ext=.pdf

40. SecureAuth IdP (2015) SecureAuth IdP for mobile. Access date: 01 Dec 2016. URL: ► https://www.secureauth.com/SecureAuth/media/Resources/SolutionBriefs/SecureAuth-IdP-for-Mobile.pdf?ext=.pdf

41. SecureAuth IdP (2015) SecureAuth IdP Office 365. Accessed date: 01 Dec 2016. URL: ► https://www.secureauth.com/SecureAuth/media/Resources/SolutionBriefs/SecureAuth-IdP-for-Office-365.pdf?ext=.pdf

42. SecureAuth IdP (2015) SecureAuth IdP authentication API. Accessed date: 01 Dec 2016. URL: ► https://www.secureauth.com/SecureAuth/media/Resources/SolutionBriefs/SA_SolutionBrief_API.pdf

43. Bill Mathers (2015) What is Azure multi-factor authentication? Accessed date: 01 Dec 2016. URL: ► https://azure.microsoft.com/en-us/documentation/articles/multi-factor-authentication/

44. Microsoft Azure (2015) Multi-factor authentication pricing. Accessed date: 01 December 2016. URL: ► https://azure.microsoft.com/en-us/pricing/details/multi-factor-authentication/

45. Collier M, Shahan R (2015) Microsoft Azure Essentials-Fundamentals of Azure. Pearson Education

46. Swivel (2015) Swivel: adaptable, active, authentication. Accessed date: 01 Dec 2016. URL: ► http://swivelsecure.com/

47. SwivelSecure (2014) Swivel secure overview. Accessed date: 01 December 2016. URL: ▶ http://hosteu.msgapp.com/uploads/96495/Documents/Data%20Sheets/1502%20DS%20Overview%20Data%20Sheet.pdf

48. SwivelSecure (2014) Risk based authentication. Accessed date: 01 Dec 2016. URL: ▶ http://hosteu.msgapp.com/uploads/96495/Documents/Data%20Sheets/1410_DS_Risk_Based_Data_Sheet.pdf

49. SwivelSecure (2014) SMS based authentication. Accessed date: 01 Dec 2016. URL: ▶ http://hosteu.msgapp.com/uploads/96495/Documents/Data%20Sheets/1410_DS_SMS_Data_Sheet.pdf

50. SwivelSecure (2014) Mobile app based authentication. Accessed date: 01 Dec 2016. URL: ▶ http://hosteu.msgapp.com/uploads/96495/Documents/Data%20Sheets/1411_DS_Mobile_App_EN.pdf

51. SwivelSecure (2014) Token based authentication. Accessed date: 01 Dec 2016. URL: ▶ http://hosteu.msgapp.com/uploads/96495/Documents/Data%20Sheets/1410_DS_Token_Data_Sheet.pdf

52. SwivelSecure (2014) PINpad. Accessed date: 01 Dec 2016. URL: ▶ http://hosteu.msgapp.com/uploads/96495/Documents/Data%20Sheets/1410_DS_PINpad_Data_Sheet.pdf

53. SwivelSecure (2014) Telephony. Accessed date: 01 Dec 2016. URL: ▶ http://hosteu.msgapp.com/uploads/96495/Documents/Data%20Sheets/1411%20Telephony%20Data%20Sheet.pdf

54. DUO Security Product Overview (2016) Accessed date: 01 Dec 2016. URL: ▶ https://duo.com/assets/pdf/Duo-Security-Product-Overview.pdf

55. DUO Security: Two-Factor Authentication Made Easy. Accessed Date: 01 Dec 2016. URL: ▶ https://duo.com/assets/pdf/Duo-Security-Product-Datasheet.pdf

56. NIST Cybersecurity whitepaper on *Best Practices for Privileged User PIV Authentication*. 21 Apr 2016. ▶ http://csrc.nist.gov/publications/papers/2016/best-practices-privileged-user-piv-authentication.pdf

57. Ferraiolo H, Cooper D, Francomacaro S, Regenscheid A, Mohler J, Gupta S, Burr W (2014) National institute of standards and technology (NIST) special publication (SP) 800-157, *Guidelines for Derived Personal Identity Verification (PIV) Credentials*. 10.6028/NIST.SP.800-157

58. RSA SECURID (2007) A comprehensive introduction to RSA SecurID® user authentication. Accessed date: 01 Dec 2016. URL: ▶ http://www.ais-cur.com/IntrotoSecurID.pdf

59. Strom D (2014) Okta verify|multifactor authentication product overview. Accessed Date: 01 Dec 2016. URL: ▶ http://searchsecurity.techtarget.com/feature/Multifactor-authentication-products-Okta-Verify

60. RSA SECURID (2010) RSA® SecurID two-factor authentication. Accessed date: 01 Dec 2016. URL: ▶ http://www.arrowecs.co.uk/ArrowECS/media/PDF-Library/Security/RSA/RSA-SecurID.pdf

Continuous Authentication

Authenticating individuals frequently during sessions to assure valid identity

©alphaspirit/Shutterstock.com

© Springer International Publishing AG 2017
D. Dasgupta et al., *Advances in User Authentication*, Infosys Science Foundation Series,
DOI 10.1007/978-3-319-58808-7_6

Users are now all time connected to the cyber world through different devices and media. It has become essential to check regularly (by monitoring the system and user behavior) whether the user who logged on with valid credential is the same person currently accessing/using resources.

Introduction

In previous chapters, authentication approaches are discussed in the context of single factor to multiple factors for validating the legitimate users once to begin the session. With the increase of usage of mobile devices and applications, one-time validation of user's identity is not a secure choice for a longer user session or period. In this case, continuous or frequent verification of a user's identity to a system is required to ensure integrity, security, and privacy of the user access throughout the session. Earlier attempts associated with continuous authentication mainly relate to locking the user's workstation after a certain period of inactivity. For smartphones, users need to provide their PIN every few minutes to resume their activities. But this type of feature sometimes annoys some users and there have been claims of a more flexible solution. This chapter discusses various approaches in frequent verification of users' authenticity in identity management; however, how often user identity needs to be verified is a deciding factor.

Continuous/Active Authentication

Continuous Authentication also referred to as Active Authentication was introduced in 2012 to address novel ways to validate users' identity rather than using typical passwords. The focus was mainly on software-based behavior biometrics to capture the session data to determine whether genuine user is using the system at a given point of time. According to the DARPA program [1, 39], a person can be authenticated at a regular interval through (as shown in ◘ Fig. 6.1):

- Physical aspects (example: fingerprint, face geometry, etc.)
- Interaction with the system (example: keystroke pattern, mouse movement, etc.)
- Existing context of the user (example: structural semantic analysis, forensic authorship, etc.)
- Experienced data usage (example: computational linguistics).

Several researches demonstrated that continuous user authentication can be done transparently as the users interact with the system and can verify their credentials if the acceptance level falls below a predefined threshold value. User nonintrusive-based authentication approaches are the plausible candidates to perform continuous authentication. For example, a user's web activities can continuously be checked for irregularities in his/her normal workflow and UI interaction patterns. Smartphone applications can regularly check the user's bio-signature to detect an imposter.

A pictorial view of continuous/active authentication is found in ◘ Fig. 6.2. Different works based on human behavioral traits (for desktop and mobile devices) such as mouse movement tracking, web browser histories, process histories, file system access, linguistic fingerprint, etc., are studied as part of active authentication, some of which are discussed later.

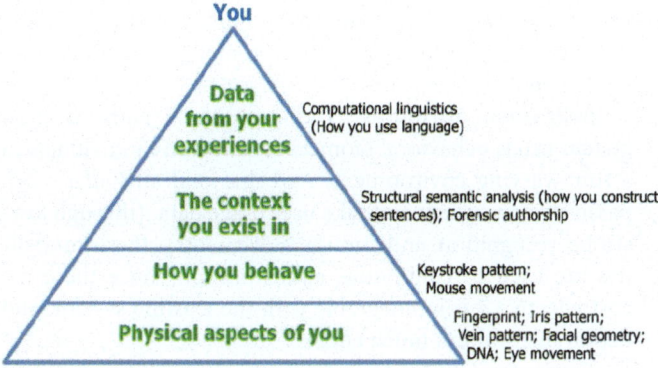

◘ **Fig. 6.1** Different perspectives for active/continuous authentication shown in pyramid [43]

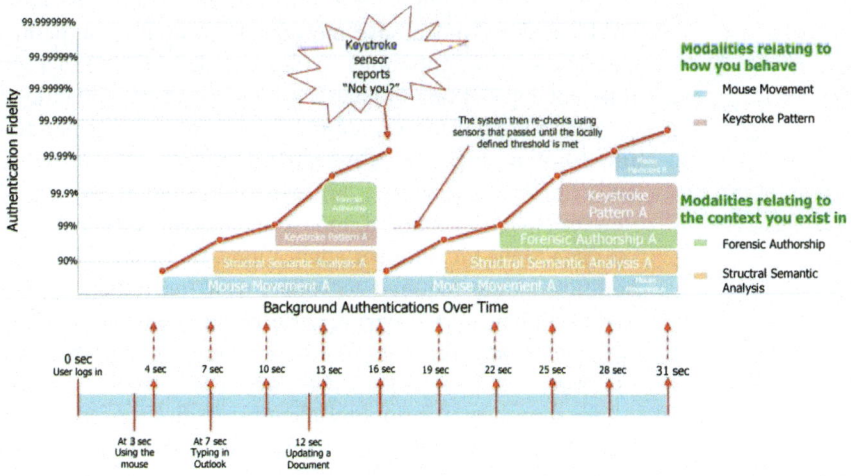

◘ **Fig. 6.2** Illustrates how active authentication concepts work [43]

Characteristics of a Good Continuous Authentication System

In general, continuous authentication systems provide an additional secured layer over the static authentication approaches. Such a strategy plays an important role in highly secured systems employing continual verification of the users' identity. A good continuous authentication system should possess the following characteristics [2]:

Non-intrusiveness: The continuous system should be able to validate users' identity in a seamless and nonintrusive manner. Intrusive approach (like passwords and PIN) requires more users' effort and time that incur significant increase in cost of authentication. Hence, a user can perform his daily tasks without any system intervention. If the system identifies an imposter, it can immediately prevent the person's access to the system and prompt for authentication verification.

Behavioral attributes: The continuous authentication system needs behavioral biometrics to identify individuals in a time-varying environment. As behavioral attributes work passively in a system and take user usage data (through keystroke recognition and mouse movements), these modalities are used in continuous authentication, since these are cost effective, easily integrable with the existing system, and deployable without much effort.

Computer System Independent: The way of continual verification of users' identity should be applicable to different applications and independent of system hardware. Keyboard, touch screen, and mouse are some input devices; when keystroke analysis and mouse movement are used to authenticate users in a continuous manner, it should work across different applications and hardware-independent modalities.

Faster Response: As continuous authentication generates the decision on the fly, it is required for faster execution of authentication modalities. Since data from different input devices (such as keyboard, mouse, etc.) are already captured by the system for different applications, there should be no additional time required to collect data for authentication and re-authentication purpose. Faster processing of identity verification also prevents imposters from gaining access to the system for a longer period of time.

In summary, continuous identity verification solutions should be faster, robust, easy-to-use, and enabled by technology that requires minimal user training.

Steps to Design Continuous Authentication System

To design a robust continuous authentication system, the following steps should be considered [2]:

1. **Choice of attributes:** It decides the behavioral attributes to be collected to design a highly accurate authentication system with the minimal processing time. These behavioral attributes should be considered as the input variables for any continuous authentication system. These variables should be easily collectible and should be limited in number to attain the minimal execution time.

2. **Choice of analysis methods:** This step mainly decides the heuristics or analytical methods to be used to build a continuous authentication system. The selected analytical algorithms should be compatible with the chosen attributes (input variables).

3. **Choice of system phases/states:** This step involves a continuous process of data collection, analysis, results, and expected behavior of the proposed system. Continuous authentication system needs to complete each of the system phases/states within a short span, due to its real-time nature.

4. **System test and design reiteration:** This step consists of rigorous testing of the newly developed system to verify its functionality, ensure robustness and performance. Design reiteration can improve the existing system by trial and error method that can be achieved through varying the number and mixture of different attributes.

Attribute Selection for Continuous Authentication System

Attributes refer to the behavioral characteristics chosen for analysis and can be divided into two major categories, namely, base and derived attributes [2]. Selection of specific base and derived attributes is a major step towards the design of continuous authentication systems. It is necessary to choose the right set of attributes for better accuracy rate of continuous authentication systems.

Base attributes are directly extracted from the raw data [2]. For example, the general speed and acceleration of the mouse in all directions and the direction-based mouse speed can be included as mouse attributes collected from the raw data. Similarly, the speed and speed/time values along with keystroke digraph latencies for 50 most frequently used alphabetical digraph combinations in English are included in the keyboard attributes. The base attributes are chosen from a large pool of attributes after these were found to exhibit promising results in preliminary studies.

Derived attributes refer to those attributes that require some computation on the base attributes [2]. For example, the ratio between mouse and keyboard operations is used in a continuous authentication system as one of the derived attributes.

Unimodal Continuous Authentication

In single modality-based continuous authentication, one attribute is used which allows the system to process data quickly to make the decision regarding the user authenticity. Some common unimodal authentication approaches in use are:

Face Recognition

Face recognition is a well-known biometric approach to authenticate users. It can be used in the areas of active or continuous authentication for smartphones [3]. As most of the smartphones have front facing camera and multiple core processors, it is more feasible to apply the existing face recognition technology in mobile phones to support continuous user authentication. Researchers (Viola and jones [4]) show that relative face detection can significantly decrease the computational overhead of face detection process during re-authentication phases. Generally, the recognition rate increases when there is no mobility of the faces on the smartphones. It is found from the dataset that the efficiency of face recognition increases as the user's face is close to the phone and thereby reduces the false positive rate. This result is supported by Fathy et al. [3] study, where it has been tested with 750 video sequences of 50 different users. With including more sophisticated features and considering the environmental conditions, its performance can be improved in future.

Temporal information is incorporated with face recognition into continuous authentication process, where body localization is considered [5]. This approach does not require pre-registration of the templates. Instead, it registers a new enrollment template every time a user logs in. Hence, this gives an option to use temporal information, like color of the user's clothing, as an enrollment template in addition to facial information. In the enrollment phase, face detection is done first, followed by body localization. Then face and body histograms are calculated and used as enrollment data for recognition. The matching between the actual user and stored template is done through comparing similarity of both face and body.

Partial Face Recognition

The partial face recognition method [6] is introduced for detecting partially cropped and occluded faces captured using a smartphone front camera mainly for continuous authentication. This recognition method consists of three phases (as shown in ■ Fig. 6.3), namely Segment Clustering, Learning SVM, and Face Detection.

In segment cluster phase, facial segments are logically clustered to estimate face regions from the training images; then in learning phase, a linear machine learning technique (SVM) is trained on the cluster set to have a cascade of classifiers; at the detection phase, those classifiers are used to identify and recognize faces from test images [6]. Experiments were performed on a dataset of 50 users (43 male, 7 female) with three different ambient lighting conditions: well-lit, dimly-lit, and natural daylight. Reported results show that the proposed method performed better than many state-of-the-art face detection methods in terms of accuracy and processing speed.

Web Browsing Behavior

Since majority of the applications are now web-based, people use the web to communicate, perform official work, and engage in online social activities; so monitoring web browsing has become an integral part of marketing and other purposes of predicting users' behavior. Web browsing patterns are also used as behavioral biometric [7] for continuously authenticating users. A scenario of browsing behavior for three users is illustrated in ■ Fig. 6.4. Global and internal session features are used where global session features capture characteristics of a user across different pageviews. A session is typically defined as a continuous stream of pageviews separated by pauses greater than 30 min [8], and a pageview refers to the URL that is visible at the time by a user. Day-of-week and day-of-time distributions [7] are part of global session features. Other global session features include total number of

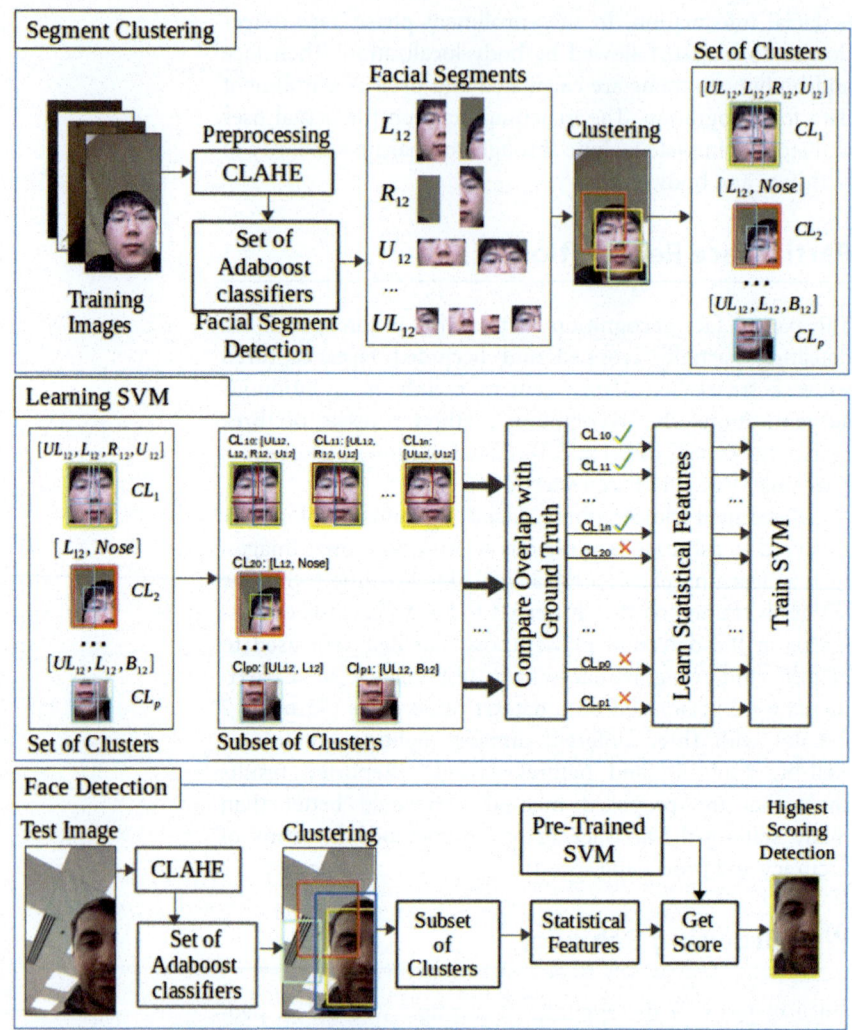

■ **Fig. 6.3** Overall method for partial face detection for continuous authentication [6]

pageviews, the average duration of pageview, and the number of unique pageview are also considered.

Moreover, for internal session features, different characteristics of pageview within a session such as burstiness, time between revisits, and genres are considered. Specifically, pauses are determined by the user's spending time on a webpage, and is calculated as the difference between the timestamp of one pageview and the timestamp of another

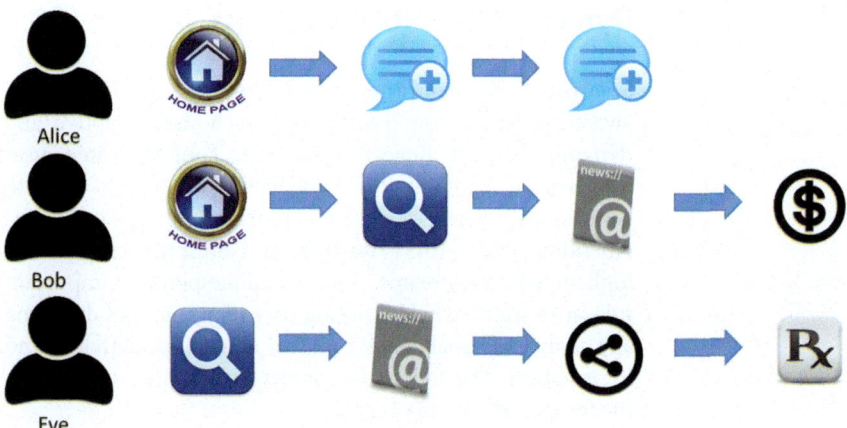

Alice

Bob

Eve

■ **Fig. 6.4** Web browsing behavior of different users

pageview. Burstiness is considered as a consequence of the human's decision process in prioritizing tasks, and is computed through the change in pause time between page views or second-order pause time. Time between revisits determines the rate at which webpages are revisited and provide a good indicator of a user's identity. Genre is considered as more indicative to distinguish different users in terms of web browsing behavior which is a mixture of style, form, and content of the webpages.

Using this approach, Kumar et al. [8] reported some results, where a mixture of these global and internal session features was used with data collected for 10 different users shown in ■ Table 6.1.

■ **Table 6.1** Performance measures (FRR and FAR values) for ensemble of the features for web browser behavior [8]

Users#	False Rejection Rate (FRR)	False Acceptance Rate (FAR)	Users#	False Rejection Rate (FRR)	False Acceptance Rate (FAR)
User 1	18 ± 2.20	7 ± 6.70	User 6	35 ± 4.28	11 ± 8.13
User 2	33 ± 7.14	7 ± 9.20	User 7	32 ± 4.11	26 ± 10.47
User 3	41 ± 4.05	10 ± 8.84	User 8	44 ± 8.99	13 ± 11.27
User 4	28 ± 3.56	5 ± 5.64	User 9	27 ± 6.11	15 ± 10.05
User 5	22 ± 3.26	19 ± 8.84	User 10	32 ± 6.61	15 ± 7.68

Dynamic Context Fingerprints

SmartAuth [9] has proposed to allow a scalable context-aware authentication framework which uses adaptive and dynamic context fingerprinting, i.e. context-related user information contains distinguishable identifying data to continuously verify the users' credentials. It basically uses Hoeffding trees [10], a learning and mining technique for high-speed data streams. This technique plays an important role in continuously identifying user identities and detecting any change of usual context-related user online activities and information. The following context fingerprints are used as the features of identifying users:

Client side: language, color depth, screen resolution, time zone, platform, plugins, etc.

Server side: IP address range, time of access, geolocation, request header, etc.

After successful login of the user, the captured fingerprints are compared with the previously stored ones. In this work [10], the authors generated dynamic context fingerprints for 10k users, and the classification accuracy of using different fingerprints is mentioned in ◻ Table 6.2.

In another work, user search patterns were investigated to identify malicious behavior. This trait identifies users through web browsing activities with the help of including semantic and syntactic session features. In particular, this study saves selected decoy files in some key directories in the system beforehand, which the legitimate users never search and access those files. This work also focuses to detect information-gathering activities by the adversaries and to prevent the imposters from accessing the service.

◻ **Table 6.2** Correct and incorrect classification of the individuals based on dynamic context fingerprints using SmartAuth [10]

Number of fingerprints considered	Correctly classified (%)	Incorrectly classified (%)
2	87	13
4	91	9
6	95	5
8	98	2
10	99	1

Keystroke Dynamics

Keystroke dynamics is a widely used nonintrusive biometric for authentication. For continuous authentication, typing behavior is used in identifying the legitimate users; for mobile devices, accelerometer and gyroscope are used to capture the alternate set of features of keystroke and typing patterns of a user. Different types of key stroke patterns (pattern name, descriptions, and examples) are shown in ◘ Fig. 6.5.

Various set of features related to keystroke are investigated to identify users in different authentication works. A summary of these features, incorporating keystroke dynamics, is mentioned below [11]:

- Digraph latency
- Keystroke time, duration, pressure or force
- Keystroke interval
- Combining key-hold and inter-key times
- Keystroke latency
- Digraph, duration time, and trigraph.

Text-independent keystroke features [12] are also considered in continuous. These features include keystroke time, tactile pressure from capacitive touchscreen with and without haptic feedback. In the work of Feng et al. [12], a dataset of 40 participants is created and performance of these selected

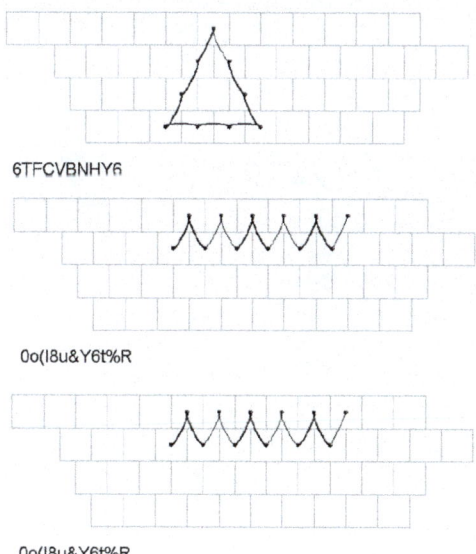

◘ **Fig. 6.5** Keystroke dynamics used in continuous authentication [43]

keystroke features is tested. Another work reported the use of typing motion behavior on a virtual keyboard in the mobile device. In addition to usual keystroke time features, it captures data from accelerometer, gyroscope, and orientation sensor to distinguish the typing behavior of individuals.

In case of static authentication, the performance of biometric systems is generally expressed in terms of FMR, FNMR, EER, and ROC curve. However, performance measures are difficult to calculate in continuous authentication. In case of keystroke dynamics for continuous authentication, the system performance is calculated by the number of keystrokes that an imposter types before the system will lock him off the system. The better the system, the lower the number of keystrokes required to detect an illegitimate user.

Cognitive Fingerprint

The concept of cognitive fingerprint [14] was coined studying keystroke dynamics, i.e., observing a user's cognitive typing rhythm as shown in ◘ Fig. 6.6. It is hypothesized that natural pauses (delays occurred between typing characters in words) are done because of cognitive factors (spelling of unfamiliar words or pause after certain syllables, for instance). These cognitive factors are considered unique and hence applied to identify individual user. Here, each feature is represented

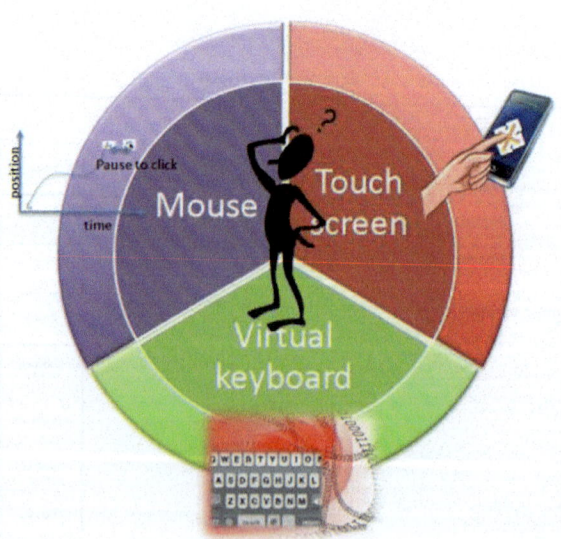

◘ **Fig. 6.6** Capturing cognitive fingerprints using mouse or touch screen and/or virtual keyboard for continuous authentication [44]

by a unique Cognitive Typing Rhythm (CTR) that contains the sequence of digraphs from a specific word usage. These features are collected for every legitimate user and build SVM classifiers for user verification. The developed system is tested with 983 participants and reported results show improved performance which is very good for behavioral biometrics. The training time for the users is 15 min per user and testing time is 0.6 s per user. These results show cognitive fingerprint is effective for authenticating users.

The linguistic fingerprint hypothesis [15, 16] states that each human uses language differently, and therefore a language fingerprint that is unique to each individual exists. The challenge is thus discovering the features of the linguistic fingerprint, i.e., a linguistic fingerprint model that would allow the accurate identification of an individual based on a sample of his language productions. The success of various features proposed to solve the authorship problem, which is about identifying the author of a set of documents, provides support for the linguistic fingerprint hypothesis (see ◘ Fig. 6.7).

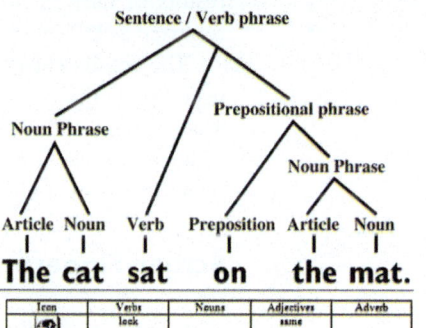

◘ **Fig. 6.7** Forensic authorship and structural semantic analysis [45]

■ **TableT 6.3**   The accuracy of keystroke behavior-based identification using stylometric features [17]

Task	Feature	No. of subject	Data size	Accuracy (%)
Identification	Writeprints feature set	79	500 characters	Exact: 55.30 In top 3: 70.30
			1000 characters	Exact: 63.98 In top 3: 79.30

Other studies show behavior biometrics captured through keyboard [17] also represents the cognitive fingerprint of users where the cognitive and linguistic features are collected from the keystroke analysis. In particular, lexical statistics of users, writing such as word length, distribution of characters, character bigrams, percentage of letters/digits/uppercase letters and numbers are analyzed [17]. Also, syntactic features like function words, part-of-speech tags, and n-grams are also computed from users' writing. In addition to writing style, other behavioral features such as domain visits, keystroke intervals and dwell times, mouse dynamics and button dwell times are calculated. In this study the dataset is created for 79 subjects for over 12 weeks, and the results of user identification tasks using these features are reported as ■ Table 6.3.

The results show that with the increase of characters in user input, the accuracy of person identification increases. The authors claim that with the incorporation of higher order cognitive features the accuracy of the identification task will significantly increase.

Screen Fingerprints

As keyboards are being replaced by touchscreen, screen fingerprint has become a new way to validate the identity of the legitimate users at a console. Screen fingerprints of user capture significant information that can be considered as a biometric for authentication. This type of biometric basically captures cognitive abilities, motor limitations, subjective preferences, and work patterns of an individual. Some user gestures for screen fingerprint including tap, double tap, drag, flick, pinch, spread, press, rotate, and some of their combinations are used as shown in ■ Fig. 6.8.

The size of text shown on the screen is used to visually capture the user cognitive ability. Motor limitations (how

Tap

Briefly touch surface
with fingertip

Double tap

Rapidly touch surface
twice with fingertip

Drag

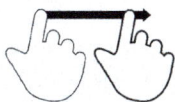

Move fingertip over
surface without
losing contact

Flick

Quickly brush surface
with fingertip

Pinch

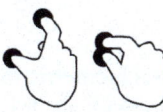

Touch surface with
two fingers and bring
them closer together

Spread

Touch surface with
two fingers and
move them apart

Press

Touch surface for
extended period
of time

Press and tap

Press surface with
one finger and briefly
touch surface with
second finger

Press and drag

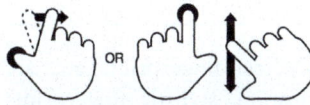

Press surface with one finger and
move second finger over surface
without losing contact

Rotate

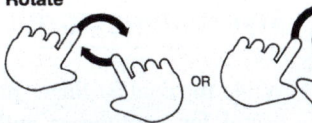

Touch surface with two fingers
and move them in a clockwise
or counterclockwise direction

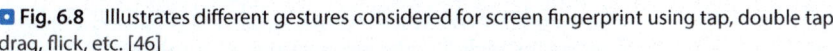

■ **Fig. 6.8** Illustrates different gestures considered for screen fingerprint using tap, double tap, drag, flick, etc. [46]

fast an operator drags a window) are captured by the amount of motion detected on the screen. Subjective preference is mainly how the user arranges multiple windows in a screen. The layout of the salient edges on the screen can visually capture this feature. Work patterns of a user signify different suites of applications that he/she uses. The distribution of application-specific visual features manifest on the screen can be used to capture this feature.

In screen fingerprints, a background program observes the computer screen and captures different screen recordings within short observation windows. These recordings are then visually analyzed to extract a screen fingerprint to verify the identity of the logged user. This is done by comparing with reference screen fingerprint of the same user.

Study results reported in [18, 19], where a dataset is created for screen recordings of 21 users. The average false acceptance rate (FAR) and false rejection rate (FRR) regarding each user interaction are shown in ■ Table 6.4.

◘ **Table 6.4**   Average FAR and FRR for each of the user's interaction and average time for each interaction [19]

Interaction	Average time (s)	Average FAR (%)	Average FRR (%)
Typing	43	37.57	36.19
Mouse movement	12	30.29	29.89
Scrolling	83	20.67	12.38
Other	2	21.85	33.82

Screen fingerprint is an emerging behavioral biometric to capture distinguishable features for different individuals. It is noted that further research in screen fingerprints to capture other richer features will improve the performance of this biometric modality.

Stylistic Fingerprint

While the context finger prints, screen fingerprints, etc., are useful for continuous authentication, stylistic fingerprints [20] can also be used to prove the legitimacy of a user. These fingerprints can be overlaid on each other and compared to detect if one person might be hiding behind a single persona or multiple people are sharing a single persona [20]. Stylistic fingerprint includes the proportions of punctuation characters, emoticons, and vocabulary measures. It mainly uses the above language traits to distinguish the language of a particular user, a set of users, or an amount of texts [20].

Metrics needed to construct this trait range from simple counts, such as the number of exclamation marks present, to more complicated measures, such as the type token ratio, a vocabulary indicator [20]. The metric calculation mainly considers text length to combine various sources where appropriate—for example, chat room texts are generally smaller than the email text.

Radar plots (shown in ◘ Fig. 6.9) of the adjusted and weighted metric scores are used here to visually represent language style fingerprints. Two plots with substantial overlap indicate the similarity of the language style, while little overlap indicates contrasting language styles.

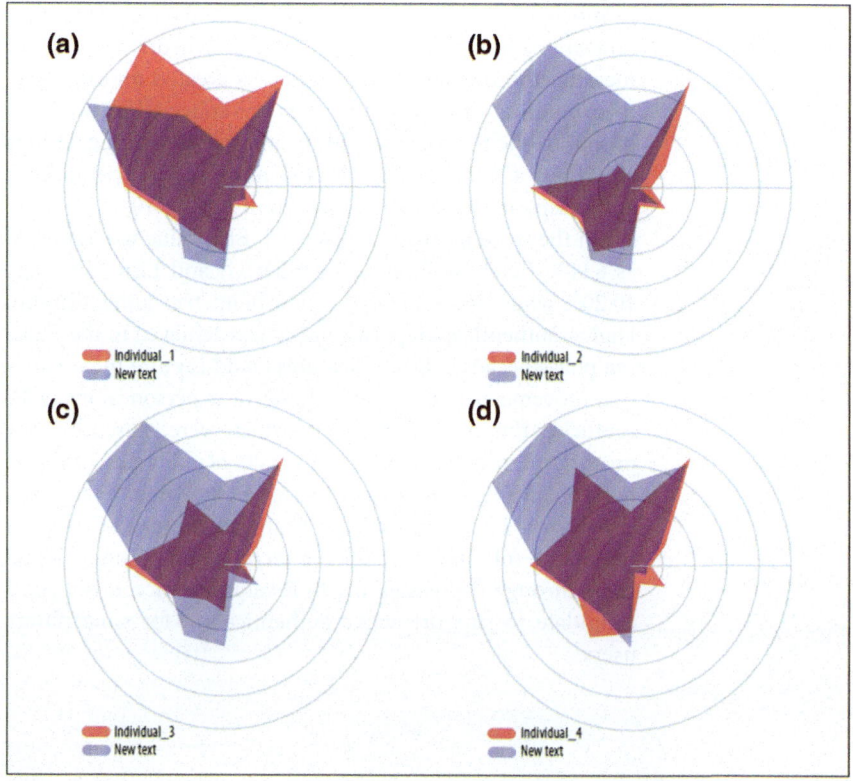

Fig. 6.9 Comparison of language style fingerprints for a new text against four individuals' fingerprints [20]

Hand Movement, Orientation, and Grasp (HMOG)

Hand movement, orientation, and grasp (HMOG) [21] is considered as a biometric modality for continuously authenticating smartphone users. This new biometric captures subtle micro-movement and orientation dynamics of the users from the pattern of his grasping, holding, and tapping on the smartphone. In this research, a total of 96 features are used to represent this biometric modality.

In this study, two types of HMOG features are considered—resistance features and stability features. Resistance features measure the micro-movements of the phone due to the force exerted by a tap gesture of the user. Stability feature, on the other hand, captures how quickly the perturbations in movement and orientation occur. The data from accelerometer, gyroscope, and magnetometer sensors are used to measure those features. The dataset is created for 100 users incorporating those features. Data are collected in two conditions: while the user was sitting and that of walking. Different types of frequently used hand movements, orientations, and grasps are considered as shown in ◘ Fig. 6.10.

In the work of Zdenka et al. [21], EER value was reported as 6.92% (while walking using HMOG) and EER value was 10.20% while sitting using this new biometric authentication. Higher authentication performance was achieved in the walking phase because HMOG features could capture the distinctive movements caused by walking of a person along with capturing the micro-movements which were caused by taps on the smartphone screen. The results of different combinations of HMOG features are reported as shown in ◘ Table 6.5.

From ◘ Table 6.5, it is clear that HMOG features play an important role in continuous authentication of smartphone users through decreasing the EER values. Hence, it is a good candidate to support active authentication for smartphone users.

◘ **Fig. 6.10**   Different hand movement, orientation, and grasp for HMOG.
©Mascha Tace/Shutterstock.com

Table 6.5 Summary of EER values with different score-level fusion of HMOG, tap, and keystroke dynamics features [21]

Score-level fusion	Sitting (%)	Walking (%)
HMOG, tap, and keystroke dynamics (KD)	10.20	6.92
HMOG and tap	11.68	8.12
Tap and KD	10.86	10.51

Multi-modal Continuous Authentication

A fusion of different behavioral biometrics produces a more robust and secure continuous authentication. The performance of multi-modal continuous authentication depends on the fusion strategies of different authentication modalities. Some of the current multi-modal continuous authentication approaches are discussed in the following sub-sections.

Keystroke and Mouse Movement, Application Usage, and System Footprint

Behaviosec [22] is working on a continuous behaviometric authentication system which captures user data on keystroke dynamics, mouse movements, application usage, and the system footprint. Keystroke and mouse data are captured along with screen resolution to get the actual picture of the user screen. CPU and RAM usage along with the list of processes running are also used to capture the system footprint. Active authentication through keystroke analysis and mouse movements are illustrated in ◻ Fig. 6.11. Again, active authentication through mouse movement with its different features is shown in ◻ Fig. 6.12.

In this approach [22], a trust model is developed to determine the trust value of the system at a given point of time. If the trust score is above a predefined threshold, trust level increases and vice versa. The trust value lies between 0 and 100. The system is started with trust value of 100 and then updated to the trust level using a formula mentioned below [22]:

$$C := \begin{cases} \text{Max } C - \frac{T-P}{100*\frac{T}{Z}} ; 0, P < T \\ \text{Min } C + \frac{P-T}{100*\frac{1-T}{Z}} ; 100, P \geq T \end{cases}$$

◘ Fig. 6.11   Keystroke analysis and mouse movement for active authentication.
©Sergey Nivens/Shutterstock.com

In the above formula,

C is the calculated trust level of a user.

T is the threshold value to determine whether the user is trusted or not, based on a single test.

P is the value that reflects the probability from the last test of the user's input against the template.

Z is a constant that controls the degree of increase or decrease of trust on each test.

The calculation to compute the increase in C value is proportional to the absolute difference between $|P - T|$ by a variable factor Z, using $(P - T)/100^* ((1 - T)/Z)$. The value of Z is generally set independently for the increase and decrease functions. The model increases the trust value of C if the probability P is equal to or greater than the threshold value of T.

The keystroke and mouse movement results for the users are shown in ◘ Table 6.6.

From ◘ Table 6.6, it is clear that using keystroke dynamics, legitimate users are not falsely rejected, and fictitious users are recognized after 38 interactions (between 20 and 25 keystrokes). These results are obtained with a small group

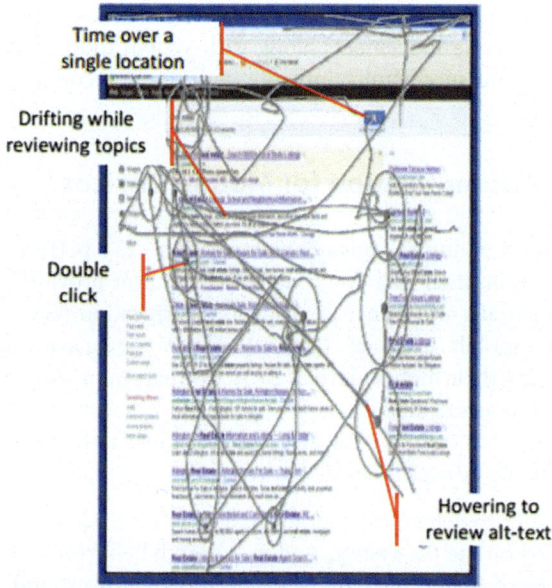

Time over a
single location

Drifting while
reviewing topics

Double
click

Hovering to
review alt-text

◻ **Fig. 6.12** Active authentication through mouse movement [48]

◻ **Table 6.6** Results of using keystroke and mouse movement of the users.
The detection times to not include the inactivity periods of more than 10 s [22]

	User (simulation hours)	Detection time (false reject/true positive)	Interactions
Mouse	Correct (1276)	823.74 min (13.73 h)	17,470
	Incorrect (4772)	4.08 min	88
Keyboard	Correct (450 h)	Never falsely rejected	Infinite (16,930)
	Incorrect (2716)	5.25 min	174 (38)
Combined	Correct (1726)	828.72 min (18.37 h)	25,028
	Incorrect (7489)	4.44 min	114

of 99 users. In case of mouse movements, it takes longer to
falsely reject the users (17,470 interactions), where incorrect users are quickly recognized (88 interactions). These

results demonstrate that with the incorporation of keystroke dynamics, mouse movements along with system footprints, it is possible to successfully authenticate users in a continuous setting.

Active Authentication for Mobile Devices

Combining multiple sensory information to perform authentication tasks in a continuous manner can provide a robust, accurate, and transparent identity verification mechanism for mobile users. ◘ Figure 6.13 shows a scenario of people using their mobile devices to access their information in a continuous manner.

Summary of various strategies used in active authentication of mobile devices is listed below:

1. In order to detect masqueraders, decoys have been placed on the file system, and user search behavior is observed to find the pattern between regular users and imposters. This is developed for desktop environment and also in mobile environment, where it includes some additional modalities like voice and image.

◘ **Fig. 6.13**   Shows people using mobile devices where they needed to continuously authenticate to protect their personal information and communication in a public place. ©Rawpixel.com/Shutterstock.com

2. Fingerprint identification from swiping patterns is another research of this category. It is being developed for mobile environment, where faster pattern recognition needs to be applied using kinematic gestures and finger images authentication.

3. Mobile perpetual authentication approach can measure the phone movement while user is in motion.

4. Type and swipe authentication looks into how a person types/swipes in the context of application for mobile environment.

5. Stylometry integrated with eye tracking for mobile environment is used to determine a user's construction thought process as well as where he/she focuses during the process.

6. Detection of heartbeat using mobile device is another category that uses the person's heartbeat.

7. The influences of power, touch, and movement in mobile device usage is also another potential research area for active authentication.

8. Human voice authentication using text-dependent verification and text-independent verification in mobile platform is another research strategy which captures how a person talks (static and continuous).

9. Stylometry-based authentication with the focus in keystroke, cogni-linguistic features and demographic classification (desktop platform) and context-aware kinetics (mobile platform) is also one of the areas of active authentication research.

10. Incorporation of covert games disguised as computer anomalies in mobile platform is also another research topic.

11. Joint physiological and behavioral authentication using fine-grained anthropometric and behavioral signatures from motion in mobile platform is another research topic. It actually detects how the phone moves when it is in use. Another research idea regarding continuous authentication is detection of natural speech and language activity performed by the user on mobile devices. The later one captures the thought process that reflects through the language usage.

12. Capturing visual images of the operator through front camera, back camera, and the screen recorder in the mobile environment is another area of research which uses passive facial recognition for continuous authentication.

13. Spatial–temporal hand micro-movements and oscillations (hand movement, device orientation, and grasping patterns) during user interaction with the touch screen is another area of research in active authentication. This actually captures the movements that occur while a person is writing/swiping.

Kryptowire [23] developed a behavior-based authentication and trust-based access control mechanisms to upgrade the security of the mobile user and the device authentication. These techniques are supposed to be deployed in Google phones and include data from available sensors like touch, pressure, movement, and other relevant information to recognize users based on the way they interact with the device and mobile applications. Through this approach, a user's phone or mobile application is embedded with a continuous authentication process that creates a model of the user's behavioral patterns like the touch pressure, way of holding the phone, etc. The user can be granted access to his/her sensitive information or resources using the level of trust established from this constant behavioral analysis. Moreover, using this technology, it is possible to identify imposters that are using sters that are using devices and stolen user credentials.

Stylometry, Application Usage, Web Browsing, and GPS Location

In a research study on mobile platforms [24], four representative authentication modalities are used to capture the behavioral traits of the users. These are stylometry (analysis of texts), application usage patterns, web browsing behavior, and physical location of the user device. In general, stylometry represents the linguistic style of a person, and can be used for author identification and verification. Web browsing behavior extracts the information regarding user behavior, habits, and interests. This study reported to achieve average FAR/FRR ratio of 0.24 on a dataset [25]. Application usage and movement patterns have also been studied to learn behavioral profiling of users in cellular networks. In this work, a dataset of 200 subjects for over a period of 5 months is collected and the following four features are tracked with 1-second resolution:

- The typed texts using soft keyboard
- Apps visited by the users

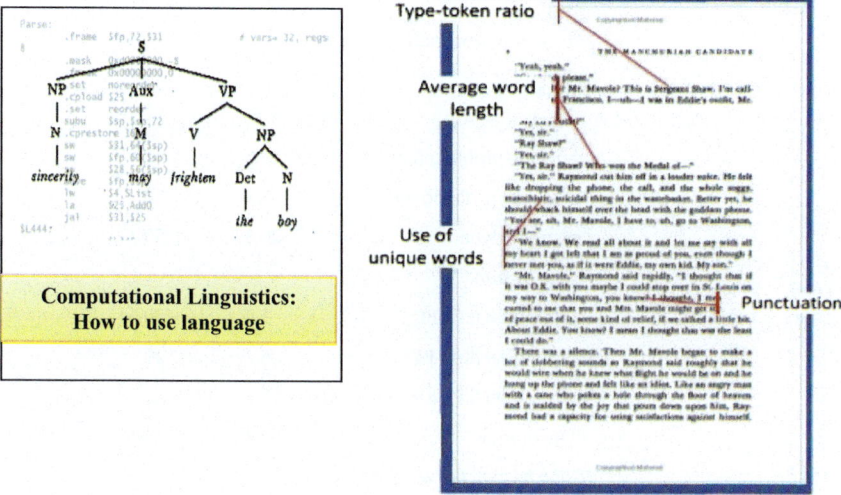

□ Fig. 6.14 Different textual features used for author identification as part of stylometry [47]

— Websites navigated by the users
— Location of the users based on GPS or Wi-Fi.

Different textual features for stylometry-based active authentication are illustrated in □ Fig. 6.14.

The reported results [25] used a binary classifier (valid or invalid) to classify 200 users based on the four above-mentioned features. Hence, there are 800 classifiers and each of those produces a probability that determines whether a user is valid or not. All these outputs are then passed through a decision fusion center (DFC) that uses these local decisions for a global decision using Chair-Varshney [26] formulation. Using this fusion approach, the proposed active authentication system could achieve an EER of 0.05 (5%) after 1 min of user interaction with the device, and an EER of 0.01 (1%) after 30 min which dominates the overall performance of the fusion-based system.

Multi-sensor Text-Based Biometric

Another multi-sensor biometric approach [27] utilizes linguistic analysis, keystroke dynamics, and behavioral profiling to design a framework to continuously authenticate users. Linguistic analysis is a behavioral trait that distinguishes people based on linguistic morphology. Using this

Table 6.7   Average case equal error rate (EER) for text-based active authentication approach [27]

Biometrics considered	Avg. equal error rate (EER)%	Worst case EER %
Linguistic profiling (LP)	12.8	40
Behavioral profiling (BP)	9.2	50
Keystroke dynamics	20.8	50.7
Fusion by sum		
Multi-modal (LP+BP)	5.8	30.6
Multi-modal (KA+BP)	6.2	20
Multi-modal (LP+KA)	11.2	45
All techniques	4.4	18.1
Fusion by matcher weighting		
Multi-modal (LP+BP)	3.6	20
Multi-modal (KA+BP)	5.3	20.2
Multi-modal (LP+KA)	8.5	44.7
All techniques	3.3	19.3

biometric trait, the performance of accuracies lies between 80 and 100%. Behavioral profiling targets to identify users based on their interaction of services on their smartphones. This approach can be used for authenticating users in mobile devices with accuracies in the range of 87–98%. Keystroke dynamics recognize users based on their typing patterns as well as their interaction with a keyboard. Common used features for this type of biometric are duration between two successive keys and hold time of a key, which are also considered here, and it is found that the average accuracy of keystroke dynamics is about 87% on mobile devices.

Next, the authors incorporated text-based biometric [27] to perform experiments with a database of 30 users, where each user's SMS, keystrokes, and text messaging activities are collected to create linguistic and behavior profiles. Results of individual biometrics and multi-modal approach are mentioned in Table 6.7.

The results (Table 6.7) demonstrate that fusion of two approaches leads to better performance than any of the individual approaches (biometrics), so the authors concluded that multi-modal biometrics have significantly lower EER values.

Gestures, Keystroke, Touch, and Body Movements

The gesture is considered as an alternative biometric modality that contains both static anatomical information and changing behavioral information. Full body gestures include a large wave of the arms, while hand gestures represent a subtle curl of the fingers and palm [28]. For authentication purpose, a specific gesture can be selected as a 'password'. In the case of compromise of this gesture, a user can easily select a new motion as a 'password'. Authors in [1] used Kinect to capture human gestures and calculate the depth and skeleton estimation as the features for this biometric [29]. To compare the registered data with verifying data, Dynamic Time Warping [30] method is used to calculate the similarity score. For user identification, average Equal Error Rate (EER) is about 4.14%, while for user authentication, the average EER is about 1.89%. Hence, this novel biometric has a good potential to verify legitimate users in time-varying environments.

In another study by Primo et al. [31], active authentication through keystrokes, touch gestures, and body movements is considered to verify user credentials in a continuous manner. In order to conduct experiments, they collected data of 74 participants for typing, 47 participants for swiping, and 11 participants for body movements [31]. Body movements are captured through accelerometer sensors while typing and swiping patterns are captured from virtual keyboard and phone screens, respectively. Using the best set of features from these modalities, a fusion-based framework is designed to integrate different modalities in a context-aware fashion. It also helps in making global decisions of the authentication and giving feedback for updating the user templates. Results of individual behavioral biometrics are shown in ◘ Table 6.8 in terms of FAR and FRR

◘ **Table 6.8** FAR and FRR values for different mobile-based biometric modalities [31]

Mobile biometric modality	FAR (%)	FRR (%)
Typing pattern through keystrokes	11.8	12.6
Touch gestures through swiping	10.3	10.3
Body movements through accelerometer sensors	3.2	3.2

values. These results demonstrate that with the use of accelerometer sensors, the error rates (FAR and FRR values) drop significantly compared to other two biometric modalities alone.

Keystroke, Mouse Movement, Stylometry, Web Browsing Behavior

Another multi-sensor continuous authentication approach uses a combination of keystroke dynamics, mouse movement, stylometry, and web browsing behavior [20, 32]. The mouse and keystroke dynamics are considered as low-level sensors while website domain frequency (part of web browsing behavior feature) and stylometry are used as high-level sensors. These features of low-level sensors and high-level sensors are listed in ◘ Table 6.9.

These four authentication biometric modalities are fused that combines the local decisions (provided by each individual modality) in order to get a global decision about authenticating legitimate users. Data are collected from 19 individuals in an office environment to verify this active authentication approach. The performance of the fusion of the different biometric modalities is shown in ◘ Table 6.10 in terms of FAR and FRR values.

From ◘ Table 6.10, it is noticed that incorporating all sensors, the overall FAR and FRR values are minimum. Removing

◘ **Table 6.9**   Features considered for mouse, keystroke, website behavior, and stylometry [32]

Sensors	Features
Mouse (M)	M1: mouse curvature angle M2: mouse curvature distance M3: mouse direction
Keystroke (K)	K1: keystroke interval time K2: keystroke dwell time
Website domain (W)	W1: website domain visit frequency
Stylometry (S)	S1: stylometry with 1000 characters and a 30-minute window S2: stylometry with 500 characters and a 30-minute window S3: stylometry with 400 characters and a 10-minute window S4: stylometry with 100 characters and a 10-minute window

■ Table 6.10 FAR and FRR values for four representative combinations of the biometric modalities [32]

Fusion of biometric modalities	FAR (%)	FRR (%)
All of the sensors (W, S, M, and K)	0.122	0.218
S, M, and K	0.210	0.27
W, M, and K	0.728	0.714
M and K	1.02	1.016

W only from consideration (second row) increases FAR and FRR values slightly. But removing S modality (third row) only from consideration, the values of FAR and FRR increase significantly. Hence, this study also shows that stylometry contributes more to reduce the error rates (EAR and EER) than web browsing behavioral biometric.

Smartphone-Based Multi-sensor Authentication

Today's mobile environment requires an identity-centric continuous authentication (both active and passive) that supports user and application identification, which is equally dynamic, flexible, and situation-aware. Your smartphone knows who you are, where you are, who you contact, what you say, what you buy, the money you have in the bank, how many steps you take in a day, whether you are sick, how you sleep—it is the best tool to authenticate you.

As shown in ■ Fig. 6.15, a smartphone has many sensors and software/hardware which are capable of identifying users in a continuous and implicit manner via many identity footprints:

- **Connectivity**: Today's mobile devices use different types of communication media and networks such as Cellular, Wireless, Bluetooth, Near Field Communication (NFC), and also remote access through Virtual Privacy Network (VPN).
- **Device and Sensors**: All mobile devices are equipped with several sensors, such as accelerometer, gyroscope, proximity, light, magnetic field, orientation, and pressure, GPS, finger scanner, camera, voice recorder, etc. in addition to MAC and other device identifiers.

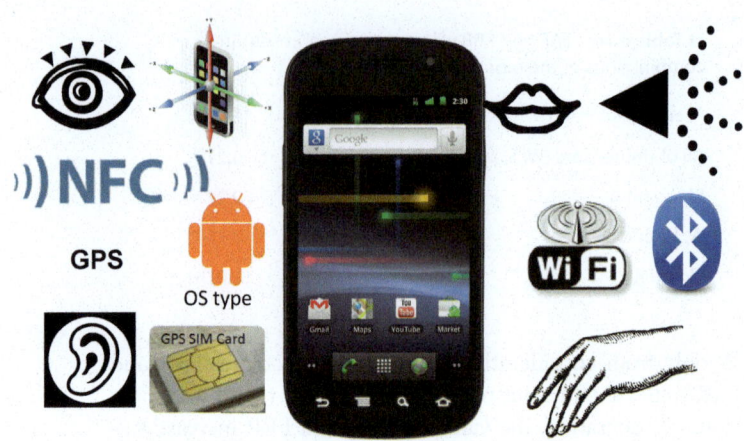

Fig. 6.15   Today's smartphone has many sensors and software/hardware identifiers for user authentication

- **OS variety**: Mobile Devices come with different desktop operating systems (Windows 7, Mac OS X, RedHat, Ubuntu, and other flavors of Linux), tablet operating systems (iOS, Android, WebOS), smart phone operating systems (Blackberry OS, iOS, Android, Symbian, Windows Mobile, WebOS).
- User identity is also critical to logging, monitoring, and audit functions that can be used to track who access what and when, for regulatory compliance or policy reasons.

Using some of the smartphone information, researchers [33] studied user behavioral patterns and then authenticated the person without interrupting his interactions with smartphone. In this work, the active authentication system creates a user profile in 10 s after capturing the sensory data and 20 s to detect an illegitimate user. It is also found that using multiple sensors, the accuracy of the system can be achieved more than 90%. A comparison among multiple sensor-based authentication approaches is mentioned in Table 6.11.

Table 6.11 highlights various approaches that use different sensors to provide user authentication for smart devices. Some of these approaches are scripted and hence they do not support implicit authentication process, while others do both implicit and continuous authentication. It is inferred from the table that, using only internal sensors of the mobile phone, user authentication can be done with good accuracy.

◼ **Table 6.11** Comparison of various sensor-based approaches for implicit and continuous authentication [33]

References	Devices	Sensors	Method	Accuracy	Detection time	Script
Lee et al. [33]	Nexus 5	Orientation, magnetometer, accelerometer	SVM	90.23%	Train: 6.07 s Test: 20 s	No
Kayacik et al. [50]	Android	Light, orientation, magnetometer, accelerometer	Temporal and spatial model	n.a.	Train: n.a Test: $\geq$122 s	No
Zhu et al. [51]	Nexus S	Orientation, magnetometer, accelerometer	n-gram language model	71.3%	n.a	Yes
Buthpitiya et al. [52]	n.a.	GPS	n-gram model on location	86.6%	Train: n.a Test: $\geq$ 30 min	No
Trojahn et al. [53]	HTC Desire	Screen	Keystroke and handwriting	FP:11% FN:16%	n.a.	Yes
Li et al. [54]	Motorola Droid	Screen	Sliding pattern	95.7%	Train: n.a Test: 0.648 s	Yes
Nickel et al. [55]	Motorola Milestone	Accelerometer	K-NN	FP: 3.97% FN: 22.22%	Train: 1.5 min Test: 30 s	Yes

■ **Fig. 6.16**   Authenticating IoT devices on a Continuous basis. *Source* Google image

Continuous Authentication for Internet of Things (IoT)

Internet of Things (IoT) has gained a lot of attention in industry as sensors, smart devices, and everyday objects now can communicate with one another using the Internet. For successful deployment of IoT technologies, these should have the following important requirements:

- Reliability, performance, quality of experience, and long-time availability are extremely high and connectivity is a critical success factor.
- Some IoT applications that require global coverage and/or mobility will use cellular technologies, but the majority of IoT devices will use non-cellular technologies.
- Need to ensure proper functionality, quality, and performance throughout the IoT product's lifetime; it is essential to test the overall communications behavior in all phases of the product life cycle (■ Fig. 6.16).

For different IoT devices to fit with Personal Area Network or Local Area Network, connect seamlessly and other media like Enhanced to Bluetooth Low Energy and Bluetooth mesh, Thread, ZigBee, Wi-Fi enhanced to Wi-Fi 802.11ah and Wi-Fi 802.11af, Low Power WAN (LP-WAN)—SIGFOX, LoRa, Weightless, and Ingenu cellular technologies (2G, 3G, 4G, 5G) with new feature enhancements like enhanced Machine Type Communications (eMTC) and Narrowband IoT (NB-IoT).

As IoT devices are connected with one another, there is an immense need to maintain the secure communication channel to transfer messages and control signals. Continuous authentication of a secure channel ensures fast and simple authentication for frequent message transmission in short time intervals. The authors in [34] mentioned a protocol for IoT applications based on secret sharing scheme.

The novelty of the protocol lies in two concepts. At first, the protocol employs the shares to both the claimer and the verifier as authentication tokens. The share token is generated by Shamir's secret sharing scheme [35]. Then, these tokens are released from the claimer to the verifier in a function of time. This single function outcome exposes portion of the secret to the recipient in order to guarantee the claimer's identity. This share is possible to be linked back to the secret, and thereby the message source can be authenticated. This solution is lightweight in terms of computation and communication cost, which justifies its efficacy to be applied in resource-constrained IoT applications.

Self-authenticable Approaches for Wearable Devices

Wearable devices are spreading at a faster rate and attaching to our body and providing seamless connectivity to the Internet or the cloud services. This new technology is an undeniable trend and many are creating new designs to satisfy a growing consumer awareness and demand for wearable products.

Secure authentication has not been explored much for different wearable devices. However, from the view of sensitivity of the information handled by wearable devices, authentication should be an essential service that any wearable device should provide. ◘ Figure 6.17 illustrates different wearable devices covering our body in different ways for various reasons.

Hence, it is essential to build an authentication protocol for wearable devices that allows a secure mutual authentication between a wearable device and any other entity such as another wearable device, a personal device (mobile phone), a remote server, or a user's application. Different entities present in the system can be authenticated using a protocol based on a digital signature scheme with a challenge-response mechanism [36]. Again, the new connectivity capabilities for wearable devices using short-range communication protocols, such as Near-Field Communication (NFC), and the need to increase their communication range to allow Internet connectivity using other personal devices as intermediate devices open up new security challenges. These intermediate devices could be malicious or adversary and may try to intercept or manipulate the information in transit during wearable devices' communications. Once connected

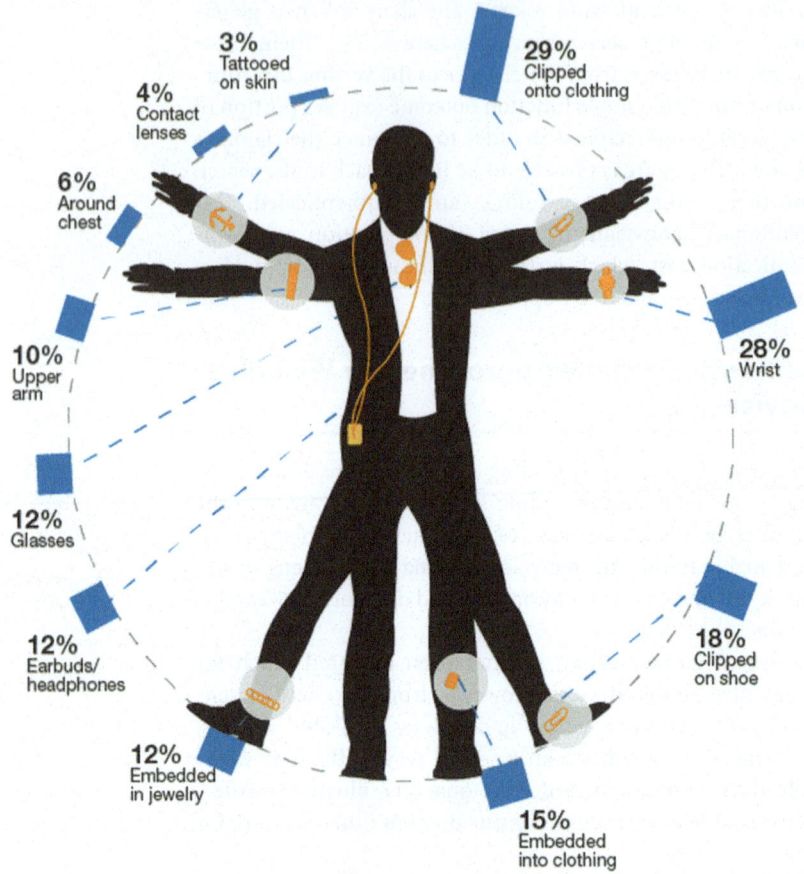

"How would you be interested in wearing/using a sensor device, assuming it was from a brand you trust, offering a service that interests you?"

3% Tattooed on skin

4% Contact lenses

29% Clipped onto clothing

6% Around chest

10% Upper arm

28% Wrist

12% Glasses

12% Earbuds/ headphones

18% Clipped on shoe

12% Embedded in jewelry

15% Embedded into clothing

Base: 4,657 US online adults (18+)
(multiple responses accepted)

Source: North American Technographics® Consumer Technology Survey, 2013

97141   Source: Forrester Research, Inc.

Fig. 6.17 Illustrate how wearable sensors will soon cover our body for monitoring and/or curing our health conditions

to the Internet, the wearable device could become an easy target for attackers without proper authentication protocols. The key functionalities of this protocol are given as follows:

1. This protocol must be secured and it can work directly over a short-range connection or use other intermediate trusted or untrusted devices for Internet access.
2. This protocol includes a secure imprinting mechanism to allow a wearable device to differentiate between trusted and untrusted intermediate devices.

3. This protocol allows to store different authentication credentials (e.g., password, certificates, etc.) within the wearable device. Hence, the holder's biometric data obtained by the wearable device itself could be used for secure authentication.
4. It allows the cryptographic and logical capabilities to be included within the wearable device, allowing it to make decisions depending on the sensitivity of information.

The current objective is to design a device-independent protocol that can exploit already proven standards, enabling its security and interoperability. Hence, the protocol proposes a Java Card based on smart card technology within the wearable device to satisfy cryptographic, logic and storage capabilities, as well as size limit. These wearable devices include NFC instead of RFID, which ensures its resilience against man in-the-middle attacks [36].

The National Strategy for Trusted Identities in Cyberspace (NSTIC)

In April 2011, NIST-NSTIC called for an Identity Ecosystem, «an online environment where individuals and organizations will be able to trust each other because they follow certain guidelines agreed upon standards to obtain and authenticate their digital identities.»

In Identity Ecosystem, individuals should be able to choose among multiple identity providers and digital credentials for convenient, secure, and privacy-enhancing transactions anywhere, anytime [37]. Identity Ecosystem solutions must be privacy enhancing and voluntary, secure and resilient, interoperable, and they must be cost effective and easy-to-use. As shown in ◘ Fig. 6.18, the proposed identity-centric infrastructure consists of three major architectural components: endpoints, policy enforcement points, and policy decision points [37, 38].

- The core of the architecture is a policy engine (the policy decision point) that allows central management of the user-centric policies for the business implementation.
- The policy decisions are made by comparing attributes (such as user identity, device type, location, time of day, application etc.) to dynamic access policies.
- Endpoints are the end-user devices but also include applications, databases, servers, and other devices connected to the network.

6

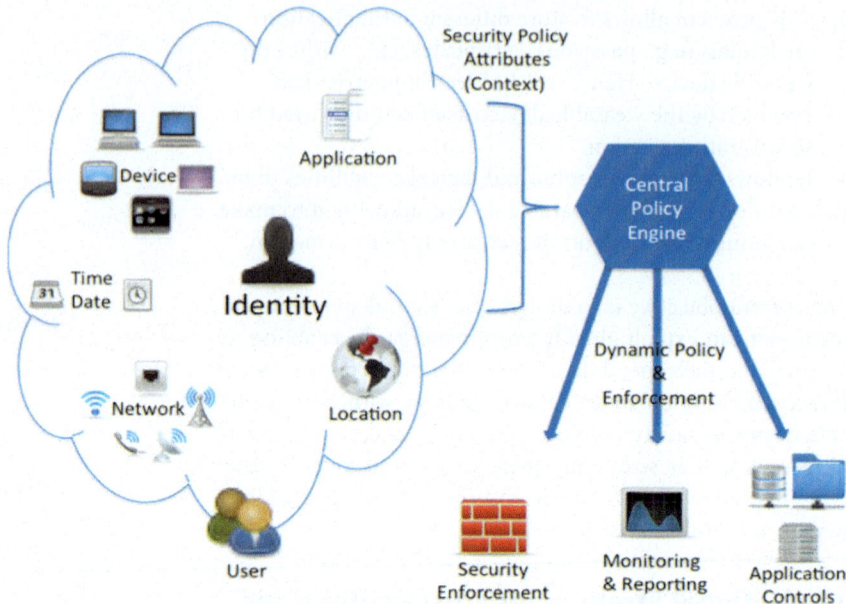

■ Fig. 6.18   NIST Identity-Management Eco-System. *Source* Nemertes Research (▶ www. nemertes.com)

    ▬ Enforcement points are various security devices or network devices (routers, switches, etc.) that can enforce access control decisions.

The Identity Ecosystem can enhance the following:
▬ **Privacy protections** for individuals, who can trust that their personal data is out of reach of the malicious users and is handled fairly and transparently.
▬ **Convenience** for individuals, who may choose to manage fewer passwords or accounts than they do today.
▬ **Efficiency** for organizations will be benefited from the reduction in paper-based and account management processes.
▬ **Ease-of-use** by automating identity solutions whenever possible and basing them on technology that is simple to operate.
▬ **Security** can be enhanced by making it more difficult for malicious users to compromise online transactions.
▬ **Confidence** that the user digital identities are adequately protected.
▬ **Innovation** can be increased by lowering the risk associated with sensitive services and by enabling service providers develop or expand their online presence.

Accordingly, Identity Ecosystem Steering Group (IDESG), a privately led governance body was formed which includes public and private sectors to collaborate and raise the level of trust associated with the identities of individuals, organizations, networks, services, and devices involved in online transactions [37]. The Identity Ecosystem Steering Group (IDESG)—an independent, non-profit organization dedicated to create the future of trusted digital identities released the first version of Identity Ecosystem Framework (IDEF) on October 15, 2015. The IDEF v.1 represents three core documents that describe the Identity Ecosystem and the requirements, best practices, and approved standards which need to be considered in compliance with it [40].

■ Figure 6.19 illustrates a multi-party trust framework built upon the foundation of the IDEF. The baseline requirements, policies, and processes ensure underlying interoperability such that credentials can be relied upon even when the participants are in different trust frameworks, which is a key NSTIC requirement [40].

The accreditation process and trustmarks are designed to foster trust among all Identity Ecosystem participants. The IDESG trustmark is a mechanism for efficiently communicating the policies and technologies that an Identity Ecosystem participant supports. For individuals, the trustmark is a simple alternative to reading documents which include

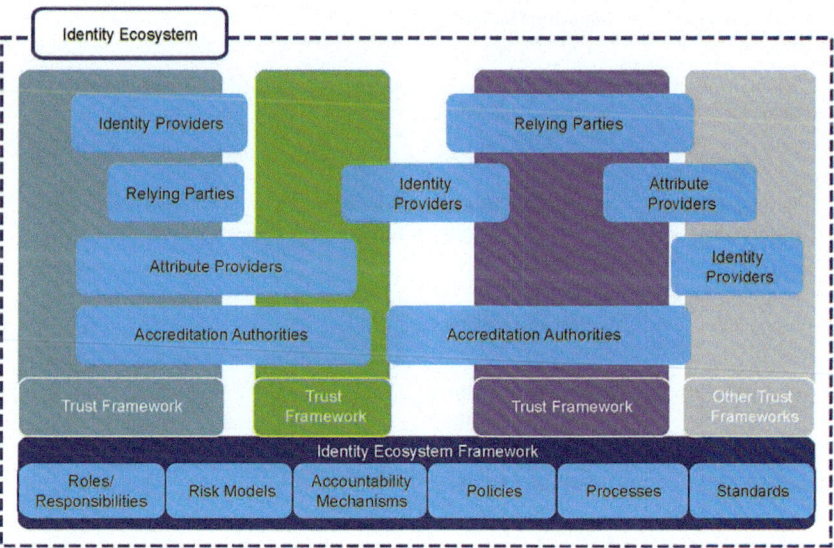

■ **Fig. 6.19**   Identity ecosystem framework highlighting different segments [49]

terms of service or detailed privacy policies [40]. It can provide easy means to identify service providers who abide by a set of uniform policies. IDESG recommends to establish a reliable third-party authentication service i.e. relying on a third party to verify the identity of the applicant [40].

Accordingly, Identity Ecosystem Framework (IDEF) provides a registry with publicly accessible listing service of entities that provide online identity services established for navigating the evolving landscape of online identity. Its purpose is to create policy foundation for strengthening privacy and security protections for organizations and consumers while asserting capabilities and responsibilities for individuals, companies, government agencies, and organizations in the identity ecosystem. There are 7 baseline requirements—interoperability, 15 privacy, 15 security and 7 usability guidelines.

□ Table 6.12 A summary comparison of continuous authentication works with their unique difference is shown in a tabular form

Continuous authentication approaches	Factor considered	Selection strategy for multi-factor	Applicability
Cognitive-centric text production and revision features [60]	Multi-factor: behavioral biometric and keystroke dynamics	Fusion of all features	All devices
Context-aware/gait-based [56, 61]	Single factor: behavioral biometric	None	Mobile devices
Typing motion behavior [13]	Single factor: behavioral biometric using statistical features	None	Mobile devices
Temporal authentication [5]	Multi-factor: face detection and body localization	Both factors individually	Fixed and portable devices
Messaging App usage [62]	Single factor: behavioral biometrics	None	Mobile devices
Touch screen gestures [58, 63]	Single factor: behavioral biometrics with finger gestures	None	Mobile devices
Keystroke dynamics [59, 64]	Single factor: behavioral biometrics	None	All devices
Behavioral biometrics [65]	Multi-factor: keyboard, mouse, and application interactions	Fusion of three in a trust model	Fixed and portable devices

A web-based company, ProctorU [41] provides student authentication services and live proctoring to validate user. They make use of web camera and microphone to continuously monitor a user throughout the session. At the very beginning of the session the registered user is authenticated by verifying his/her state ID and cross-verified if it is the same person who registered to take the test. Activities such as looking away from the scene, talking to other person, or getting up and walking away are sensed. Users engaged in such activities are flagged. All the flagged videos are then viewed by security personals and appropriate action is taken. The web camera and audio feeds are analyzed using computer vision and pattern recognition algorithms. Another company Vproctor [42] provides a similar service but the only difference is that in Vproctor, a software application can be installed at the client.

This chapter discusses different approaches for continuous authentication of users and ▪ Table 6.12 summarizes some of the works published so far.

Chapter Summary

This chapter discusses on several aspects of continuous authentication to verify legitimate users during session in a periodic manner. Active authentication is one way of continuously authenticating users mainly with the help of several behavioral and cognitive traits. Active authentication becomes an emerging trend of modern day authentication research, as one-time validation of user's identity is not a smart choice to secure user's credentials for a longer period. A good active authentication system should be fast in execution, nonintrusive, application independent, hardware independent, and it should use different behavioral attributes, like mouse movements, keystroke dynamics, temporal information, cognitive and screen fingerprint, etc. Active authentication can be unimodal (using only one behavioral attribute); however, several behavioral attributes can be combined to form multi-modal active authentication system. Many recent research works show the applicability of active authentication on desktop and mobile environments. Moreover, interesting research continues on developing cyber identity-eco systems which may require modification and refinement in continuous authentication systems.

Review Questions

Question 1:
What is continuous authentication? Discuss four characteristics for good continuous authentication system.
Question 2:
What are the features to be considered for Keystroke dynamics? Explain them.
Question 3:
How web browsing behavior distinguishes the users in continuous authentication systems?
Question 4:
Describe the key features of cognitive fingerprint and how is it used in continuous authentication?
Question 5:
What type of temporal information is used in continuous authentication? Briefly discuss.
Question 6:
What is meant by HMOG? Describe this feature for continuous authentication for mobile devices.
Question 7:
What is the difference between multi-modal and uni-modal continuous authentication?
Question 8:
Describe a multi-modal continuous authentication where keystroke and mouse are one of the modalities for authentication.
Question 9:
Describe the text-based biometrics as multi-modal behavioral biometrics.
Question 10:
What features of application usage and screen fingerprints are used for continuous authentication?

Multiple-choice Questions

Question 1
Which of the following(s) is not considered as cognitive fingerprint for active authentication?
A. User search
B. Stylometry
C. Covert games
D. How people talks

Question 2

What is the most common behavioral biometric used for both static and continuous authentication?

A. Keystroke and Mouse
B. Fingerprint
C. Face recognition in mobile device
D. Voice recognition

Question 3

Which of the following characteristics is required for a good continuous authentication system?

A. Physiological attributes
B. Application-oriented
C. Hardware-oriented
D. Non-intrusiveness

Question 4

Which characteristic is NOT required for a good continuous authentication system?

A. Non-intrusiveness
B. Application-independent
C. Hardware-dependent
D. Fast

Question 5

Which metric is better for continuous authentication using keystroke dynamics?

A. EER
B. number of keystrokes before detecting imposter
C. FAR
D. FNMR

References

1. Guidorizzi RP (2013) Security: active authentication. IT Prof 15(4):4–7
2. Jagadeesan H, Hsiao MS (2011) Continuous authentication in computers. Continuous Authentication Using Biometrics: Data, Models, Metrics: Data, Models, Metrics (2011):40
3. Fathy ME, Patel VM, Chellappa R (2015) Face-based active authentication on mobile devices. In: 2015 IEEE international conference on acoustics, speech and signal processing (ICASSP). IEEE, pp 1687–1691
4. Viola P, Jones MJ (2004) Robust real-time face detection. Int J Comput Vision 57(2):137–154

5. Niinuma K, Jain AK (2010) Continuous user authentication using temporal information. In: SPIE Defense, Security, and Sensing. International Society for Optics and Photonics, pp 76670L–76670L

6. Mahbub U, Patel VM, Chandra D, Barbello B, Chellappa R (2016) Partial face detection for continuous authentication. ►arXiv:1603.09364

7. Abramson M, Aha DW (2013) User authentication from web browsing behavior. Naval Research Lab, Washington DC

8. Kumar R, Tomkins A (2010) A characterization of online browsing behavior. In: Proceedings of the 19th international conference on World Wide Web. ACM, pp 561–570

9. Preuveneers D, Joosen W (2015) SmartAuth: dynamic context fingerprinting for continuous user authentication. In: Proceedings of the 30th annual ACM symposium on applied computing. ACM, pp 2185–2191

10. Domingos P, Hulten G (2000) Mining high-speed data streams. In: Proceedings of the sixth ACM SIGKDD international conference on knowledge discovery and data mining. ACM, pp 71–80

11. Karnan M, Akila M, Krishnaraj N (2011) Biometric personal authentication using keystroke dynamics: a review. Appl Soft Comput 11(2):1565–1573

12. Feng T, Zhao X, Carbunar B, Shi W (2013) Continuous mobile authentication using virtual key typing biometrics. In: 12th IEEE international conference on trust, security and privacy in computing and communications (TrustCom). IEEE, pp 1547–1552

13. Gascon H, Uellenbeck S, Wolf C, Rieck K (2014) Continuous authentication on mobile devices by analysis of typing motion behavior. Sicherheit, pp 1–12

14. Chang JM, Fang CC, Ho KH, Kelly N, Wu PY, Ding Y, Chu C, Gilbert S, Kamal AE, Kung SY (2013) Capturing cognitive fingerprints from keystroke dynamics. IT Prof 15(4):24–28

15. Stamatatos E (2009) A survey of modern authorship attribution methods. J Am Soc Inform Sci Technol 60(3):538–556

16. Olsson J (2008). Forensic linguistics, 2nd edn. London: Continuum. ISBN: 978-0826461094

17. Juola P, Noecker JI, Stolerman A, R MV, Brennan P, Greenstadt Rachel (2013) Keyboard-behavior-based authentication. IT Prof 15(4):8–11

18. Fathy ME, Patel VM, Yeh T, Zhang Y, Chellappa R, Davis LS (2014) Screen-based active user authentication. Pattern Recogn Lett 42:122–127

19. Patel VM, Yeh T, Fathy ME, Zhang Y, Chen Y, Chellappa R, Davis L (2013) Screen fingerprints: a novel modality for active authentication. IT Prof 15(4):38–42

20. Rashid A, Baron A, Rayson P, May-Chahal C, Greenwood P, Walkerdine J (2013) Who am I? Analyzing digital personas in cybercrime investigations. Computer 4:54–61

21. Sitova Z, Sedenka J, Yang Q, Peng G, Zhou G, Gasti P, Balagani K (2015) HMOG: a new biometric modality for continuous authentication of smartphone users. arXiv: 1501.01199

22. Deutschmann I, Nordstrom P, Nilsson L (2013) Continuous authentication using behavioral biometrics. IT Prof 15(4):12–15

23. Snapshots: DHS S&T, DARPA co-leading development of higher-level mobile device authentication methods. ▶https://www.dhs.gov/science-and-technology/news/2016/07/20/snapshots-dhs-st-darpa-co-leading-development-higher-level. Accessed on 20 July 2016

24. Fridman L, Weber S, Greenstadt R, Kam M (2015) Active authentication on mobile devices via stylometry. Appl Usage, Web Browsing, GPS Location. ▶arXiv:1503.08479

25. Li F, Clarke N, Papadaki M, Dowland P (2014) Active authentication for mobile devices utilising behaviour profiling. Int J Inform Secur 13(3):229–244

26. Chair Z, Varshney PK (1986) Optimal data fusion in multiple sensor detection systems. IEEE Trans Aerosp Electron Syst 1(1986):98–101

27. Saevanee H, Clarke N, Furnell S, Biscione V (2014) Text-based active authentication for mobile devices. In: ICT systems security and privacy protection. Springer, Berlin, pp 99–112

28. Wu J, Ishwar P, Konrad J (2016) Two-stream CNNs for authentication and identification: learning user gesture style. In: IEEE conference on computer vision and pattern recognition (CVPR) biometrics workshop

29. Shotton J, Sharp T, Kipman A, Fitzgibbon A, Finocchio M, Blake A, Cook M, Moore R (2013) Real-time human pose recognition in parts from single depth images. Commun ACM 56(1):116–124

30. Myers CS, Rabiner LR (1981) A comparative study of several dynamic time-warping algorithms for connected-word recognition. Bell Syst Tech J 60(7):1389–1409

31. Primo A, Phoha VV, Kumar R, Serwadda A (2014) Context-aware active authentication using smartphone accelerometer measurements. In: IEEE conference on computer vision and pattern recognition workshops (CVPRW), IEEE, pp 98–105

32. Fridman A, Stolerman A, Acharya S, Brennan P, Juola P, Greenstadt R, Kam M (2013) Decision fusion for multimodal active authentication. IT Prof 4:29–33

33. Lee WH, Lee RB (2015) Multi-sensor authentication to improve smartphone security. In: Conference on information systems security and privacy, pp 1–11

34. Bamasag OO, Youcef-Toumi K (2015) Towards continuous authentication in internet of things based on secret sharing scheme. In: Proceedings of the WESS'15: workshop on embedded systems security. ACM, p 1

35. Shamir A (1979) How to share a secret. Commun ACM 22(11):612–613

36. Diez FP, Touceda DS, Cámara JMS, Zeadally S (2015) Toward self-authenticable wearable devices. IEEE Wireless Commun 37

37. ▶www.nist.gov

38. ▶http://www.ffiec.gov/pdf/authentication_guidance.pdf

39. DARPA's Active Authentication (2014) Date accessed 15 Nov 2015. ▶http://www.smartcardalliance.org/wp-content/uploads/karime.pdf

40. The Identity Ecosystem Steering Group (IDESG) developed identity ecosystem framework (IDEF). ▶www.IDESG.org

41. ▶http://www.proctoru.com/safety.php

42. ▶http://vproctor.com/

43. Guidorizzi R (2012) Active authentication moving beyond passwords. Second round table: from biometric to augmented human recognition, 10 May 2012
44. DARPA's Active Authentication: moving beyond passwords. ►http://www.connectidexpo.com/creo_files/expo2014-slides/connectID%20slides.pdf. Accessed on 17 Mar 2014
45. ►https://www.pinterest.com/explore/sentence-structure/
46. Bank C (2014) Gestures and animations: the pillars of mobile design. 1351. ►https://uxmag.com/articles/gestures-animations-the-pillars-of-mobile-design. Accessed on 1 Dec 2014
47. Grant TD (2007) Quantifying evidence for forensic authorship analysis. Int J Speech Lang Law 14(1):1–25
48. Chen MC, Anderson JR, Sohn MH (2001) What can a mouse cursor tell us more? Correlation of eye/mouse movements on web browsing. In: CHI'01 extended abstracts on human factors in computing systems. ACM, pp 281–282
49. The National Strategy for Trusted Identities in Cyberspace (NSTIC or Strategy), 15 April 2011. ►https://www.whitehouse.gov/sites/default/files/rss_viewer/NSTICstrategy_041511.pdf
50. Kayacik HG, Just M, Baillie L, Aspinall D, Micallef N (2014) Data driven authentication: on the effectiveness of user behaviour modelling with mobile device sensors. ►arXiv:1410.7743
51. Zhu J, Wu P, Wang X, Zhang J (2013) Sensec: mobile security through passive sensing. In: International conference on computing, networking and communications (ICNC). IEEE, pp 1128–1133
52. Buthpitiya S, Zhang Y, Dey AK, Griss M (2011) N-gram geo-trace modeling. In: International conference on pervasive computing. Springer, Berlin, pp 97–114
53. Trojahn M, Ortmeier F (2013) Toward mobile authentication with keystroke dynamics on mobile phones and tablets. In: 27th international conference on advanced information networking and applications workshops (WAINA). IEEE, pp 697–702
54. Li L, Zhao X, Xue G (2013) Unobservable re-authentication for smartphones. In: NDSS
55. Nickel C, Wirtl T, Busch C (2012) Authentication of smartphone users based on the way they walk using k-NN algorithm. In: Eighth international conference on intelligent information hiding and multimedia signal processing (IIH-MSP). IEEE, pp 16–20
56. Guidorizzi RP (2014) DARPA's active authentication: moving beyond passwords
57. Serwadda A, Balagani K, Wang Z, Koch P, Govindarajan S, Pokala R, Goodkind A, Brizan DG, Rosenberg A, Phoha VV (2013) Scan-based evaluation of continuous keystroke authentication systems. IT Prof 4:20–23
58. Chang JM, Fang CC, Ho KH, Kelly N, Wu PY, Ding Y, Chu C, Gilbert S, Kamal AE, Kung SY (2013) Capturing cognitive fingerprints from keystroke dynamics. IT Prof 15(4):24–28
59. Zhong Y, Deng Y, Jain AK (2012) Keystroke dynamics for user authentication. In: IEEE computer society conference on computer vision and pattern recognition workshops (CVPRW), IEEE, pp 117–123

60. Locklear H, Sitova Z, Govindarajan S, Goodkind A, Brizan DG, Gasti P, Rosenberg A, Phoha V, Balagani KS (2014) Continuous authentication with cognition-centric text production and revision features. In: International joint conference on biometrics (IJCB), Clearwater, Florida
61. Primo A, Phoha VV, Kumar R, Serwadda A (2014) Context-aware active authentication using smartphone accelerometer measurements. In: The IEEE conference on computer vision and pattern recognition (CVPR) workshops
62. Klieme E, Engelbrecht KP, Möller S (2014) Poster: towards continuous authentication based on mobile messaging app usage
63. Feng T, Liu Z, Kwon KA, Shi W, Carbunar B, Jiang Y, Nguyen N (2012) Continuous mobile authentication using touchscreen gestures. In: IEEE conference on technologies forhomeland security (HST). IEEE, pp 451–456
64. Serwadda A, Wang Z, Koch P, Govindarajan S, Pokala R, Goodkind A, Brizan DG, Rosenberg A, Phoha VV, Balagani K (2013) Scan-based evaluation of continuous keystroke authentication systems. IT Prof 4(15):20–23
65. Deutschmann I, Lindholm J (2013) Behavioral biometrics for DARPA's active authentication program. In: Biometrics Special Interest Group (BIOSIG). IEEE, pp 1–8

Adaptive Multi-factor Authentication

Bring dynamicity in multi-factor authentication process

©Bloomua/Shutterstock.com

© Springer International Publishing AG 2017

D. Dasgupta et al., *Advances in User Authentication*, Infosys Science Foundation Series,

DOI 10.1007/978-3-319-58808-7_7

Introduction

With the advancements of modern technology, most user activities rely upon various online services, which need to be trusted and secured to prevent the thorny issue of illegal access. Authentication is the primary defense to address the growing need of authentications, though a single-factor (user id and password, for example) is suffering from some significant pitfalls as mentioned in earlier chapters. As a consequence, authentication through different factors is an ongoing trend to provide secure, resilient, and robust access verification to legitimate users of computing systems and, at the same time, makes it harder for intruders to gain unauthorized access. Rapidly developing mobile technology continues to increase user's access to online services. Hence, checking the authenticity of the registered users on a continuous basis is of utmost importance to protect any sensitive user data. This enhances the need to move toward the multi-factor authentication system and provide different choices to the users during authenticating to verify their identity. Multi-factor Authentication (MFA) also comes with a fail-safe feature in case of compromising any authentication factor as users are authenticated utilizing the other existing non-compromised modalities. Chapters 5 and 6 discussed regarding various multi-factor authentication approaches and continuous authentication approaches, respectively.

To design a resilient MFA solution, one of the critical issues is to choose the better set of authentication factors out of all possible choices in any given operating environment of the user. The selection of a good set of authentication factors determines the performance of the MFA as a whole. Surprisingly, recent MFA products do not address this issue while providing solutions to the end users [1–6]. Additionally, using the same factors or a random selection of factors for MFA to validate users across all possible operating environments (devices, media, and surroundings factors) is not reasonable as the selected sets are predictable, exploitable (for the set of same factors), and less trustworthy (for random selection, all the selected sets of authentication factors are not equally trusted). Hence, there is an increasing need for adaptive selection of authentication factors (to validate the users at any given time) sensing the existing operating conditions. In this chapter, an adaptive multi-factor authentication system is described, where the system is composed of following two components:

1. Developing optimization strategies to calculate trust-worthy values for various authentication factors.
2. Developing an adaptive selection framework for MFA using a nonlinear multi-objective optimization approach with the probabilistic objective.

Different authentication modalities are used for a multi-factor authentication system. Some of them could be classified as physiological biometrics like face, fingerprint, voice, etc., while some others fall into behavioral biometrics like keystroke dynamics, mouse movement, gait, etc. Also, some non-biometric modalities like SMS (for PIN code), password, and passphrases are also popular for user authentication.

Adaptive-MFA (A-MFA) Framework

An MFA system is designed and implemented as a client-server model-based service that covers all of the abovementioned categories by offering a wide variety of authentication modalities and information as shown in ◘ Table 7.1.

In this work, an adaptive multi-factor-based authentication system, shown in ◘ Fig. 7.1, is designed and implemented to address some of the current shortcomings of available multi-factor authentication solutions. The A-MFA system chooses the best set of authentication modalities for a user, sensing the connected device, communicating medium, and surrounding conditions (light, noise, motion, location, etc.). The consideration of these environment factors into

◘ **Table 7.1** Categorization of authentication modalities supported by the A-MFA system

Knowledge-based modalities	Password, passphrase Security challenge questions
Possession-based modalities	SMS code TOTP code
Biometric modalities	Face recognition Fingerprint recognition Voice recognition Keystroke recognition
Location-based modalities	GPS, IP address, MAC address Wi-Fi triangulation Cellular triangulation

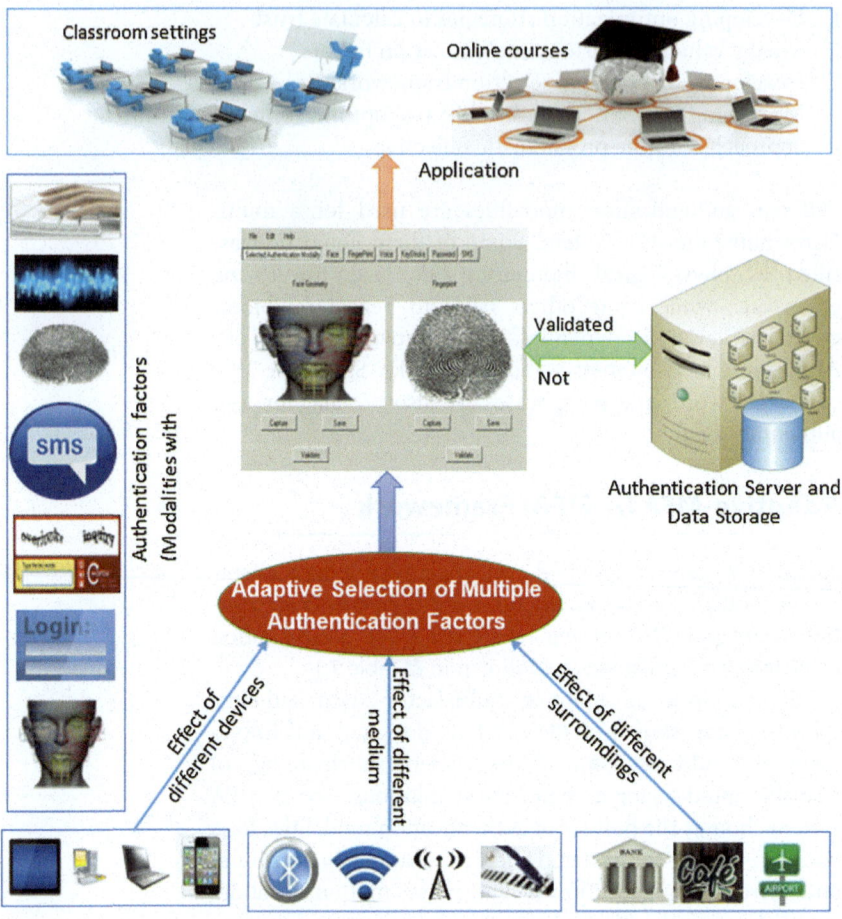

Fig. 7.1 Conceptual diagram highlighting the overall adaptive multi-factor authentication (A-MFA) approach

selection decision assures to capture the dynamic nature of the environment and to provide a better set of authentication factors suitable for that environment settings. In order to design a more secure solution for selection of MFA, the selection decision (set of authentication factors) should not be predicted beforehand. If there is any predefined pattern of selection for a particular user in any particular environment, it opens a window for attackers to sabotage the authentication decision. To address this mission-critical issue, the Adaptive Multi-factor solution considers the previous selection history of a particular user and it ensures that the same set of authentication factor will not be used to challenge the authenticity of that user in successive reauthentication attempts. The design of authentication factor selection process is discussed below.

Details of the concept, design, and implementation of Adaptive-MFA solution are discussed later in this chapter.

Authentication Factor

An authentication factor is redefined in this chapter to include authentication modalities along with their features. This helps in increasing the total number of possible authentication factors with a given list of authentication modalities which assists to choose:

1. Single feature of an authentication modality; or,
2. Any combination of features of an authentication modality; or,
3. Combination of multiple features of different authentication modalities.

Let $M_{k\in\mathbb{Z}^+}$ be the k-th authentication modality and $\{M_k{:}f_{k,i}\}$ be its i-th feature. Then, $\{\{M_k\}{:}\{f_{k,i}\}_{i\in\mathbb{Z}^+}\}_{k\in\mathbb{Z}^+}$ can be defined as the i-th features of different combinations of $\{M_k\}_{k\in\mathbb{Z}^+}$. For example, $\{M_1{:}f_{1,1}\}$ and $\{M_2{:}f_{2,1}\}$, the first features of M_1 and M_2, respectively, can be considered as two authentication factors (as shown in scenario 1). Moreover, $\{M_1{:}f_{1,1},f_{1,2}\}$, combinations of $\{M_1{:}f_{1,1}\}$ and $\{M_1{:}f_{1,2}\}$, can also be considered as an authentication factor (scenario 2). Additionally, the combination of $\{M_1{:}f_{1,1}\}$ and $\{M_2{:}f_{2,1}\}$, i.e., $\{M_1, M_2{:}f_{1,1},f_{2,1}\}$, can also be considered as one authentication factor (scenario 3). This form of defining authentication factor can significantly increase the available options for choosing the better set of authentication factors, and make the selection decision in a wise manner by ensuring diversity and randomness in nature. Different biometric and non-biometric authentication modalities with their features and usability are discussed in Chaps. 1 and 2.

Considered Surrounding Conditions for MFA

The adaptive selection of MFA provides the selection decision sensing the existing surrounding conditions. For this research, light and noise conditions are considered. For light, luminance (light intensity) is a measure to represent the lighting conditions and decides whether any particular modality can be chosen (for example, face recognition) or not. ◼ Table 7.2 shows different lighting conditions in

◻ **Table 7.2**   Different lighting conditions and their corresponding luminance values [39]

Luminance	Surfaces illuminated by
0.0001 lx	Moonless, almost overcast night sky
0.002 lx	Moonless clear night sky with airflow
0.27–1.0 lx	Full moon on a clear night
3.4 lx	Dark limit of civil twilight under a clear sky
50 lx	Family living room with regular lights
80 lx	Office building hallway/toilet lighting, dull lighting condition
100 lx	Extremely dark and overcast day with bad light
320–500 lx	Office lighting mainly
400 lx	Sunrise or sunset on a clear day
1000 lx	Overcast day; typical TV studio lighting with dull light
10,000–25,000 lx	Full daylight
32,000–100,000 lx	Direct sunlight or bright day light

everyday life and the luminance values. The acceptable range of luminance value lies between 100 and 500 lx.

◻ Table 7.3 shows the details of various surrounding conditions and the acceptable ranges of different conditions and which authentication modalities are affected by the surrounding conditions.

In Adaptive-MFA approach, various types of devices, media, and surrounding conditions are incorporated to provide a better set of authentication factors to identify the legitimate users in various environment settings. The A-MFA approach considers three types of devices, namely, fixed device (FD), portable device (PD), and handheld device (HD). Similarly, three different types of media are considered, namely, wired medium (WI), wireless (WL), and cellular (CL). The light and noise are considered as the surrounding conditions. However, this framework can easily

◘ **Table 7.3** Effect of various surrounding conditions in different authentication factors and the acceptable ranges

Surrounding conditions	The acceptable range of surrounding conditions	Affected authentication modalities
Light (L)	The luminance of the light levels varies with indoor and outdoor conditions. The typical light level falls in the range of 100–1000 lx. The range of light is generally categorized as human visible range and human invisible range. The wavelength of the visible range varies from 380 to 750 nm. A typical human eye will respond to the visible range of wavelengths The nearest wavelength of visible light is infrared light, which has its wavelength in the range of 750 nm–1 mm. Most of the thermal radiations emitted by objects near room temperature are infrared	Face recognition
Sound and background noise (N)	The reasonable frequency range of audible sound is 20 Hz–20 k Hz. Humans cannot hear any sound with a frequency less than 20 Hz and above 20 kHz In general, the threshold level of hearing is 30–90 dB. Within this range, any noise below 10 dB is considered as very faint noise. Faint is considered as the level of noise that falls between 1035 dB. Quiet to moderate level noise falls into 36–70 dB range. Any noise level that falls within 70–90 dB is considered as loud noise	Voice recognition

be extended with other surrounding conditions and other types of device and media. ◘ Figure 7.2 shows different types of device and media and frequency of the authentication required for different environment settings. Using this information, required authentication factors can be adjusted to fit with various operating conditions.

Device Type	Connection Media	Environment	Frequency of Authentication	No. of Modalities Activate
Fixed Device	Wired	Private	Static	small
		Public with Login	Semi-static	medium
	Wireless	Public with Login	Semi-static	medium
Portable Device	Wired	Private	Static	small
		Public with Login	Semi-static	medium
	Wireless	Public Hotspot	Dynamic	large
		Public with Login	Semi-static	medium
Handheld Device	Wireless	Public Hotspot	Dynamic	large
		Public with Login	Semi-static	medium
	Cellular	Public with Login	Semi-static	medium
Internet of Things	Wired	Private	Static	medium
	Wireless	Public with Login	Semi-static	large

Terminology:
Device:
Fixed Device: Desktop, server, workstation
Portable Device: Laptop, notebook
Handheld Device: Tablet, Smart phone, PDA
Internet of Things: Any electronic device with internet connection.
Types of Network Connection:
Wired: connected with Ethernet cable
Wireless: connected using Wi-Fi network
Cellular: connected using mobile operator's network
Environment:
Private: Home or business network
Public Hotspot: Star bucks, McDonald
Public with Login: Airport, Hotel, Station.
Frequency of Authentication:
Static: Once in a given day
Semi-static: After a fixed duration (like after 3 hours) or after inactivity of use for a fixed time.
Dynamic: frequently (like every hour) or when usage type changes (use of banking app, credit card app, using corporate application app)
No. of Modality Considered:
Small: 2 Medium: 3 Large: 4.

Fig. 7.2 Criteria for selecting authentication factors

Trustworthy Framework for Authentication Factors

One of the challenges in designing the adaptive selection of different authentication factors is to come out with a metric to compare different authentication factors in different operating conditions. The operating conditions are varied in different types of devices and media used by the users. In this chapter, the trustworthy factor is defined to measure the effectiveness of different authentication factors among the various devices and media. As of now, there is no approach available to calculate the trustworthy values among different authentication factors. First, a formulation of trustworthy model for various modalities with using the pairwise comparative preference information is introduced. It is assumed that all the required pairwise comparative preference information is available beforehand to compute the trustworthy model. Second, another approach with the probabilistic formulation of the trustworthy values from the pairwise comparative preference is discussed later.

First Approach

In this approach [7–10], trustworthy value function for different modality–media–device combinations is done by utilizing pairwise comparative preference information. For instance, the following type of question/answer were collected:

For a Wired Media, which of the two Devices—Fixed or Portable—is more trustworthy?

The answer could be any one of the following three cases: [8]:

Case 1: A Fixed Device is more trustworthy than a Portable Device.

Case 2: A Portable Device is more trustworthy than a Fixed Device.

Case 3: Both Devices are almost equally trustworthy or non-trustworthy.

Interestingly, since an optimization method is used to build the model, not all pairwise combinations of modalities, media, and devices are needed to be considered.

The representation of deterministic trustworthy value is T_{ij}^m for the m-th modality (of M options), i-th device (of N options), and j-th media (of P options). Thus, for a particular pairwise comparison involving i-th and k-th devices for a fixed (j-th) media and fixed (m-th) modality, it was desired to come up with T_{ij}^m values, such that $T_{ij}^m > T_{kj}^m$, if Case 1 is the answer; $T_{ij}^m < T_{kj}^m$, if Case 2 is the answer; and $T_{ij}^m \approx T_{kj}^m$, if Case 3 is the answer. Based on these three cases, the following optimization problem can be formed to find a set of normalized T_{ij}^m values:

Maximize $\in = \in_1 - \in_2$
Subject to:

$$\left| T_{ij}^m - T_{kj}^m \right| < \in_1, \quad \text{for each } j = 1, 2, 3 \text{ and } i, k = 1, 2, 3, \ i \neq k \tag{7.1}$$

$$\left| T_{ij}^m - T_{kj}^m \right| < \in_2, \quad \text{for each } j = 1, 2, 3 \text{ and } i, k = 1, 2, 3, \ i \neq k \tag{7.2}$$

$$0 \leq T_{ij}^m \leq 1, \quad \text{for all } i, j = 1, 2, 3. \tag{7.3}$$

The above optimization problem considers $\in$ as an additional variable, and it makes a total of $N \times P + 1$ variables for each authentication modality. The first three constraints enforce prescribed preferences. The fourth and fifth constraints together ensure that all T_{ij} values lie in [0, 1] with the condition that the highest T_{ij} value is exactly one. This condition acts as a normalization factor. It will be useful when the trustworthy values for different modalities will be combined later to construct the overall trustworthy function value. The fourth constraint can be relaxed with each $T_{ij}^m \leq 1$ and split the third constraint set into two inequality constraint sets, so that the resulting problem becomes a linear programming (LP) problem. The resulting LP problem can then be solved using an LP solver (like Matlab's) and the optimal T-matrix can be obtained [8].

If the optimal solution to the above problem causes the objective function value ϵ^* to be strictly positive, a set of acceptable T-matrix is then found honoring all prescribed preferences. The T-matrix can then be used to compute the overall trustworthy function value (trustworthy factor). The detailed steps of forming the optimization problem using the preference information and the calculations of trustworthy factors for a modality are shown below.

This example only considers the pairwise comparative preference information as a metric to calculate the trustworthy function values. An example of face recognition modality (M_1) with three devices ($i = 1$ for fixed device, 2 for portable device, 3 for handheld device) and three media ($j = 1$ for wired media, 2 for wireless media, and 3 for cellular media) were considered, thereby making a total of nine trustworthy values. The preference conditions for M_1 are shown below:

(i) For wired media (WI), a portable device (PD) is more trustworthy than a handheld device (HD), meaning $T_{21}^1 - T_{31}^1 \geq \epsilon_1$.

(ii) Irrespective of media, a fixed device is more trustworthy than a portable device, $T_{FD} > T_{PD}$, meaning $(T_{11}^1 + T_{12}^1 + T_{13}^1) - (T_{21}^1 + T_{22}^1 + T_{23}^1) \geq \epsilon_1$.

(iii) Irrespective of media, a fixed device is more trustworthy than a handheld device, $T_{FD} > T_{HD}$, meaning $(T_{11}^1 + T_{12}^1 + T_{13}^1) - (T_{31}^1 + T_{32}^1 + T_{33}^1) \geq \epsilon_1$.

(iv) A fixed device (irrespective of media) and wired media (irrespective of devices) are almost equally trustworthy, $T_{FD} \approx T_{WI}$, meaning $|(T_{11}^1 + T_{12}^1 + T_{13}^1) - (T_{11}^1 + T_{21}^1 + T_{31}^1)| \leq \epsilon_2$.

The LP solution for this optimization problem (*obtained using Matlab's linprog() routine*) is given by the following T-matrix.

$$\begin{bmatrix} T_{11}^1 & T_{12}^1 & T_{13}^1 \\ T_{21}^1 & T_{22}^1 & T_{23}^1 \\ T_{31}^1 & T_{32}^1 & T_{33}^1 \end{bmatrix} = \begin{bmatrix} 1 & 0.636 & 0.364 \\ 1 & 0 & 0 \\ 0 & 0.268 & 0.139 \end{bmatrix},$$

and $\in_1 = 1$, $\in_2 = 0$, $\in^* = 1$. Since $\in^* > 0$, above solution is acceptable. The trustworthy factors are as follows:

$$T_{FD} = 1 + 0.636 + 0.364 = 2,$$
$$T_{PD} = 1 + 0 + 0 = 1,$$
$$T_{HD} = 0 + 0.268 + 0.139 = 0.407,$$
$$T_{WI} = 1 + 1 + 0 = 2,$$
$$T_{WL} = 0.636 + 0 + 0.268 = 0.904,$$
$$T_{CL} = 0.364 + 0 + 0.139 = 0.503,$$

All the above conditions (7.1)–(7.3) are satisfied appropriately. Many other T-matrices would satisfy all four conditions, but the one found by above T-matrix maximizes the difference between desired and undesired trustworthy values.

From the current literature, the error rate (FAR, FRR, or AER) is available for different modalities. For naïve or advanced users, the more accuracy rate (the lesser error rate)

■ **Table 7.4** The error rates available for the seven authentication modalities

	Modality	FAR (%)	FRR (%)		Modality	AER (%)
M_1	Face	1	10	M_3	Password	<3.77
M_2	Fingerprint	2	2	M_4	CAPTCHA	<20
M_6	Voice	2	10	M_5	SMS	5.3
M_7	Keystroke	7	0.10			

■ **Table 7.5** Trustworthy values for various devices and media using the first approach

Modality	Trustworthiness factor of each modality for a particular device			Trustworthiness factor of each modality for a particular Media		
	FD	PD	HD	WI	WL	CL
M_1	2	1	0.407	2	0.904	0.503
M_2	2.3	1.82	1.6	2.3	1.75	1.67
M_3	1.76	1.26	1.13	1.76	1.43	1.26
M_4	1.44	0.87	0.87	1	0.5	0.5
M_5	0.63	0.78	0.63	1.76	1.64	1.48
M_6	1	0.5	0.5	1.52	1.43	1.35
M_7	0.82	0.63	0.52	0.82	0.52	0.46

of a modality is, the more trustworthy it is to them. ◻ Table 7.4 shows a comparative analysis of the considered modalities. For biometric modalities, false acceptance rate (FAR) and false rejection rate (FRR) are shown [11], while for others average error rate (AER) is shown [12–14].

The calculated trustworthy values are adjusted based on their error rates. In the current calculation of trustworthy values, the lower the error rate is, the higher the trustworthy value of that modality. ◻ Table 7.5 shows the trustworthy values calculated from these error rates and pairwise preference information.

Second Approach

To extend this trustworthy model into authentication modalities along with their features, the second approach uses a probabilistic optimization model to calculate the trustworthy values. As there is no state-of-art study that guarantees the supremacy of one authentication factor over the other in a particular device or medium, it cannot be formulated using deterministic approach to calculating the trustworthy values. The pairwise preference information regarding the authentication factors are also the prime factors in designing the optimization model for this approach. The detailed generic formulation [15, 16] of the trustworthy model is illustrated below.

In this subsection, a nonlinear programming problem with probabilistic constraints is formulated for calculating the trustworthy values of individual features of different authentication modalities. Let there be s number of authentication modalities, l number of features for each modality, d number of different classes of devices, and e number of different classifications of media available for the present problem.

Now, let us assume that an authentication modality (with a set of features), M_i; $(i = 1, 2 \ldots s)$ is more (or less or equally) trusted for a user in a device D_j; $(j = 1, 2, \ldots, d)$ rather than in device D_k; $(k = 1, 2, \ldots, d; k \neq j)$ for a particular medium Me_l; $(l = 1, 2 \ldots e)$.

Accordingly, these pairwise comparisons have been done among the devices in a fixed medium, and all the corresponding outcomes are equally likely. No prior assumption regarding the trustworthiness of the devices and media is made because of that. For example, it can never be confirmed if the face recognition authentication modality in fixed devices (desktops, workstations, etc.) is more trusted than in portable devices (laptops, tablets, etc.) in wired medium (WI) , as there are equal chances of it being equal and less trusted. The corresponding representation for the pair-wise comparative trustworthy preference is: $T_{ij}(M_s{:}f_{s,l})$ for the m-th modality (s options) with features $\{M_s{:}f_{s,l}\}$ (taking one feature at a time), i-th device (d options) and the j-th medium (e options). Thus, for a particular pairwise comparison involving i-th and k-th devices for a fixed (j-th) media and fixed (m-th) modality, the following conditions will occur:

$T_{ij}(M_s{:}f_{s,l}) >=< T_{kj}(M_s{:}f_{s,l}); i \neq k;$ and in this research, it is considered that they are equally likely:

$$\{P(T_{ij}(M_s{:}f_{s,l}) > T_{kj}(M_s{:}f_{s,l})); i \neq k\}$$
$$= \{P(T_{ij}(M_s{:}f_{s,l}) < T_{kj}(M_s{:}f_{s,l})); i \neq k\}$$
$$= \{P(T_{ij}(M_s{:}f_{s,l}) = T_{kj}(M_s{:}f_{s,l})); i \neq k\} = \frac{1}{3}$$

These comparisons only involve the trustworthiness of different devices with respect to the various features of a particular authentication modality in a single medium. Consequently, the random variable $\{T_{ij}(M_s{:}f_{s,l})\}$ can be constructed to determine the comparisons of the trustworthiness of a specific feature of a particular authentication modality (with a set of features) in different devices in a fixed medium. Similarly, the comparison can also be done among the trustworthiness of a particular authentication modality (with a set of features) in different media, keeping the device selection static.

In a similar way, let a particular feature of one authentication modality $\{M_s{:}f_{s,l}\}$ be more (or less or equally) trusted for a user in a medium $Me_j(j = 1, 2 \ldots e)$ rather than in medium $Me_k(k = 1, 2 \ldots e; k \neq j)$ for a particular device $D_l; (l = 1, 2 \ldots d)$.

All the above cases are equally likely as no prior assumptions regarding the trustworthiness of devices and media have been made. Thus, for a particular pairwise comparison involving i-th devices, j-th and k-th media, the following conditions will occur:

$T_{ij}(M_s{:}f_{s,l}) >=< T_{ik}(M_s{:}f_{s,l}); j \neq k;$ and

$P\{T_{ij}(M_s{:}f_{s,l}) >=< T_{ik}(M_s{:}f_{s,l}); j \neq k\} = \dfrac{1}{3},$ as all the three cases are equally likely.

These comparisons involve the trustworthiness of different devices, keeping the medium selection static.

Consequently, the random variable $\{T_{ij}(M_s{:}f_{s,l})\}$ can be constructed to determine the comparisons of the trustworthiness of different media, keeping the device selection static.

Based on the above cases, the following nonlinear programming problem with probabilistic constraints [16, 17] has been formed to find a set of $T_{ij}(M_s{:}f_{s,l})$ values.

$$\textbf{Maximize} \sum_{j} \sum_{i} \sum_{\substack{k \\ i \neq k}} \left| \frac{T_{ij}(M_s{:}f_{s,l}) - T_{kj}(M_s{:}f_{s,l})}{\varepsilon_1} \right|$$

$$+ \sum_{i} \sum_{j} \sum_{\substack{k \\ j \neq k}} \left| \frac{T_{ij}(M_s{:}f_{s,l}) - T_{ik}(M_s{:}f_{s,l})}{\varepsilon_2} \right| \tag{7.4}$$

Subject to:

$$P\{T_{ij}(M_s{:}f_{s,l}) \geq T_{kj}(M_s{:}f_{s,l}); \quad \text{for each } j = 1,2,3$$
$$\text{and } i,k = 1,2,3; i \neq k; s,l \in \mathbb{Z}^+\} \geq 1 - \varepsilon_1; \tag{7.5}$$

$$P\{T_{kj}(M_s{:}f_{s,l}) > T_{ij}(M_s{:}f_{s,l}); \quad \text{for each } j = 1,2,3$$
$$\text{and } i,k = 1,2,3; i \neq k; s,l \in \mathbb{Z}^+\} \geq \varepsilon_1; \tag{7.6}$$

$$P\{T_{ij}(M_s{:}f_{s,l}) \geq T_{ik}(M_s{:}f_{s,l}); \quad \text{for each } j = 1,2,3$$
$$\text{and } i,k = 1,2,3; j \neq k; s,l \in \mathbb{Z}^+\} \geq 1 - \varepsilon_2; \tag{7.7}$$

$$P\{T_{ik}(M_s{:}f_{s,l}) > T_{ij}(M_s{:}f_{s,l}); \quad \text{for each } j = 1,2,3$$
$$\text{and } i,k = 1,2,3; j \neq k; s,l \in \mathbb{Z}^+\} \geq \varepsilon_2; \tag{7.8}$$

$$0 \leq T_{ij}(M_s{:}f_{s,l}) \leq 1; \quad \forall i,j = 1,2,3; \ l \in \mathbb{Z}^+; \varepsilon_1, \varepsilon_2 \in (0,1) \tag{7.9}$$

In this work, ten authentication modalities have been considered (face, fingerprint, voice, keystroke, iris, hand geomerty, password, passphrase, SMS, TOTP), three types of devices (fixed device, handheld device , and portable device) , and three different media (wired, wireless, and the cellular medium) to calculate the trustworthy values of the authentication features. However, the method can be applied for any number of devices, media, and authentication modalities with various features. The primary objective of Nonlinear programing problem with probabilistic constraints is to find the trustworthy value of every single feature of any authentication modality (authentication factors). The given formulation for calculating the trustworthy values is generalized, although it is more applicable for the biometric authentication modalities. The reason is the error rates are available for most of the biometric authentication modalities with different sets of features. However, it is worth trying to investigate how the trustworthy framework can produce a better visualization of trust values for different non-biometric authentication modalities. In this work, the range of s can be decided as $1 \leq s \leq 10$. The process is shown to calculate trustworthy values of those features of the authentication modalities for which the FAR, FRR, EER, FMR, and FNMR can be found.

As a consequence, the random variables, $\left\{T_{ij}\left(M_s{:}f_{s,l}\right)\right\}_{j=1}^{e}$ and $\left\{T_{ij}\left(M_s{:}f_{s,l}\right)\right\}_{i=1}^{d}$ are converted to $\left\{T_{ij}\left(M_s{:}f_{s,l}\right)\right\}_{j=1}^{3} \sim N(0,1)$ (media changed, i.e., $\left\{T_{i1}\left(M_s{:}f_{s,l}\right)_{s,l}, T_{i2}\left(M_s{:}f_{s,l}\right)_{s,l}, T_{i3}\left(M_s{:}f_{s,l}\right)_{s,l}\right\}$) and $\left\{T_{ij}\left(M_s{:}f_{s,l}\right)\right\}_{i=1}^{3} \sim N(0,1)$ (device changed, i.e., $\left\{T_{1j}\left(M_s{:}f_{s,l}\right)_{s,l}, T_{2j}\left(M_s{:}f_{s,l}\right)_{s,l}, T_{3j}\left(M_s{:}f_{s,l}\right)_{s,l}\right\}$) respectively.

$\varepsilon_1 \in (0,1)$ is the critical region for $P\{T_{ij}\left(M_s{:}f_{s,l}\right) \geq T_{kj}\left(M_s{:}f_{s,l}\right)$; for each $j = 1,2,3$ and $i,k = 1,2,3; i \neq k; l \in \mathbb{Z}^+\}$ as well as the acceptance region of $P\{T_{kj}\left(M_s{:}f_{s,l}\right) > T_{ij}\left(M_s{:}f_{s,l}\right)$; for each $j = 1,2,3$ and $i,k = 1,2,3; i \neq k; l \in \mathbb{Z}^+\}$. If $\varepsilon_1 \to 1$, then the probability of accepting the hypothesis given in Eq. (7.5) decreases, increasing the probability of accepting the hypothesis presented in Eq. (7.6). Similarly, $\varepsilon_2 \in (0,1)$ is the rejection region of $P\{T_{ij}\left(M_s{:}f_{s,l}\right) \geq T_{ik}\left(M_s{:}f_{s,l}\right)$; for each $j = 1,2,3$ and $i,k = 1,2,3; j \neq k; l \in \mathbb{Z}^+\}$; however, it can be considered as the acceptance or critical region for $P\{T_{ik}\left(M_s{:}f_{s,l}\right) > T_{ij}\left(M_s{:}f_{s,l}\right)$; for each $j = 1,2,3$ and $i,k = 1,2,3; j \neq k; l \in \mathbb{Z}^+\}$.

The objective of the proposed NLPPPC is to observe the influence of the size of the critical regions on the difference of the trustworthy values of a particular feature of a given modality, i.e., if the difference becomes significant, the rejection regions corresponding to the constraints $P\{T_{ij}\left(M_s{:}f_{s,l}\right) \geq T_{kj}\left(M_s{:}f_{s,l}\right)$; for each $j = 1,2,3$ and $i,k = 1,2,3; i \neq k; l \in \mathbb{Z}^+\}$ and $P\{T_{ij}\left(M_s{:}f_{s,l}\right) \geq T_{ik}\left(M_s{:}f_{s,l}\right)$; for each $j = 1,2,3$ and $i,k = 1,2,3; j \neq k; l \in \mathbb{Z}^+\}$ reduce. For example, if the difference between $T_{12}\left(M_1{:}f_{1,1}\right)$ and $T_{22}\left(M_1{:}f_{1,1}\right)$ (different instances of the random variables) is very significant, i.e., if, $T_{12}(M_1{:}f_{1,1}) \gg T_{22}(M_1{:}f_{1,1})$ or $T_{22}(M_1{:}f_{1,1}) \gg T_{12}(M_1{:}f_{1,1})$, the rejection region corresponding to $P\{T_{ij}(M_s{:}f_{s,l}) \geq T_{ik}(M_s{:}f_{s,l})$; for each $j = 1,2,3$ and $i,k = 1,2,3; i \neq k; l \in \mathbb{Z}^+\}$ decreases. Similar conclusion can be made in case of the trustworthy value of a particular modality in a device and on different media. Now, combining both the cases the objective function of the proposed NLPPPC has been constructed. The nature of the objective function (Eq. (7.4)) is convex.

Now considering the modality as M_1 with feature $f_{1,1}$, Eq. (7.5) can also be written as follows:

Now the first constraint can be formed in the following way:

$$\mathrm{Pr}\{T_{ij}(M_1{:}f_{1,1}) \geq T_{kj}(M_1{:}f_{1,1})\} \geq 1 - \varepsilon_1,$$

$$\text{where } i,j,k = 1,2,3 \text{ and } i \neq k$$

$$\Rightarrow \mathrm{Pr}\{T_{ij}(M_1{:}f_{1,1}) \geq x \geq T_{kj}(M_1{:}f_{1,1})\}$$

$$\geq 1 - \varepsilon_1; x \sim N(0,1)$$

$$\Rightarrow \int_{T_{kj}(M_1{:}f_{1,1})}^{T_{ij}(M_1{:}f_{1,1})} \frac{1}{\sqrt{2\pi}} e^{\frac{-x^2}{2}} \, dx \geq 1 - \varepsilon_1,$$

where the mean is 0 and standard deviation is 1.

$$\Rightarrow \int_{T_{kj}(M_1{:}f_{1,1})}^{T_{ij}(M_1{:}f_{1,1})} e^{\frac{-x^2}{2}} \, dx \geq \sqrt{2\pi}(1 - \varepsilon_1).$$

The left part of the equation is actually an inappropriate integral and we can write the above term as

$$T_{ij}(M_1{:}f_{1,1}) - T_{kj}(M_1{:}f_{1,1}) \geq 2.508(1 - \varepsilon_1)$$

$$\Rightarrow T_{ij}(M_1{:}f_{1,1}) - T_{kj}(M_1{:}f_{1,1}) \tag{7.10}$$

$$+ 2.508\varepsilon_1 - 2.508 \geq 0$$

Second, third, and fourth of constraints of optimization problem can be formed in a similar manner:

$$T_{kj}(M_1{:}f_{1,1}) - T_{ij}(M_1{:}f_{1,1}) - 2.508 \geq 0 \tag{7.11}$$

$$T_{ij}(M_1{:}f_{1,1}) - T_{kj}(M_1{:}f_{1,1}) + 2.508\varepsilon_2 - 2.508 \geq 0 \tag{7.12}$$

$$T_{ik}(M_1{:}f_{1,1}) - T_{ij}(M_1{:}f_{1,1}) - 2.508 \geq 0 \tag{7.13}$$

Next, from Eqs. (7.10) through (7.13) the following T-matrix could be constructed. The construction of the T-matrix is similar to the first approach of trustworthy value calculation.

$$\begin{bmatrix} T_{11}(M_1{:}f_{1,1}) & T_{12}(M_1{:}f_{1,1}) & T_{13}(M_1{:}f_{1,1}) \\ T_{21}(M_1{:}f_{1,1}) & T_{22}(M_1{:}f_{1,1}) & T_{23}(M_1{:}f_{1,1}) \\ T_{31}(M_1{:}f_{1,1}) & T_{32}(M_1{:}f_{1,1}) & T_{33}(M_1{:}f_{1,1}) \end{bmatrix}$$
$$= \begin{bmatrix} 0.99 & 0.761 & 0.415 \\ 0.502 & 0.856 & 0.473 \\ 0.594 & 0.434 & 0.047 \end{bmatrix}$$

Along with $\varepsilon_1 = 0.4429$ and $\varepsilon_2 = 0.2392$.

The calculated trustworthy values of $(M_1{:}f_{1,1})$ for different settings of device and medium using the abovementioned approach are shown below:

$$T_{FD}(M_1{:}f_{1,1}) = 2.17, \quad T_{PD}(M_1{:}f_{1,1}) = 1.83,$$
$$T_{HD}(M_1{:}f_{1,1}) = 1.07, \quad T_{WI}(M_1{:}f_{1,1}) = 2.08,$$
$$T_{WL}(M_1{:}f_{1,1}) = 2.05, \quad T_{CL}(M_1{:}f_{1,1}) = 0.92.$$

Next, the EER for M_1; $f_{1,1}$ is 6.8%. Hence, for 93.2% cases [6], this feature performs well. The values of trustworthy factor mentioned above do not consider the error rates. With the incorporation of the EER, we can scale the abovementioned trustworthy values.

$$T_{FD}(M_1{:}f_{1,1}) = 2.02$$
$$T_{PD}(M_1{:}f_{1,1}) = 1.67.$$
$$T_{HD}(M_1{:}f_{1,1}) = 1.002.$$
$$T_{WI}(M_1{:}f_{1,1}) = 1.95.$$
$$T_{WL}(M_1{:}f_{1,1}) = 1.92.$$
$$T_{CL}(M_1{:}f_{1,1}) = 0.87.$$

In a similar manner, the trustworthy values for other authentication modalities with their various features in different settings of devices and media can be calculated. Some very interesting and realistic results regarding the trustworthy values of various features of different authentication modalities are tabulated and are shown in ◘ Table 7.5. For example, from ◘ Table 7.5, it can be found that eye $(M_1{:}f_{1,2})$ (one of the features of the face recognition modality) is more trusted than lip $(M_1{:}f_{1,1})$ (another feature of the face recognition modality), which is supported by the trustworthy values found in ◘ Table 7.6. These trustworthy values can be used to calculate the trustworthy values for combined features, which is described in the next subsection. Same types

□ **Table 7.6** Trustworthy values of different authentication factors with individual features

Modalities	Features	Trustworthy values					
		FD	PD	HD	WI	WL	CL
(M_1)	$(M_1:f_{1,1})$	2.02	1.67	1.00	1.95	1.92	0.87
	$(M_1:f_{1,2})$	2.15	1.79	1.07	2.08	2.04	0.93
	$(M_1:f_{1,3})$	1.84	1.52	0.92	1.78	1.74	0.8
	...	...	...	...	...	...	...
(M_2)	$(M_2:f_{2,i}; i \in \mathbb{Z}^+)$	2.17	1.79	1.07	2.09	2.05	0.94
(M_3)	$(M_3:f_{3,i}; i \in \mathbb{Z}^+)$	2.12	1.76	1.05	2.04	2.01	0.92
(M_4)	$(M_4:f_{4,1})$	2.07	1.71	1.03	1.99	1.96	0.89
	$(M_4:f_{4,2})$	2.05	1.70	1.02	1.98	1.94	0.89
	$(M_4:f_{4,3})$	2.09	1.72	1.03	2.01	1.97	0.90
(M_5)	$(M_5:f_{6,i}; i \in \mathbb{Z}^+)$	1.76	1.26	1.13	1.76	1.43	1.26
(M_6)	$(M_6:f_{6,i}; i \in \mathbb{Z}^+)$	1.44	0.87	0.87	1.00	0.5	0.5
(M_7)	$(M_7:f_{8,i}; i \in \mathbb{Z}^+)$	0.63	0.78	0.63	1.76	1.64	1.48
(M_8)	$(M_8:f_{9,i}; i \in \mathbb{Z}^+)$	1.95	1.61	0.97	1.88	1.85	0.84
(M_9)	$(M_9:f_{9,i}; i \in \mathbb{Z}^+)$	1.84	1.52	0.91	1.77	1.74	0.80
(M_{10})	$(M_{10}:f_{10,i}; i \in \mathbb{Z}^+)$	0.99	0.96	0.92	0.92	0.82	0.86

of arguments can be made for different features of remaining modalities.

Computation of Trustworthy Values for Different Combinations of Authentication Factors

The previous subsection discusses the calculation of trustworthy values for individual features (authentication factors) incorporating their respective error rates. Quite on the contrary, the present subsection discusses the calculation of trustworthy values of authentication factors combining individual features. In this research, we assume that the influence of the individual trustworthy value of different authentication factors on their combined trustworthy values follows the standard normal distribution ($N(0, 1)$). The motivation behind this assumption is as follows:

The error rates (EER, ERR, FAR, etc.) [6] for the combined authentication factors are not currently available in the literature. Again, the error rates of different individual features are already incorporated in calculating their trustworthy values. Hence, it is required to propagate the influence of individual error rates into the trustworthy value of the combined authentication factor.

Let us define two continuous random variables $X_{d\bar{m}pj}$ and $X_{\bar{d}mpj}$ in the following way:

$X_{d\bar{m}pj} \sim N(0,1)$ The influence of $T_{d\bar{m}}\left(M_p{:}\{f_{p,j}\}_j\right)_p$ on $T_{d\bar{m}}\left(\{M_p\}{:}\{f_{p,j}\}_j\right)_p; \forall d, m, p, j \in \mathbb{Z}^+,$ in a particular medium irrespective of devices, where $T_{d\bar{m}}\left(M_p{:}\{f_{p,j}\}_j\right)_p$ and $T_{d\bar{m}}\left(\{M_p\}{:}\{f_{p,j}\}_j\right)_p$ can be expressed as $T_{d\bar{m}}\left(M_p{:}f_{p,1}, f_{p,2}, \ldots\right)$ and $T_{d\bar{m}}\left(M_1, M_2, \ldots {:} f_{p,1}, f_{p,2}, \ldots\right)$, respectively.

$X_{\bar{d}mpj} \sim N(0,1)$ The influence of $T_{\bar{d}m}\left(M_p : (f_{p,j})_j\right)_p$ on $T_{\bar{d}m}\left(\{M_p\}{:}\{f_{p,j}\}_j\right)_p; \forall d, m, p, j \in \mathbb{Z}^+,$ in a particular device irrespective of medium. These notations can be explained as follows:

$\mathbf{T}_{d\bar{m}}\left(\mathbf{M}_p{:}\{f_{p,j}\}_j\right)_p$ Trustworthy value of $\left(M_p{:}\{f_{p,j}\}_j\right)_p$ on a specific medium irrespective of devices

$\mathbf{T}_{\bar{d}m}\left(\mathbf{M}_p{:}\{f_{p,j}\}_j\right)_p$ Trustworthy value of $\left(M_p{:}\{f_{p,j}\}_j\right)_p$ on a specific device irrespective of media

$\mathbf{T}_{d\bar{m}}\left(\{\mathbf{M}_p\}{:}\{f_{p,j}\}_j\right)_p$ Trustworthy value of $\left(\{M_p\}{:}\{f_{p,j}\}_j\right)_p$ on a specific medium irrespective of devices

$\mathbf{T}_{\bar{d}m}\left(\{\mathbf{M}_p\}{:}\{f_{p,j}\}_j\right)_p$ Trustworthy value of $\left(\{M_p\}{:}\{f_{p,j}\}_j\right)_p$ on a specific device irrespective of media

These notations are used to calculate the trustworthy values that have been described in the next subsections.

Computation of Trustworthy Values in a Specific Medium

In order to calculate the trustworthy value of a particular combination of authentication factors in a specific medium, the following two quantities are needed to combine:
1. The influences of trustworthy values of the authentication factors involved in the previously mentioned combination irrespective of devices.

2. The influence of individual trustworthy value of the previously mentioned authentication factors in that particular medium. Detailed descriptions regarding the calculation of the influences of a particular combination of authentication factors in a specific medium are given as follows:

Let us assume that the highest and lowest trustworthy values of an authentication factor $\left(\left(M_{p_k}:f_{p_k j}\right)_{k \in \mathbb{Z}+}\right)$ on different devices are $b_{k \in \mathbb{Z}+}$ and $a_{k \in \mathbb{Z}+}$, respectively. Now, the trustworthy value of $\left\{M_{p_k}:f_{p_k j}\right\}_{k \in \mathbb{Z}+}$ in a specific medium can be calculated as follows:

$$
\mathbf{T}_{d\bar{m}}\left(\{\mathbf{M}_{p_k}\} : \{f_{p_k j}\}_j\right)_k = E\{a_1 \leq X_{d\bar{m}p_1 j} \leq b_1; a_2
$$
$$
\leq X_{d\bar{m}p_2 j} \leq b_2; \ldots; a_n \leq X_{d\bar{m}p_n j} \leq b_n\}
$$
$$
= \int_{a_1}^{b_1} \int_{a_2}^{b_2} \cdots \int_{a_n}^{b_n} X_{d\bar{m}p_1 j} * X_{d\bar{m}p_2 j} * \ldots * X_{d\bar{m}p_n j}
$$
$$
* \psi_{X_{d\bar{m}p_1 j}, X_{d\bar{m}p_2 j}, \ldots, X_{d\bar{m}p_n j}}(x_{d\bar{m}p_1 j}, x_{d\bar{m}p_2 j}, \ldots, x_{d\bar{m}p_n j})
$$
$$
dx_{d\bar{m}p_1 j} dx_{d\bar{m}p_2 j} \ldots dx_{d\bar{m}p_n j}
$$
$$
= \int_{a_1}^{b_1} \int_{a_2}^{b_2} \cdots \int_{a_n}^{b_n} X_{d\bar{m}p_1 j} * X_{d\bar{m}p_2 j} * \ldots * X_{d\bar{m}p_n j}
$$
$$
* \frac{1}{\sqrt{2\pi}} e^{-\frac{1}{2}X_{d\bar{m}p_1 j}^2} \frac{1}{\sqrt{2\pi}} e^{-\frac{1}{2}X_{d\bar{m}p_2 j}^2} \cdots \frac{1}{\sqrt{2\pi}} e^{-\frac{1}{2}X_{d\bar{m}p_n j}^2}
$$
$$
dx_{d\bar{m}p_1 j} dx_{d\bar{m}p_2 j} \ldots dx_{d\bar{m}p_n j}
$$

Again, the random variables $X_{d\bar{m}p_1 j}, X_{d\bar{m}p_2 j}, \ldots, X_{d\bar{m}p_n j}$ are independent to one other as different authentication modalities along with their several features are independent. For example, eye $(M_1:f_{1,1})$ and lip $(M_1:f_{1,2})$ (two different features of face recognition authentication modality (M_1)) are independent of each other. Now, the contributions of all the features of an authentication modality have been considered for calculating its trustworthy value. Consequently, the above expression can be written as follows:

Next, to calculate the trustworthy value of $\left(\{M_{p_k}\}:\{f_{p_k j}\}_j\right)_k$ in a particular medium, the influence of the individual trustworthy values of $\{M_{p_k}:f_{p_k j}\}_{k \in \mathbb{Z}+}$ in that medium (which follows the standard normal distribution with the combined trustworthy value) have been added with $T_{\bar{d}m}\left(\{M_{Pk}\}:\{f_{Pkj}\}_j\right)_k$ as follows:

$$T_{d\bar{m}}\left(\{M_{Pk}\}:\{f_{Pkj}\}_j\right)_k + \frac{1}{\sqrt{2\pi}}\sum_k\sum_j e^{-\frac{1}{2}\left\{T_{m_i}\left(M_{Pk}:f_{Pkj}\right)\right\}^2}$$

$$= E\left\{a_1 \le X_{d\bar{m}p1j} \le b_1; a_2 \le X_{d\bar{m}p2j} \le b_2; \dots a_n \le X_{d\bar{m}pnj} \le b_n\right\}$$

$$+ \frac{1}{\sqrt{2\pi}}\sum_k\sum_j e^{-\frac{1}{2}\left\{T_{m_i}\left(M_{Pk}:f_{Pkj}\right)\right\}^2},$$

where $T_{m_i}\left(M_{Pk}:f_{Pkj}\right)$ is the trustworthy value of $\left(M_{Pk}:f_{Pkj}\right)$ on the i-th medium.

For better and clearer understanding a sample scenario has been pictorially represented in ◻ Fig. 7.3, which shows the influence of the individual trustworthy values of $M_1:f_{1,1}; M_2:f_{2,1}\dots M_n:f_{n,1}$ for a device (irrespective of any medium) on the combined trustworthy value of $\left\{M_1:f_{1,1}, M_2:f_{2,1}\dots M_n:f_{n,1}\right\}$.

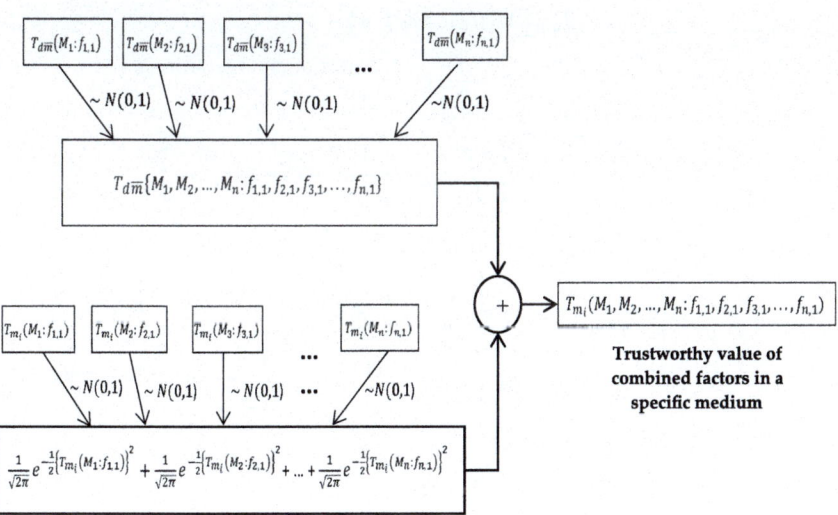

◻ **Fig. 7.3** Pictorial representation of calculation strategy of trustworthy values of combined factor from individual trustworthy values in a particular medium

Now the above methodology for calculating trustworthy value of combined authentication factors is described with the following example, which calculates the combined trustworthy value of $\{M_1{:}f_{1,1}, M_1{:}f_{1,2}\}$ in the wired medium (WI).

$$T_{d\bar{m}}\{M_1{:}f_{1,1}, M_1{:}f_{1,2}\}$$
$$= E\{1.00 \leq X_{d\bar{m}11} \leq 2.02;\ 1.07 \leq X_{d\bar{m}12} \leq 2.15\}$$

$$= \int_{1.00}^{2.02}\int_{1.07}^{2.15} x_{d\bar{m}11}\frac{1}{\sqrt{2\pi}}e^{-\frac{1}{2}x_{d\bar{m}11}^2}x_{d\bar{m}12}\frac{1}{\sqrt{2\pi}}e^{-\frac{1}{2}x_{d\bar{m}12}^2}\,dx_{d\bar{m}11}\,dx_{d\bar{m}12}$$

$$= \left(\int_{1.00}^{2.02}x_{d\bar{m}11}\frac{1}{\sqrt{2\pi}}e^{-\frac{1}{2}x_{d\bar{m}11}^2}\,dx_{d\bar{m}11}\right)\left(\int_{1.07}^{2.15}x_{d\bar{m}12}\frac{1}{\sqrt{2\pi}}e^{-\frac{1}{2}x_{d\bar{m}12}^2}\,dx_{d\bar{m}12}\right) \approx 2.67$$

Next, to get the trustworthy value of $\{M_1{:}f_{1,1}, M_1{:}f_{1,2}\}$ in WI, the influence of the individual trustworthy values of $\{M_1{:}f_{1,1}\}$ and $\{M_1{:}f_{1,2}\}$ in WI (which follows the standard normal distribution with the combined trustworthy value) has been incorporated as follows:

$$T_{d\bar{m}}\{M_1{:}f_{1,1};\ M_1{:}f_{1,2}\}$$
$$+ \frac{1}{\sqrt{2\pi}}e^{-\frac{1}{2}\left(T_{WI}(M_1{:}f_{1,1})^2 + T_{WI}(M_1{:}f_{1,2})^2\right)}$$

$$= \left(\int_{1.00}^{2.02}x_{d\bar{m}11}\frac{1}{\sqrt{2\pi}}e^{-\frac{1}{2}x_{d\bar{m}11}^2}\,dx_{d\bar{m}11}\right)$$

$$\times \left(\int_{1.07}^{2.15}x_{d\bar{m}12}\frac{1}{\sqrt{2\pi}}e^{-\frac{1}{2}x_{d\bar{m}12}^2}\,dx_{d\bar{m}12}\right)$$

$$+ \left\{\frac{1}{\sqrt{2\pi}}e^{-\frac{1}{2}1.95^2} + \frac{1}{\sqrt{2\pi}}e^{-\frac{1}{2}2.08^2}\right\} \approx 2.78.$$

Similarly, the combined trustworthy values of other authentication factors in a particular medium can be calculated; some of them are listed in ◘ Table 7.7.

Computation of Trustworthy Values in a Specific Device

In order to calculate the trustworthy value of a particular combination of authentication factors in a specific device (◘ Fig. 7.4), the following two quantities are needed to combine:

◻ **Table 7.7** The calculation of trustworthy values of combined authentication factors from individual trustworthy values

Features	Trustworthy values					
	FD	PD	HD	WI	WL	CL
$(M_1{:}f_{1,1}, M_1{:}f_{1,2})$	2.72	2.81	3.1	2.78	2.79	3.21
$(M_1{:}f_{1,1}, f_{1,2}, f_{1,3})$	3.15	2.99	2.97	2.57	2.71	0.99
...	...	...	...	...	...	...
$(M_1{:}f_{1,1}, f_{1,2}, f_{1,3}, f_{1,4})$	3.57	3.79	3.19	2.61	2.75	0.99
$(M_5{:}f_{5,1}, f_{5,2})$	3.95	3.94	3.97	3.73	2.78	0.97
$(M_5{:}f_{5,1}, f_{5,3})$	3.90	2.89	3.79	3.79	3.69	3.79
$(M_5{:}f_{5,1}, f_{5,2})$	3.95	3.94	3.97	3.73	2.78	0.97
...	...	...	...	...	...	...

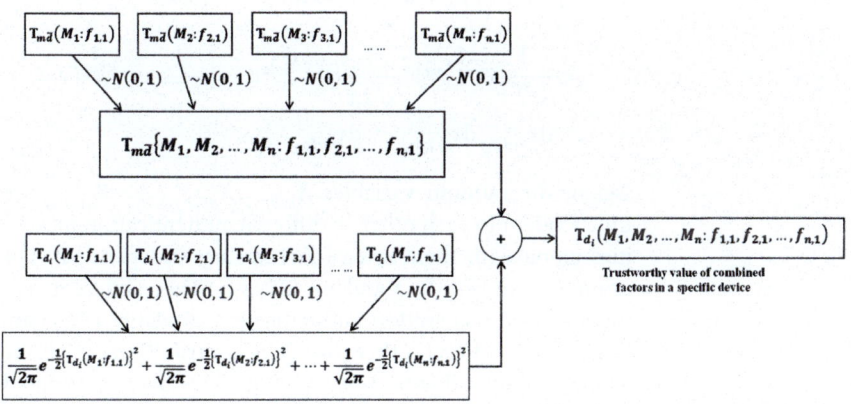

◻ **Fig. 7.4** Pictorial representation of calculation strategy of trustworthy values of combined authentication factors from individual trustworthy values in a particular device

1. The influences of trustworthy values of the authentication factors involved in the previously mentioned combination irrespective of medium.
2. The influences of individual trustworthy values of the previously mentioned authentication factors in that particular device. Detailed descriptions regarding the calculation of the influences of a particular combination of authentication factors in a specific device are given as follows:

Let us assume that the lowest and highest trustworthy values of an authentication factor $((M_{p_k}{:}f_{p_kj})_{k\in\mathbb{Z}^+})$ on different media are $c_{k\in\mathbb{Z}^+}$ and $d_{k\in\mathbb{Z}^+}$, respectively. Now, the trustworthy value of $\{M_{p_k}{:}f_{p_kj}\}_{k\in\mathbb{Z}^+}$ in any device can be calculated as follows:

$$T_{m\bar{d}}\left(\{M_{p_k}\}{:}\{f_{p_kj}\}_j\right)_k = E\Big\{c_1 \le X_{m\bar{d}p_1j} \le d_1; c_2 \le X_{m\bar{d}p_2j}$$

$$\le d_2; \dots; c_n \le X_{m\bar{d}p_nj} \le d_n\Big\}$$

$$= \int_{c_1}^{d_1}\int_{c_2}^{d_2}\cdots\int_{c_n}^{d_n} x_{m\bar{d}p_1j} * x_{m\bar{d}p_2j} * \cdots * x_{m\bar{d}p_nj}$$

$$* \psi_{X_{m\bar{d}p_1j},X_{m\bar{d}p_2j},\dots,X_{m\bar{d}p_nj}}(x_{m\bar{d}p_1j},x_{m\bar{d}p_2j},\dots,x_{m\bar{d}p_nj})$$

$$dx_{m\bar{d}p_1j}dx_{m\bar{d}p_2j}\cdots dx_{m\bar{d}p_nj}$$

$$= \int_{c_1}^{d_1}\int_{c_2}^{d_2}\cdots\int_{c_n}^{d_n} x_{m\bar{d}p_1j} * x_{m\bar{d}p_2j} * \cdots * x_{m\bar{d}p_nj}$$

$$* \frac{1}{\sqrt{2\pi}}e^{-\frac{1}{2}x_{m\bar{d}p_1j}^2}\frac{1}{\sqrt{2\pi}}e^{-\frac{1}{2}x_{m\bar{d}p_2j}^2}\cdots\frac{1}{\sqrt{2\pi}}e^{-\frac{1}{2}x_{m\bar{d}p_nj}^2}$$

$$dx_{m\bar{d}p_1j}dx_{m\bar{d}p_2j}\cdots dx_{m\bar{d}p_nj}$$

Again, the random variables $X_{m\bar{d}p_1j}, X_{m\bar{d}p_2j}, \dots, X_{m\bar{d}p_nj}$ are independent to each other as different authentication modalities along with their several features are independent. For example, eye ($M_1{:}f_{1,1}$) and lip ($M_1{:}f_{1,2}$) (two different features of face recognition authentication modality (M_1)) are independent of each other. Now, the contributions of all the features of an authentication modality have been considered for calculating its trustworthy value. Consequently, the above expression can be written as follows:

$$T_{d\bar{m}}\left(\{M_{p_k}\}{:}\{f_{p_kj}\}_j\right)_k$$

$$= \int_{a_1}^{b_1} x_{m\bar{d}p_1j} \frac{1}{\sqrt{2\pi}}e^{-\frac{1}{2}x_{m\bar{d}p_1j}^2} dx_{m\bar{d}p_1j}$$

$$\int_{a_2}^{b_2} x_{m\bar{d}p_2j} \frac{1}{\sqrt{2\pi}}e^{-\frac{1}{2}x_{m\bar{d}p_2j}^2} dx_{m\bar{d}p_2j}\cdots$$

$$\int_{a_n}^{b_n} x_{m\bar{d}p_nj} \frac{1}{\sqrt{2\pi}}e^{-\frac{1}{2}x_{m\bar{d}p_nj}^2} dx_{m\bar{d}p_nj}$$

Next, to calculate the trustworthy value of $\left(\{M_{p_k}\} : \{f_{p_k j}\}_j \right)_k$ in a particular device, the influence of the individual trustworthy values of $\{M_{p_k} : f_{p_k j}\}_{k \in \mathbb{Z}+}$ in that device (which follows the standard normal distribution with the combined trustworthy value) has been added with $T_{d\bar{m}} \left(\{M_{p_k}\} : \{f_{p_k j}\}_j \right)_k$ as follows:

$$T_{m\bar{d}} \left(\{M_{p_k}\} : \{f_{p_k j}\}_j \right)_k$$

$$+ \frac{1}{\sqrt{2\pi}} \sum_k \sum_j e^{-\frac{1}{2} \left\{ T_{d_i} \left(M_{p_k} : f_{p_k j} \right) \right\}^2}$$

$$= E\left\{ a_1 \leq X_{d\bar{m}p_1 j} \leq b_1; a_2 \leq X_{d\bar{m}p_2 j} \right.$$
$$\left. \leq b_2; \dots a_n \leq X_{d\bar{m}p_n j} \leq b_n \right\}$$

$$+ \frac{1}{\sqrt{2\pi}} \sum_k \sum_j e^{-\frac{1}{2} \left\{ T_{d_i} \left(M_{p_k} : f_{p_k j} \right) \right\}^2},$$

where $T_{d_i} \left(M_{p_k} : f_{p_k j} \right)$ is the trustworthy value of $\left(M_{p_k} : f_{p_k j} \right)$ on the i-th medium.

For better and clearer understanding a sample scenario has been pictorially represented in ◘ Fig. 7.8, which shows the influence of the individual trustworthy values of $M_1 : f_{1,1}; M_2 : f_{2,1} \dots M_n : f_{n,1}$ for a device (irrespective of any medium) on the combined trustworthy value of $\{M_1 : f_{1,1}, M_2 : f_{2,1} \dots M_n : f_{n,1}\}$.

The above methodology for calculating trustworthy value of a combination of combined authentication factors in a specific device is described in the following example, which calculates the combined trustworthy value of $\{M_1 : f_{1,1}, M_1 : f_{1,2}\}$ in the fixed device (FD).

$$T_{m\bar{d}} \{M_1 : f_{1,1}, M_1 : f_{1,2}\}$$

$$= E\left\{ 0.87 \leq X_{m\bar{d}11} \leq 1.95; 0.93 \leq X_{m\bar{d}12} \leq 2.08 \right\}$$

$$= \int_{0.87}^{1.95} \int_{0.93}^{2.08} X_{m\bar{d}11} \frac{1}{\sqrt{2\pi}} e^{-\frac{1}{2} x^2_{m\bar{d}11}} X_{m\bar{d}12}$$

$$\frac{1}{\sqrt{2\pi}} e^{-\frac{1}{2} x^2_{m\bar{d}12}} dx_{m\bar{d}11} dx_{m\bar{d}12}$$

$$= \left(\int_{1.00}^{2.02} X_{m\bar{d}11} \frac{1}{\sqrt{2\pi}} e^{-\frac{1}{2} x^2_{m\bar{d}11}} dx_{m\bar{d}11} \right)$$

$$\left(\int_{1.07}^{2.15} X_{m\bar{d}12} \frac{1}{\sqrt{2\pi}} e^{-\frac{1}{2} x^2_{m\bar{d}12}} dx_{m\bar{d}12} \right)$$

Next, to get the trustworthy value of $\{M_1{:}f_{1,1}, M_1{:}f_{1,2}\}$ in FD, the influence of the individual trustworthy values of $\{M_1{:}f_{1,1}\}$ and $\{M_1{:}f_{1,2}\}$ in FD (which follows the standard normal distribution with the combined trustworthy value) has been incorporated as follows:

$$
\begin{aligned}
&T_{m\bar{d}}\{M_1{:}f_{1,1}; M_1{:}f_{1,2}\} \\
&\quad + \frac{1}{\sqrt{2\pi}}e^{-\frac{1}{2}\left(T_{WI}(M_1{:}f_{1,1})^2 + T_{WI}(M_1{:}f_{1,2})^2\right)} \\
&\quad = \left(\int_{0.87}^{1.95} x_{m\bar{d}11} \frac{1}{\sqrt{2\pi}} e^{-\frac{1}{2}x_{m\bar{d}11}^2} dx_{m\bar{d}11} \right) \\
&\qquad \left(\int_{0.93}^{2.08} x_{m\bar{d}12} \frac{1}{\sqrt{2\pi}} e^{-\frac{1}{2}x_{m\bar{d}12}^2} dx_{m\bar{d}12} \right) \\
&\qquad + \left\{ \frac{1}{\sqrt{2\pi}}e^{-\frac{1}{2}1.95^2} + \frac{1}{\sqrt{2\pi}}e^{-\frac{1}{2}2.08^2} \right\} \approx 2.72,
\end{aligned}
$$

which can be found in the fourth row of ◻ Table 7.7.

These trustworthy values are used in the selection procedure of the adaptive-MFA discussed in next section.

Adaptive Selection Approach for MFA

Trustworthy framework mentioned in the previous section provides a metric to design the selection procedure for adaptive multi-factor authentication. This section focuses on the step-by-step procedure for designing the objective functions to choose a better set of authentication factors in different time triggering events. The selection procedure considers different devices, media, and surrounding conditions to accurately tailor the appropriate constraints to get the better set of authentication factors.

Moreover, the following criteria should be considered to design an adaptive selection approach for Multi-factor Authentication:
The performance of the selected authentication factors.
The total trustworthy values of all the selected authentication factors.
The cardinality of the selected authentication factors.
The selection of previous authentication factors in the same operating conditions.

In the design of the adaptive selection procedure, it is to be guaranteed that the selected set should have higher total trustworthy values with the lower number of authentication factors. This assures the selected set of authentication factors has higher overall performance and considers the users' burden into account (because more authentication factors bring more burden to the users). In addition, the selection of the authentication factors should not follow any predictable pattern that can be easily exploited by the attackers. Hence, the selected set of authentication factors should be different from previously selected authentication factors if the device, medium, and surrounding conditions remain the same.

To achieve these criteria, a multi-objective nonlinear quadratic optimization problem with probabilistic constraints has been formulated. The effects of surrounding conditions are considered in the selection of the set of authentication factors as some of them may have a dependence on the light (for example, face recognition), sound (for example, voice recognition), etc. These surrounding conditions are formulated as the constraints of the multi-objective optimization problem. It is worth noting the formulation is generic enough to capture the variabilities of the device, media, and surrounding conditions. Other surrounding conditions can be added based on the settings of the system and the availability of the sensor data from the client application. In this current implementation of this MFA, light, noise, and motion are considered as surrounding conditions.

The formulation of the multi-objective optimization is given below:

Objectives:

$$\text{Maximize} \quad \left(WD_l * D_l + \sum_j WM_j * Me_j \right)$$

$$* \left\{ TD_l(\{M_k\}\colon \{f_{k,i}\}) * \sum_j TMe_j(\{M_k\}\colon \{f_{k,i}\}) \right\}^{-|(\{M_k\}\colon \{f_{k,i}\})|} \tag{7.14}$$

$$\text{Minimize} \quad P(X_t | X_{t-1}, X_{t-2}, X_{t-3} \ldots X_1) \tag{7.15}$$

Subject to:

$$\left(D_l - TD_l(\{M_k\}\colon \{f_{k,i}\}_i)_k \right)^2$$

$$+ \sum_j \left(Me_j - TMe_j(\{M_k\}\colon \{f_{k,i}\}_i)_k \right)^2 \le U_{max}; \tag{7.16}$$

where $(a_1 \, lx \le L \le a_2 \, lx; \, a_3 \, Hz \le S \le a_4 \, Hz;$
motion within a sustainable range; $\{a_i\}_{i \in \mathbb{N}})$

$$\left(D_l - TD_l\big(\{M_k\}\!:\!\{f_{k,i}\}_i\big)_{k\neq 8} \right)^2$$

$$+ \sum_j \left(Me_j - TMe_j\big(\{M_k\}\!:\!\{f_{k,i}\}_i\big)_{k\neq 8} \right)^2 \leq U_{max}; \quad (7.17)$$

where $(a_3\ Hz \leq S \leq a_4\ Hz;$

motion within a sustainable range)

$$\left(D_l - TD_l\big(\{M_k\}\!:\!\{f_{k,i}\}_i\big)_{k\neq 1,2,8} \right)^2$$

$$+ \sum_j \left(Me_j - TMe_j\big(\{M_k\}\!:\!\{f_{k,i}\}_i\big)_{k\neq 1,2,8} \right)^2 \leq U_{max}; \ (7.18)$$

(where motion is within a sustainable range)

The present multi-objective programming problem with probabilistic constraints has two objectives. Generally, the multi-objective optimization could be considered as an area of multiple criteria decision making that is concerned with mathematical optimization problems involving more than one objective function to be optimized simultaneously. In mathematical terms, a k-objective optimization problem can be formulated as follows [16]:

$$\text{Max } f(x) = \{f_1(x), f_2(x), \ldots f_k(x)\}$$
$$\text{Subject to } \quad x \in X,$$

where $f(x)$ is a k-dimensional objective vector ($k \geq 2$), $f_i(x)$ is the i-th objective to be maximized ($i = 1, 2, \ldots, k$), x is a decision vector, and X is the set of all feasible decision vectors (i.e., X is the feasible region in the decision space) or the set of constraints. The feasible set is defined by some constraint functions (can also be probabilistic in nature). Moreover, the vector-valued objective function is often defined as $f:X \rightarrow \mathbb{R}^k$, then $f(x) = (f_1(x), f_2(x), \ldots f_k(x))^T$. The space in which the objective vector belongs is called the objective space, and the image of the feasible set under f is called the attained set.

In the present formulation, first objective contains two quantities—trustworthy values and cardinality of the authentication factors. The second objective ensures that same set of authentication factors are less likely to be selected. Two objectives are detailed described below:

In first objective term, $TD_l\big(\{M_k\}\!:\!\{f_{k,i}\}\big)$ means the trustworthy value of $\{M_k\}\!:\!\{f_{k,i}\}$ in the selected device. Similarly, $TMe_j\big(\{M_k\}\!:\!\{f_{k,i}\}\big)$ means the trustworthy value of $\{M_k\}\!:\!\{f_{k,i}\}$ in the selected medium. These two terms are considered together to compute the combined trustworthy values of specific device and medium. Two different weights are also included in the first objective to show the effect of trustworthy values. They are WD_l and WM_j and they are calculated in the following way:

$$WD_l = \frac{\text{Total trustworthy value of different features of activated modalities of the } l^{th} \text{ device}}{\text{Total trustworthy value of all the features of different modalities in the } l^{th} \text{ device}}$$

$$WM_j = \frac{\text{Total trustworthy value of different features of the activated modalities of the } j^{th} \text{ medium}}{\text{Total trustworthy value of all the features of different modalities in the } j^{th} \text{ medium}}$$

The variables D_l and Me_j are determined by solving the appropriate constraints mentioned in above equations. The cardinality term, $\left|(\{M_k\}:\{f_{k,i}\})\right|$, counts the total number of individual authentication factors to be considered at any given time. The main goal is to achieve higher trustworthy values with lesser number of authentication factors. In order to have the joint effect of two contrasting criteria, the cardinality term is included as negative exponent term of total trustworthy values. Hence, maximizing the terms means minimizing the cardinality value of the selected set of authentication factors, which serves the required purpose. In the current proposal of the objective function, the coefficient of the negative exponent is set as 1. Further exploration of this formulation will be future work for this research.

The second objective captures the necessity to reduce the possibility of the same set of authentication factors to be selected in successive authentication triggering events. Here, X_t can be expressed as the selection of authentication factors at tth triggering event. Mathematically, $\{X_t\}$ is the random variable that can be defined as follows:

$$X_t = \left\{\left\{TD_l(\{M_k\}:\{f_{k,i}\}_i)_k\right\} * \left\{TMe_j(\{M_k\}:\{f_{k,i}\}_i)_k\right\}\right\}_t^{-\left|(\{M_k\}:\{f_{k,i}\})\right|_t}$$

and its $(t-1)$th occurrence can be defined as

$$X_{t-1} = \left\{\left\{TD_l(\{M_k\}:\{f_{k,i}\}_i)_k\right\} * \left\{TMe_j(\{M_k\}:\{f_{k,i}\}_i)_k\right\}\right\}_{t-1}^{-\left|(\{M_k\}:\{f_{k,i}\})\right|_{t-1}},$$

Here, X_t is the tth observation of the random variable $\{X_t\}$. The observation of tth and $(t-1)$th authentication triggering events is expressed in terms of trustworthy values and cardinality of the authentication factors. This approach of formulation provides a better way to calculate the conditional probability as mentioned in Eq. (7.15). Equation (7.15) can be simplified to only consider the last authentication event to calculate the conditional probability. Hence Eq. (7.15) becomes Minimize $P(X_t|X_{t-1})$ and the distribution of $\{X_t\}$ has been considered as exponential (due to memory-less property of exponential distribution). This second objective guarantees that the successive selected authentication factors in the same environment are not exactly same. This option provides a good feature of unpredictability to the attackers. The detailed illustrations regarding constraints of multi-objective optimization are given below.

Equation (7.16) is constructed to calculate the values of D_l and Me_j if all the surrounding conditions are within the acceptable range that will allow considering all the authentication modalities. U_{max} is defined as the highest trustworthy value of any possible combination of the authentication factors. Previous chapter's trustworthy value of combined factors provides the highest value of 3.97 as trustworthy value which is considered the value of U_{max}. Hence, U_{max} can take the highest value of the considered authentication factors for any given scenario of multi-factor authentication.

Equations (7.17) and (7.18) are similar as an earlier constraint but it will be applicable under certain surrounding conditions. These types of constraints can be extended as other surrounding sensors can provide data regarding the existing environment condition of the client side. The surface plots for the given constraints are shown in ◧ Fig. 7.5a for fixed device (FD) and media as wired connection (WI). 7.6B shows the top view of the surface plot of the given constraints.

(a)   **(b)**

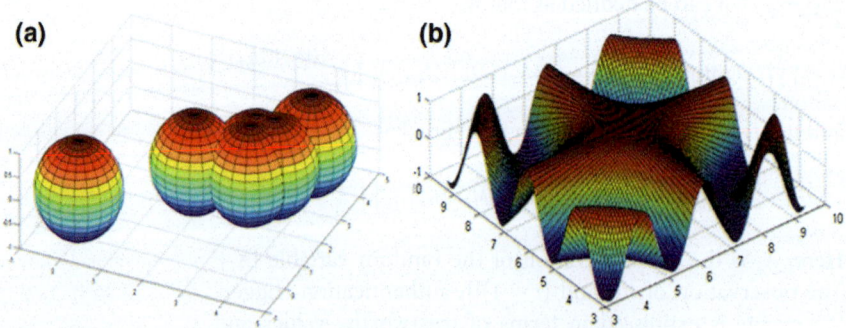

◧ **Fig. 7.5** Surface plots of the constraints of adaptive selection procedure

Now, a sample scenario is described to explain the constraints and in details. Let us consider, fixed device (FD) and wired connection (WI) are used. M_1, M_2, M_3, and M_5 (see ◻ Table 7.6) are the usable set of authentication modalities along with their various features. It is also assumed that all the surrounding environmental conditions, i.e., light (L), noise (N), motion (M), etc., are favorable to the user device and single device and medium are employed. Consequently, Eq. (7.16) will be selected and its different forms are listed below:

$$(FD - 3.57)^2 + (WI - 2.61)^2 + (WL - 2.75)^2 \leq 3.97;$$
$$(FD - 2.72)^2 + (WI - 2.74)^2 + (WL - 2.75)^2 \leq 3.97;$$
$$(FD - 3.90)^2 + (WI - 3.79)^2 + (WL - 3.69)^2 \leq 3.97, \text{ etc.}$$

Here, 3.57, 2.72, and 3.90 are the trustworthy values of M_1, $(M_1{:}f_{1,1},f_{1,2})$ and $(M_5{:}f_{5,1},f_{5,2})$ in fixed device (FD), respectively. Again, 2.61, 2.74, and 3.79 are the trustworthy values of M_1, $(M_1{:}f_{1,1},f_{1,2})$ and $(M_5{:}f_{5,1},f_{5,2})$ in the wired medium (WI) , respectively. Lastly, 2.75, 2.75, and 3.69 are the trustworthy values of M_1, $(M_1{:}f_{1,1},f_{1,2})$, and $(M_5{:}f_{5,1},f_{5,2})$ in the wireless medium (WL), respectively, (from Table 7.6).

If the surrounding light condition lies inside the region of visibility (given in ◻ Table 7.2), the background noise is not too high and surrounding motion is not affecting the selection procedure (as mentioned in ◻ Table 7.3), then Eq. (7.16) will be selected as a constraint for the selection mechanism. Similarly, other constraints can be chosen sensing the current surrounding conditions data from the client device.

Implementation of Adaptive-MFA System

This section highlights the implementation details of the previously mentioned Adaptive-MFA system. This authentication system is implemented using seven authentication modalities along with different features. The prototype system is developed to justify the applicability of the adaptive selection process of the multi-factor authentication system. The seven authentication modalities are face, fingerprint, voice, keystroke, password, SMS, and Time-enabled One Time Password (TOTP). The light and noise conditions are captured as part of surrounding conditions in this implementation of MFA system.

The adaptive-MFA system is implemented where the client application captures all the required authentication

7

modalities and sends the data to the authentication server. The server then extracts all the required features from the captured authentication data and stores the features into the database as part of user registration phase. During authentication, the client sends the surrounding conditions, device, and media information to the server. The server then runs the selection algorithm based on the information and chooses the best set of authentication factors to push it back to the client for verifying the user. The captured user data for selected authentication factors are later sent back to server, and server calculates the matching score for each of these factors. If all these factors are matched with the stored user template, the successful authentication decision is sent to the client machine. Otherwise, unsuccessful authentication decision is sent to the client and asked for reauthentication. Details of the implementation steps and different client-side and server-side components are described in later subsections.

Surrounding Conditions (Light and Noise)

Surrounding conditions play a big role in biometric-based authentication modalities. The performance of face- and voice-based authentication strongly depends on light and noise condition of the surroundings, respectively. In this research, both light and noise conditions of the surrounding are captured at the time of authentication triggering and the selection algorithm uses this information to provide a better set of authentication factors to validate users.

To capture the surrounding conditions, Arduino MEGA 2560 [18] based microcontroller is used to capture the light and sound sensor data and sends the data to the connected device using USB port. The sensor data is captured in every 100 milliseconds (ms) and hence, it provides the current surrounding conditions to the system. Arduino Mega has 54 digital input/output pins, 16 analog inputs, and 4 UARTs, which provide great flexibility to interface with a good number of sensors and get the surrounding data easily in real time to the system. ▪ Figure 7.6 shows the Arduino MEGA 2560 Microcontroller and its different parts. This is a widely used microcontroller to get the input of a good variety of sensors and output the signal back to the connecting computing system.

◘ **Fig. 7.6** Arduino MEGA 2560 microcontroller

Arduino-compatible Mini Luminance light sensor and Diymall Microphone Noise Decibel sound sensor are used to capture the surrounding conditions. The light sensor detects the surrounding luminance or light and delivers the output in terms of voltage. It has adjustable sensitivity and stable performance to capture the variability of different lighting conditions. The sound sensor detects the surrounding noise and delivers the decibel output in terms of voltage. The values are needed to be calibrated to fit with actual decibel values. The calibration is achieved by comparing their values with actual known decibel values in different existing noise settings.

Arduino Mega 2560 microcontroller is used to send the sensor values to the desktop or laptop. It is connected through USB port and is programmed using Visual C#. NET to capture the sensor values in the native application. ◘ Figure 7.7 shows the connections of the light and noise sensors with Arduino MEGA microcontroller.

Database to Store Authentication Factors

In this implementation, the database server is created to store all the user registration data along with the configuration parameter information for the server applications. MySQL Server version 5.7 is used for MFA implementation. The database stores all the extracted features of authentication modalities. All different authentication modalities are

■ **Fig. 7.7** The light and noise sensors are connected in a breadboard and are communicated with the client device through an Arduino MEGA microcontroller device

stored in separate tables, and the email address of the user is used as the primary key for all the modality tables. Additionally, configuration parameters are stored in separate tables that are used to run different algorithms to extract and match the required features on the server side. To keep track of various types of device and media information, database stores all the mac address of the client machines and addresses of Ethernet adapter and wireless LAN adapters.

Configuration of A-MFA System

The configurations of different types of devices including the client and server machines that are used to implement the A-MFA are shown in ■ Table 7.8.

The device types are listed in the first column while settings of those devices are mentioned in the second column

◘ **Table 7.8** Different types of devices used in current MFA implementation

Different devices used in implementing A-MFA	Settings
Authentication Server (storing database, calculating trustworthy scores, features' values, and adaptive selection decision)	**Dell PowerEdge R710** • RAM: 32 GB • OS: Windows Server 2012 • Processors: Intel Xeon X5650 with 12 cores at 2.67 GHz. • Database: MySQL 5.7
Client machine 1	**Dell XPS Laptop** • RAM: 8 GB • OS: Windows 10 Professional 64 bit • Processors: Intel i7-3517U with 4 cores at 2.4 GHz
Client machine 2	**Dell OptiPlex 740** • RAM: 8 GB • OS: Windows 7 Enterprise 64 bit • Processors: AMD Athlon 64 X2 Dual Core at 2.70 GHz
Fingerprint device	**Digital Persona U. are. U 4000B Sensor and 4500 Sensor** • Cross match fingerprint driver to access the fingerprint device • Supported OS: Windows 7, 8 and 10
Web camera	**Any Windows supported Web camera** • Integrated and external web camera
Recording and playback device	**Any Windows supported microphone and speakers** • To record the voice and playback the recorded audio
Keyboard	**Any Windows supported PS/2 or USB Keyboard** • To record the keystroke information of the user
Mobile phone	**Any mobile phone** • Capable of receiving SMS for a temporary passcode

of ◘ Table 7.8. Also, some devices required for capturing the biometric data are also mentioned in the table. These are fingerprint device, web camera, recording and playback device, keyboard, and cellular phone to receive passcode through SMS.

7

Client Application

In this implementation, seven different authentication modalities are used to create the prototype of Adaptive Multi-factor authentication. Of them, four are biometric modalities (face, fingerprint, voice, and keystroke) and three are non-biometric modalities (password, SMS, and TOTP). Also two different variations of passwords namely, passphrase and security questions are also integrated into this implementation. The client application is developed using Visual C#.NET platform with .NET Framework 4.5. The first window of launching A-MFA client application is shown in ◘ Fig. 7.8. A user needs to provide his/her email address to start the registration process. The server will check whether the email address is already registered or not. If this email address is not registered, it prompts the user to start the registration process. Otherwise, it will ask the user to authenticate using A-MFA to access the application.

Registration Process

To start with the registration process, a user needs to provide basic account information including the first and last name. Then the user needs to set up his/her password, passphrase,

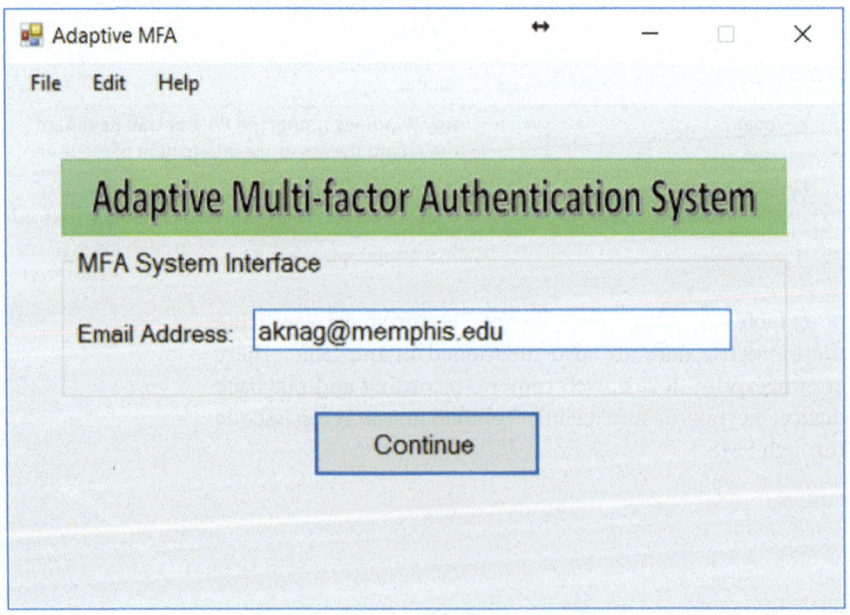

◘ **Fig. 7.8**   The welcome screen of adaptive multi-factor authentication client application

and security questions as part of password-based authentication modality. Later, the user needs to provide his/her telephone number along with carrier as part of registration data. The phone number will be used to send one-time password code (OTP) to verify user identity. As part of registration, a one-time code is sent to the user and user needs to type the given code to verify his/her phone. Once the user provides all this required information and saves them, the user account will be created with *incomplete registration* flag in the database. The screenshot of the client application is shown in ▢ Fig. 7.9. These fields are checked for validation as the user enters

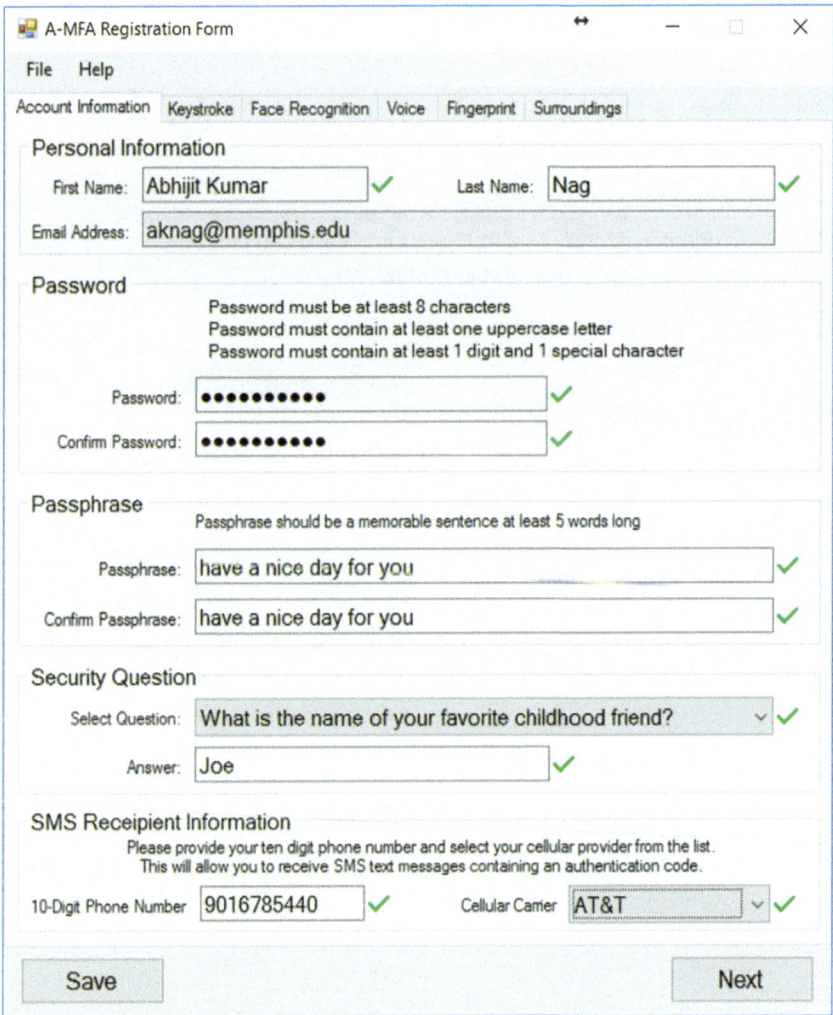

▢ **Fig. 7.9** The first window of user registration for A-MFA application

information. This reduces the time required for the server to validate the registered data and provide the registration status back to the client in shorter time. The «*Next*» button shown in ▣ Fig. 7.9 is used to navigate among different screens.

The next screen is for registering the keystroke. Here, the user needs to type the text that he/she provided as his/her first and last name. To better train the keystroke algorithm, five keystroke samples are captured through this process and dwell time and flight time are recorded for every entered key. The screenshot of captured keystroke biometrics is shown in ▣ Fig. 7.10.

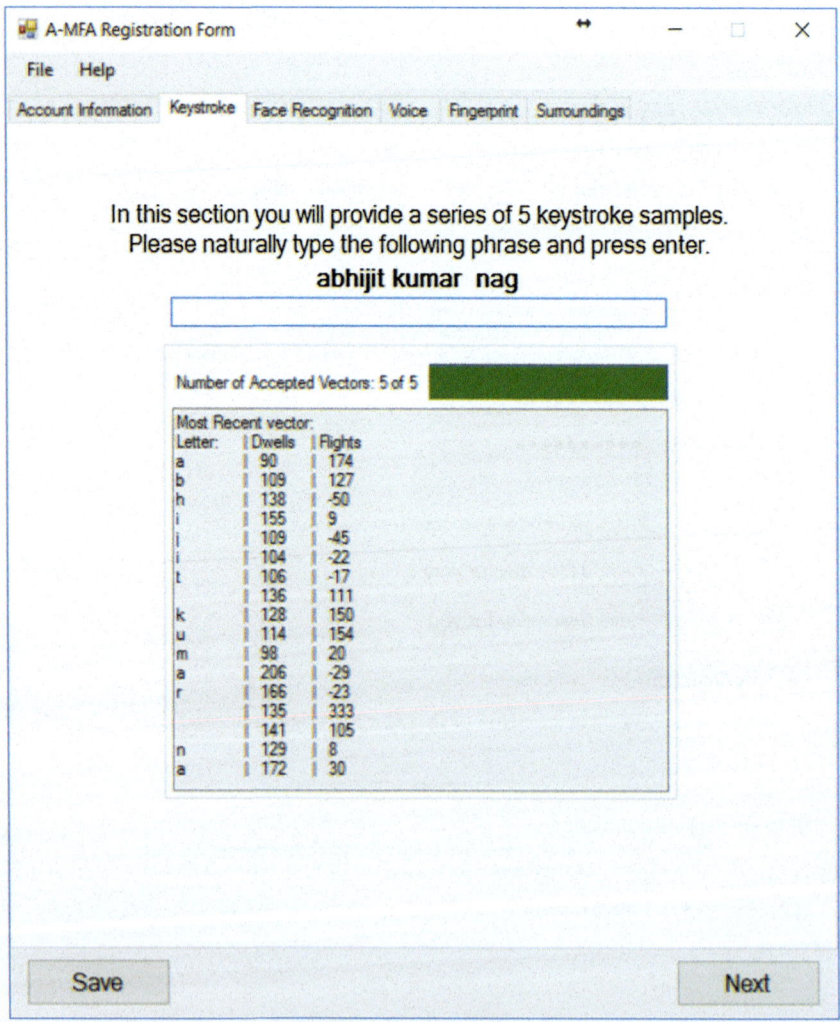

▣ **Fig. 7.10**   Keystroke capture using A-MFA application

To capture face data for registration, a user needs to choose the web camera from the available options and click on *Start Camera*. Face detector program then detects the face from the video and draws a rectangle to identify the region. Upon pressing «*Capture*» ten images are recorded and the capture status is shown completed. The user then can press *Next* button to move to next tab or can press *Save* button to store the facial information in the database. The screenshot for capturing face biometric is shown in ◘ Fig. 7.11. The user

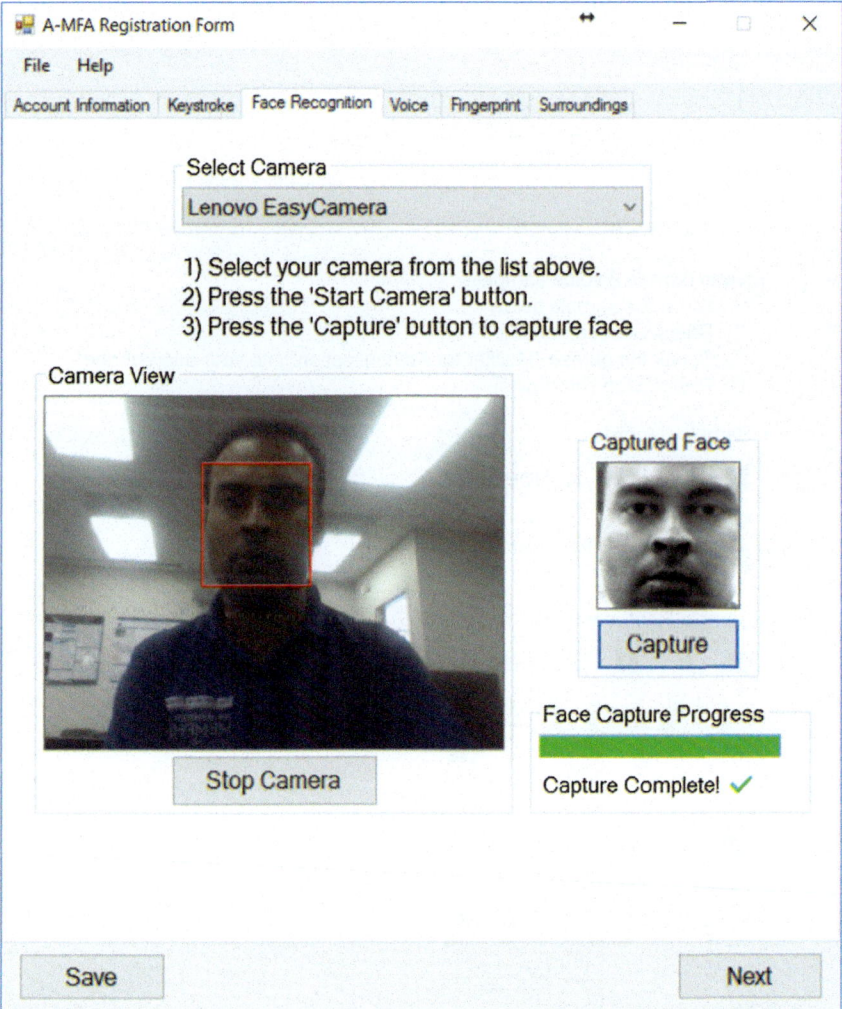

◘ **Fig. 7.11**   Face biometric obtained using A-MFA application as part of registration

is allowed to take as much time as possible to capture his image before saving the information to the server.

To capture voice data for registration, a user needs to provide three voice samples for a predefined phrase («Multi-factor Authentication»). The user interface provides flexibility to the users to playback the recorded voice and if needed, it can be recorded multiple times. A real-time visualization of the voice signal is also provided for every voice sample as it is being recorded. The screenshot for capturing voice biometric is shown in ◘ Fig. 7.12.

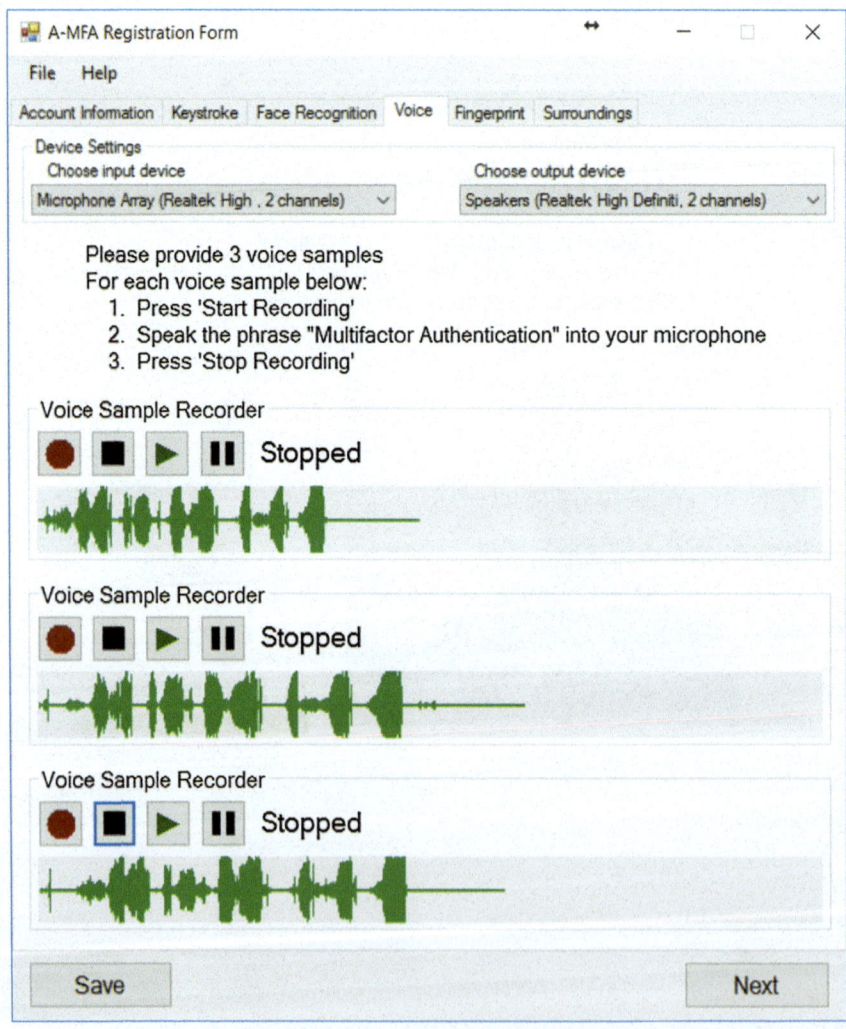

◘ **Fig. 7.12** Voice biometric obtained using A-MFA application as part of registration

To capture the fingerprint biometric, the user needs to provide three samples of the same thumb to create the registration profile. The GUI keeps tracks of the fingerprint captured and provides feedback to the user at a real time. In this implementation, two different fingerprint devices are supported as mentioned in ■ Table 7.8. Upon completion of the fingerprint capture, the user can save the registered data and move to see the surrounding conditions in the next tab. ■ Figure 7.13 provides a screenshot of the Fingerprint GUI of A-MFA system.

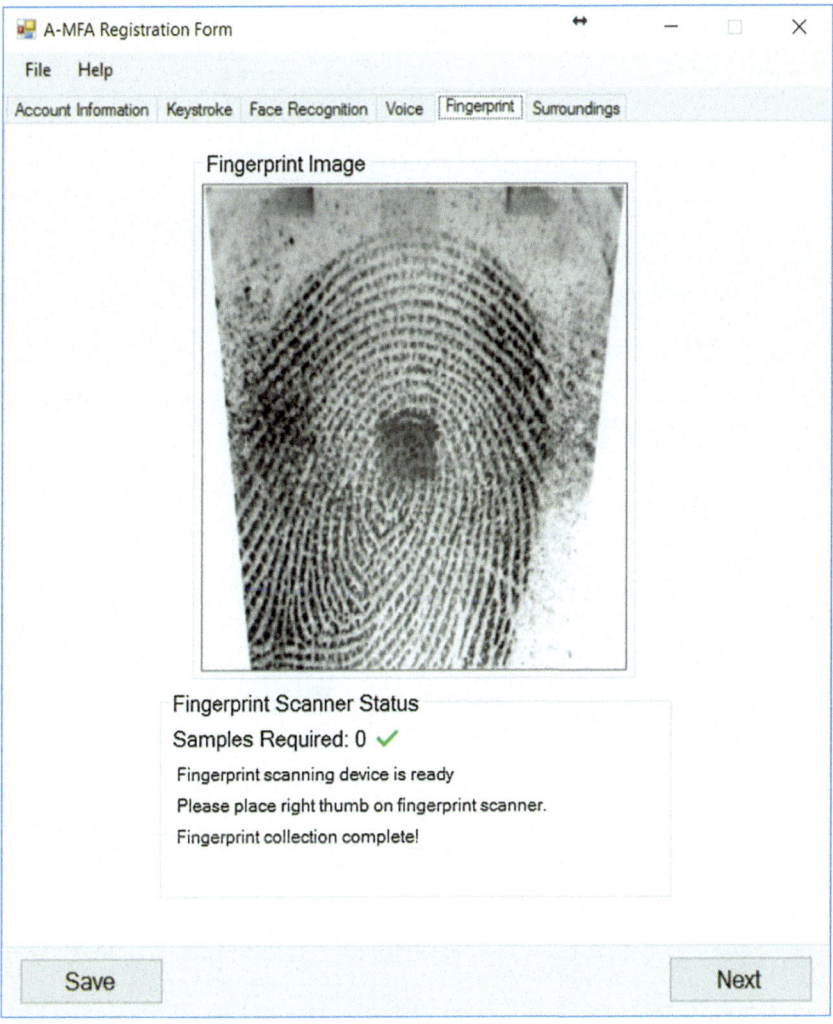

■ **Fig. 7.13** Fingerprint data is captured through A-MFA application

The surrounding conditions (light and noise) are captured in the client application as well (■ Fig. 7.14). The available ports are displayed for Arduino MEGA microcontroller and a visual scale is also shown to quantify the sensor values. Light condition is shown in the unit of Luminance (lux), and noise condition is shown in the unit of a decibel. The availability of light and noise sensor data will be stored as part of the registration template, which allows capturing these data during the authentication phase to correctly choose the best

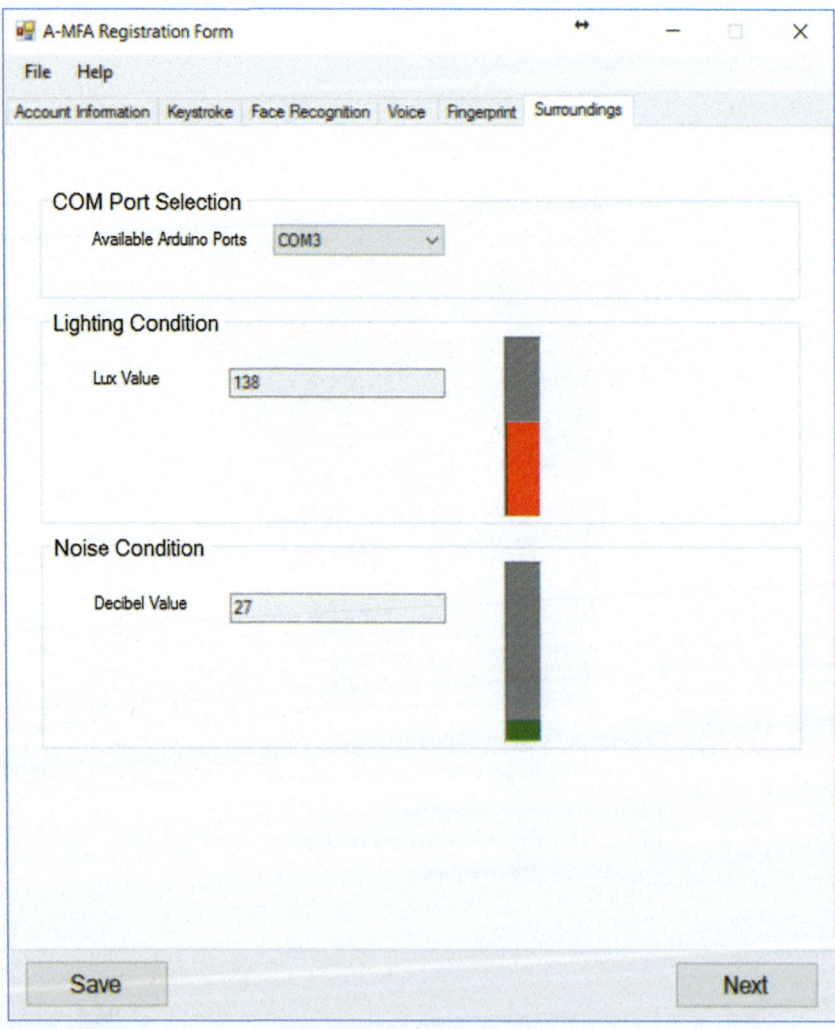

■ **Fig. 7.14** Surrounding conditions obtained in client application through Arduino microcontroller

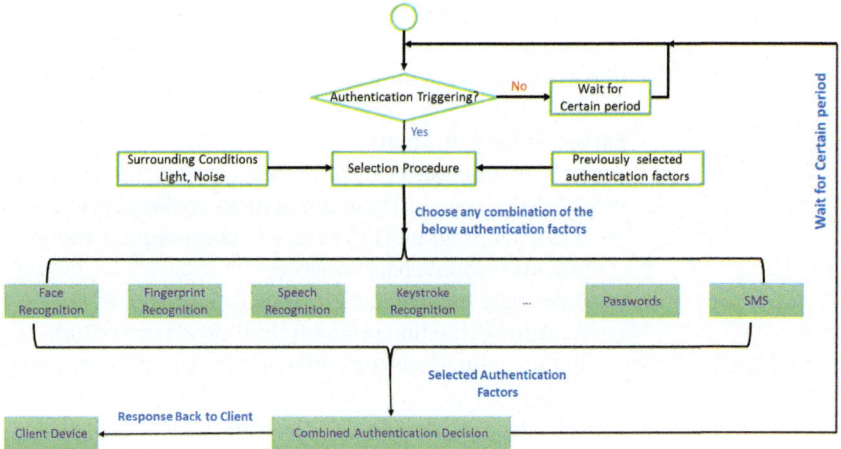

○ **Fig. 7.15** Flow diagram for A-MFA authentication triggering process

set of authentication factors at any given authentication triggering event.

In this implementation, incomplete registration is allowed. If the user tries to finish registration process using the same device (MAC address), the system will ask for a password to verify identity. If the user tries to complete registration using a different device, the system will verify the user's identity through SMS code. This is one of the novel features of the A-MFA system supporting the incomplete registration. Also, the registered data are stored in the memory buffer on the client side and it will pass to the server through secure HTTPS connection as a JSON object. No biometric and non-biometric data is stored in the client device during the registration and authentication phase. The flowchart of A-MFA authentication triggering events is included in ○ Fig. 7.15.

According to ○ Fig. 7.15, authentication triggering can happen in various time intervals and the selection algorithm chooses the best set of authentication factors considering surrounding conditions, the previous history, and currently selected device and medium. The selected factors are sent back to the client application to challenge against the claimed identity.

Authentication Process

During authentication, some of the chosen authentication factors are challenged to the users to verify them using the same GUI as registration phase. The client application stores the server address and port number to connect to the authentication server and then send the captured authentication information for verification.

Server-Side Application

Server-side application includes five components to implement A-MFA system. These components are listed below:

- A listener program (LP) to receive client request and provide response back to server
- A database implemented in MySQL 5.7 database
- A feature extraction program (FEP) to extract features of all the captured authentication modalities and store into database
- The selection procedure (SP) to run adaptive approach to generate set of authentication factors sensing device, media, and surrounding conditions
- A Matching Program (MP) to compute score by comparing registered user data (stored in database) with the captured authentication data and calculate the authenticate decision for the client

LP in server side is implemented in Visual C#.NET. It listens to all incoming requests from clients and accepts JSON objects. Parsing the type field of the JSON object, it can classify the clients and different request types and then, address them accordingly. FEP is launched from LP program and processes all the authentication modalities data to extract the features. If the extracted features' qualities are not good, FEP sends a reply back to the listener program, which eventually will forward back to the client reporting bad user data. If the features' qualities are good, the features are stored in the database server and the *success* message is sent to the client application to acknowledge the user accordingly.

LP module triggers SP module when an authentication request comes to the server. SP generates the appropriate authentication factors and sends the result to the LP. The selected factors for that user are then stored in the database for records. LP then forwards the selected authentication factors to the client to capture them. MP calls the FEP to extract features for the selected authentication factors during the authentication phase. These captured feature values are then compared with the stored feature values in the database. The final score (along with authentication decision) computed by MP will then be forwarded to LP, which is eventually delivered to appropriate client device.

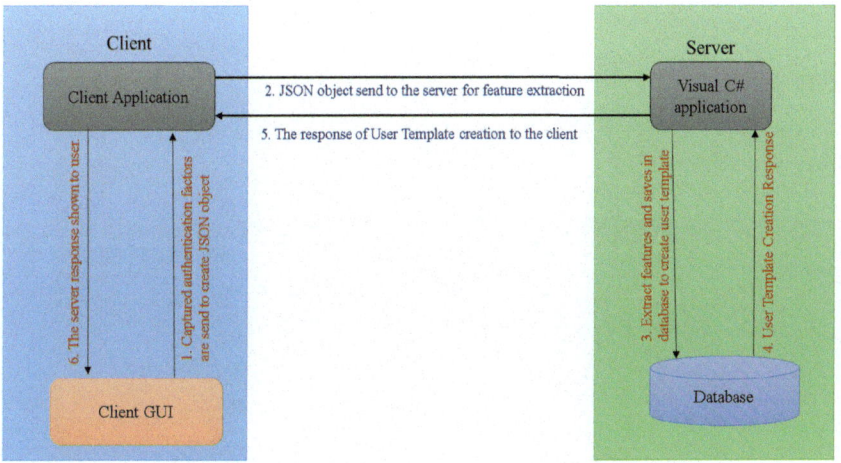

Fig. 7.16 Overall process of user registration in the A-MFA system

The required steps for user registration process between a client application and server machine is shown in ■ Fig. 6.15. According to ■ Fig. 7.16, the client application captures the required authentication data and forms a JSON object. The JSON object is then sent to the server-side application. The server then deserializes the JSON object and gets the individual authentication modality data. The server-side Visual C# application then extracts feature values from each authentication modality and saves the feature values as part of user registration profile in MySQL database. After successful completion, «registration was successful» message is sent back to the client to acknowledge the user.

In this implementation, processing of an individual authentication modality is done using a thread to increase the extraction performance and reduce the time required for completing the registration process. Again, to handle multiple registration requests regarding storing the captured information in the database, a queue is maintained on the server side to avoid deadlock and better handling the waiting periods of the client applications for registration response.

The undertaking steps between a client application and server machine to do the authentication process are shown in ■ Fig. 7.17. This figure clearly shows all the process and communication steps conducted between the server machine and client application to successfully authenticate a user. According to ■ Fig. 7.17, an application developed in Visual C# on the server side listens to all incoming requests from the clients. This application then communicates with

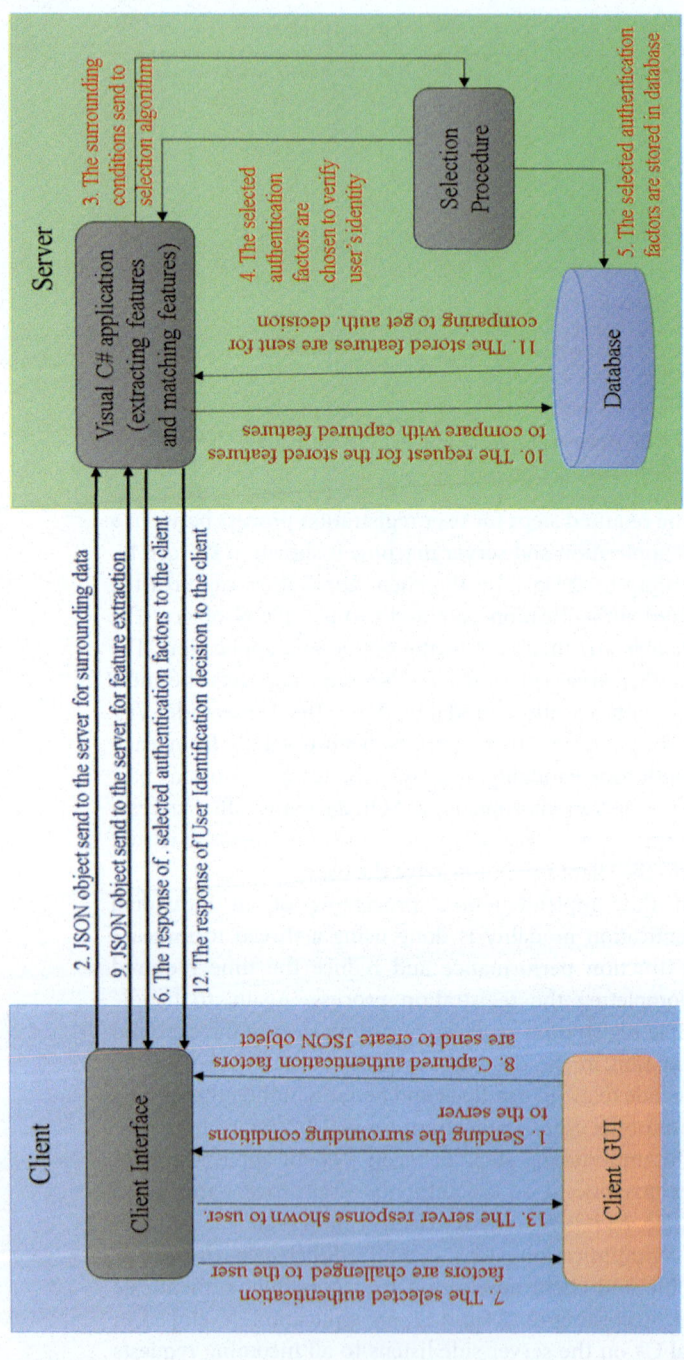

■ **Fig. 7.17** Overall process of user authentication in the A-MFA system

database and selection procedure to get the best set of authentication factors and push back these decisions back to the client application. The client application then captures the required data for the selected factors and sends back to the server for verification. The server then runs the matching algorithms and takes authentication decision. The decision is then forwarded to the client to either successfully authenticate or deny the user to the system.

In this implementation, three different approaches used for every biometric modality are considered. Every approach uses a given set of features and has their individual error rates. In the case of password-based authentication modality, password, passphrase, and security question are added as three possible choices. In addition, two out-of-band authentication modalities are implemented to provide a diverse set of authentication factors. They are SMS- and Time-based one-time password (TOTP). The details about biometric and non-biometric approaches are listed below:

Biometric Algorithms Used in A-MFA System

Each biometric modality that the A-MFA system uses can be divided into two distinct processes: enrollment (◘ Fig. 7.18) and verification (◘ Fig. 7.19). Enrollment involves the collection of biometric samples from the user, extracting a set of features from these raw samples, creating a template from the extracted features, and finally storing the feature templates in the database. Verification is a similar process to enrollment; however, in this case the feature template may not be stored in the database but instead is used as a query template in the matching algorithm, where stored feature templates are retrieved and used as target templates in the matching algorithm. The matching algorithm evaluates the similarity of the two templates and returns a similarity score. If this similarity score is within a predefined threshold value, the query template is accepted as a genuine identifier of the stored user. If the score is not within the threshold, the query template is rejected.

Face Recognition: Face Recognition is the first biometric-based authentication modality that is implemented in the A-MFA framework. In this prototype application, three different feature-based authentication approaches are considered. They are Eigenface, Fisherface, and Local Binary Pattern Histogram (LBPH)-based approaches. Eigenface approach [19] uses Principal Component Analysis (PCA) to find the set of vectors that best describes the distribution of

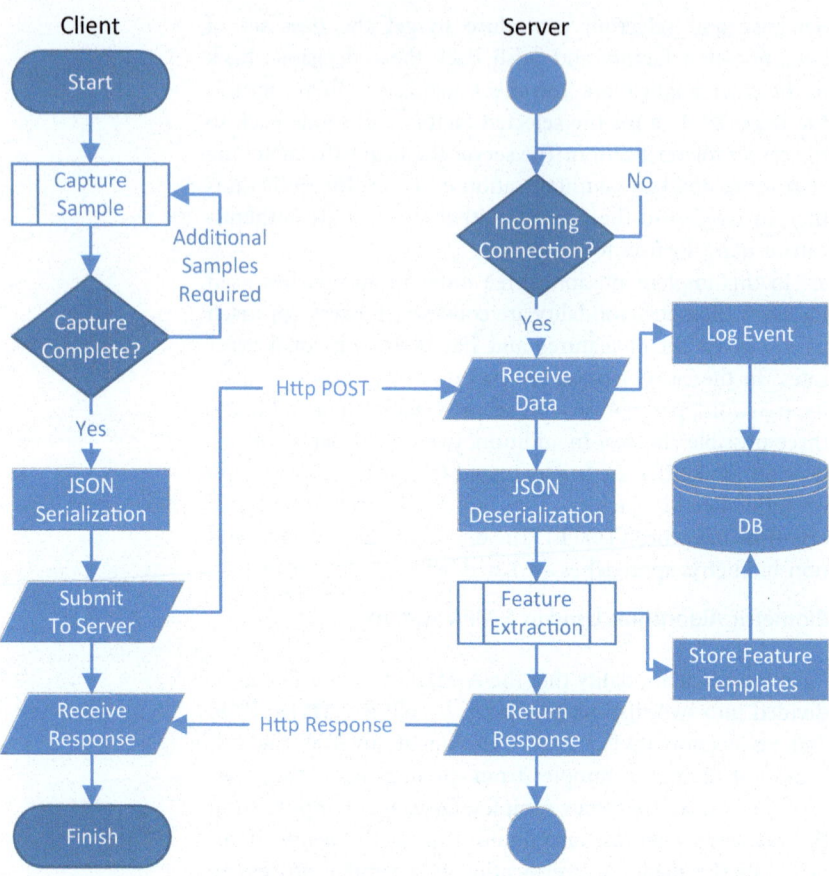

Client

Server

■ **Fig. 7.18**   Process flow of biometric enrollment and matching

face images within the entire face image space. Because the vectors are N × N linear combinations which represent the eigenvectors of the covariance matrix of the original images and resemble actual faces, they are called «Eigenfaces» Eigenfaces [19]. These Eigenfaces are considered as one of the features of face recognition. The Fisherface method is based on Fisher's Linear Discriminant method [20]. Like the Eigenface method, Fisherface uses a global model and performs dimensionality reduction by projecting 2D points down to 1D, though intra-class scatter is reduced and classification is, therefore, simplified. Fisherface exercises more discriminative power than Eigenface with a significantly lower error rate, though under admittedly idealized conditions [21]. LBPH approach [22] uses spatial information of face to do the face recognition. This approach divides the

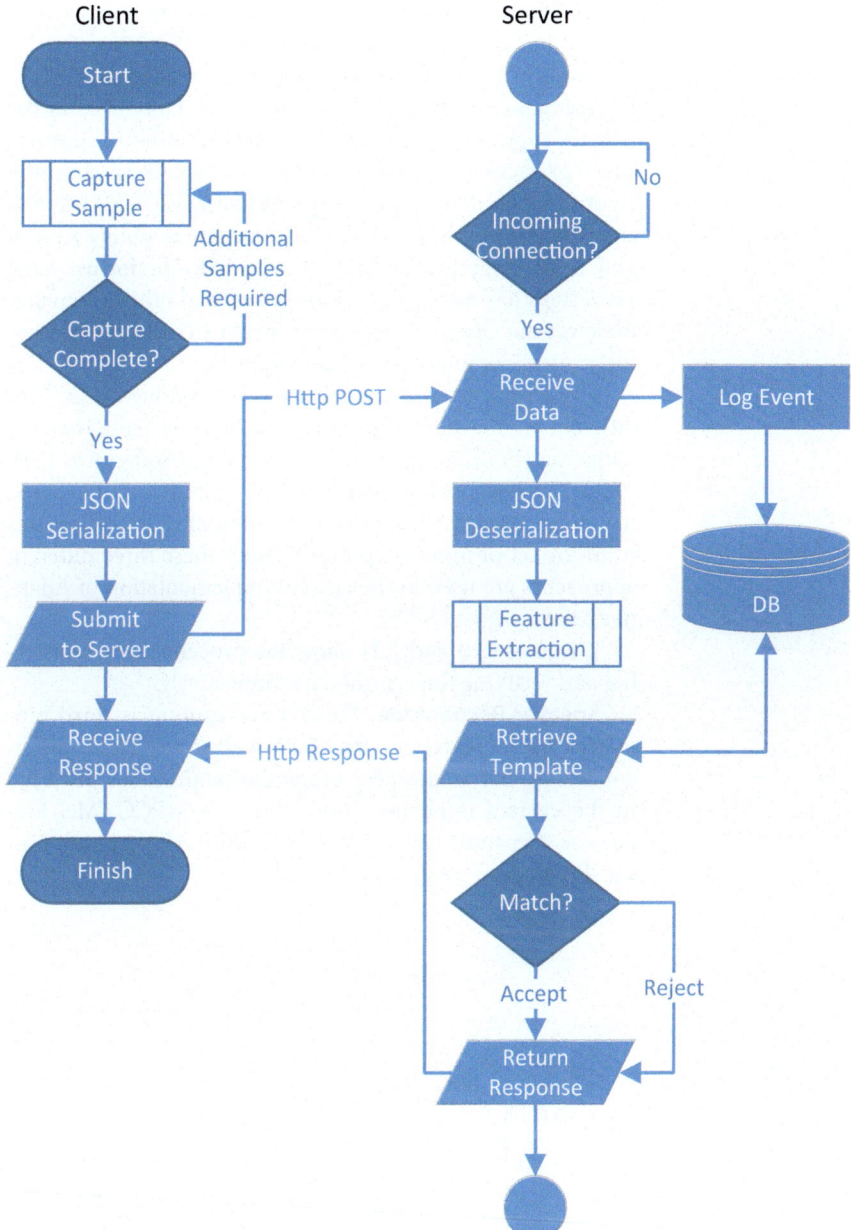

Fig. 7.19 Process flow of biometric matching algorithm

LBP image into 'm' local regions and extracts histogram from each region. The spatially enhanced feature vector is then obtained by concatenating the local histograms. In this

7

current implementation, Emgu OpenCV library is used to extract those mentioned features for face detection.

Fingerprint Recognition: Fingerprint recognition is the second biometric-based authentication modality used in the A-MFA approach. In this implementation, minutiae feature-based extraction methods are used. The two most common algorithms used are Ratha [23] and SourceAFIS [24]. Ratha algorithm attempts to remove false minutiae points from a gray-scale fingerprint image. SourceAFIS performs some extra steps to remove gaps, dots, spikes, and other anomalies to detect the false minutiae points better in order to increase the accuracy of fingerprint recognition.

Fingerprint feature matching process is done using three different approaches in the Adaptive-MFA System. They are Jiang-Yau (JY) [25], Medina-Perez [26], and SourceAfis [24]. These three algorithms are based on minutiae point-based matching, though they each derive a unique set of features from the set of minutiae points. Hence, these three different approaches are used in the current implementation of Adaptive MFA.

■ Figures 7.20 and 7.21 show the procedure for registering and verifying fingerprint data, respectively.

Speaker Recognition: Speaker recognition is third biometric implemented in the A-MFA system. Three different feature-based speaker recognition approaches are used in the current implementation. They are MFCC (Mel Frequency Cepstral Coefficient) [27], PLP (Perceptual Linear Prediction) [28], and Prosody [29]. MFC coefficients

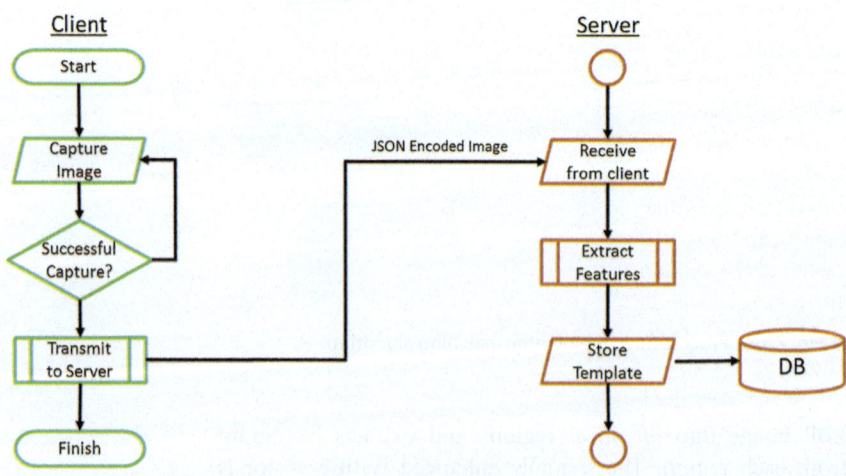

■ **Fig. 7.20**   Fingerprint registration process flowchart

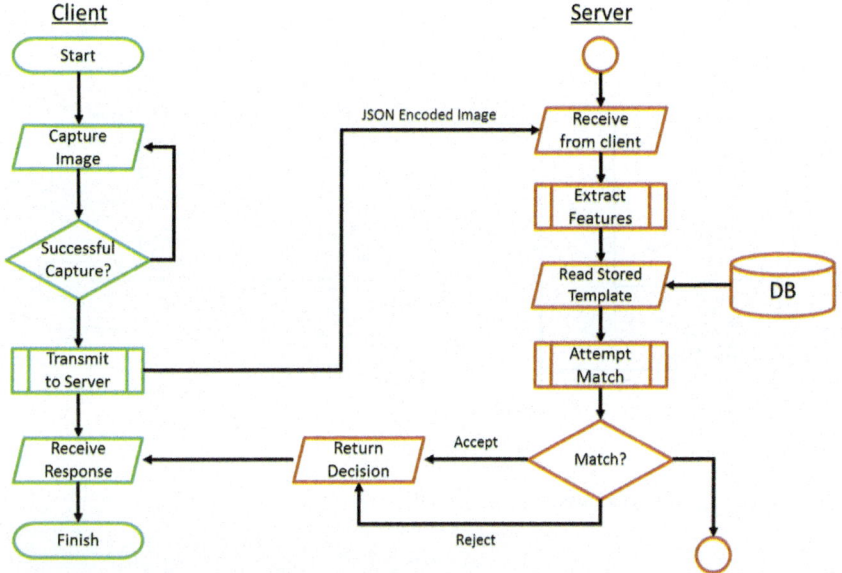

☐ **Fig. 7.21**   Fingerprint verification process flowchart

result from a discrete cosine transform of the logarithm of the short-term energy spectrum on a Mel Frequency scale, which more closely approximates the human auditory sense and captures important phonetic characteristics in a speech [27]. PLP uses psychophysical concepts of hearing to derive an estimate of the auditory spectrum [28]. MFCC and PLP both represent features that capture short-term representations of human speech. Prosody features are concerned with longer term aspects of human speech such as intonation, stress, tempo, and rhythm, all of which convey information about the speaker [29]. To compute the similarity to find the speaker, Dynamic Time Warping (DTW) [30] is used. DTW measures the similarity between two voice samples that are misaligned in time of speed. The preprocessing to the captured voice signal (trimming the beginning and ending silence period) is done in the developed prototype application and three different feature sets are stored in the database for every registered user. ☐ Figures 7.22 and 7.23 illustrate the processes of speaker registration and verification, respectively.

Keystroke Recognition: Keystroke recognition is the fourth biometric authentication modality (passive biometric) used in this Adaptive-MFA prototype. It is used to identify an individual and is based on the typing pattern, the rhythm,

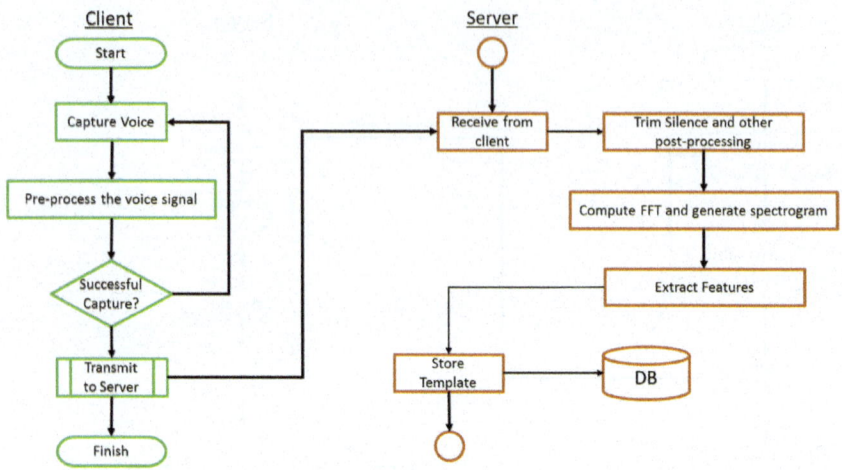

☐ **Fig. 7.22** Speaker data registration process flowchart

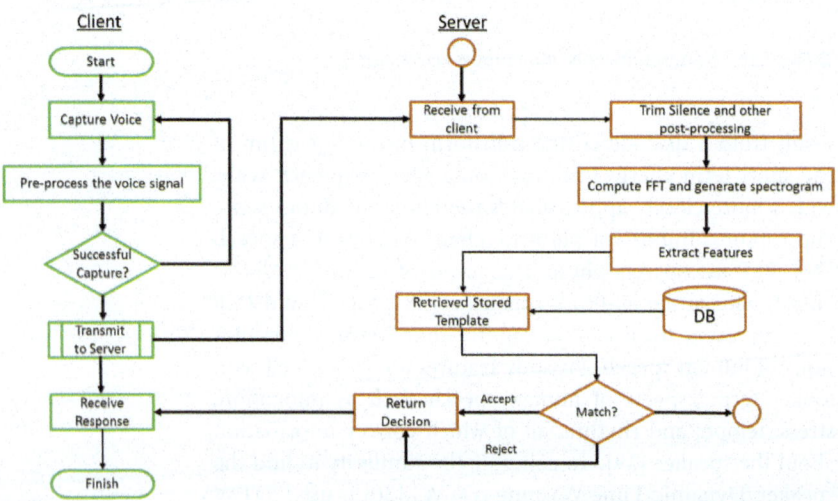

☐ **Fig. 7.23** Speaker verification process flowchart

and the speed of typing on a keyboard. The raw features for keystroke dynamics are dwell time and flight time [31]. Dwell time is the time duration that a key is pressed. Flight time is the time duration in between releasing a key and pressing the next key. In A-MFA implementation, the first name and last name of the user are selected to type to capture the keystroke data. When typing a series of characters, the time the subject needs to find the right key (flight time) and the time he holds down a key (dwell time) is specific to

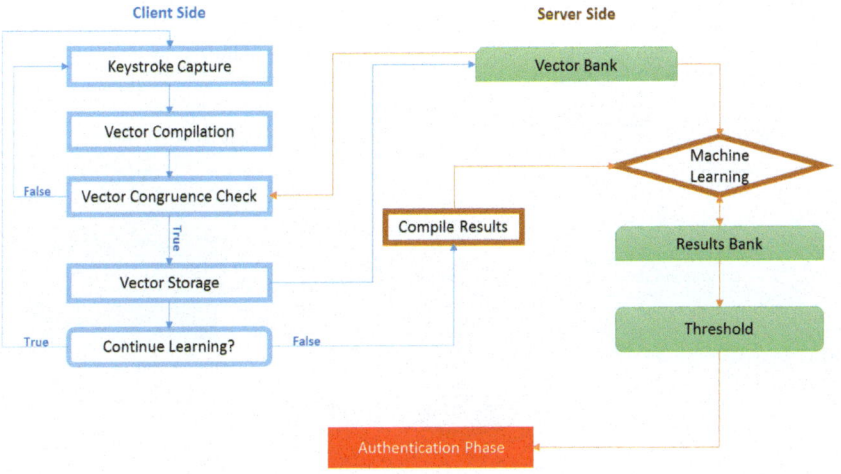

Fig. 7.24 Keystroke vector learning phase process flowchart

that subject, and can be calculated in such a way that it is independent of overall typing speed. The rhythm with which some sequences of characters are typed can be very person dependent. Three different approaches are used to calculate the similarity score of the user's keystroke samples. They are Manhattan scaled, Mahalanobis, and Nearest Neighbor Mahalanobis metrics [31]. In future, neural network-based approach will be used along with two and three letter combinations (considered as features), which provides better accuracy for continuous authentication. A user needs to provide at least five samples of the same phrase to make a good keystroke data model. The keystroke learning phase and verification phase are illustrated in ◘ Figs. 7.24 and 7.25, respectively.

The client and server-side component model is shown in ◘ Figs. 7.26 and 7.27, respectively, and discussed in details.

According to ◘ Fig. 7.27, the client device provides an interface to the user for enrollment into the system and verification for access to the system. All data sent from the client to the server is stored in a custom data container. The client container defines a request type, which is used to determine the client's intent and the server's response action. It also contains data fields for all user information, biometric data, device information, network information, and environmental information. The client data container is serialized as a JavaScript Object Notation (JSON) string and is posted to the server using HTTPS. Because HTTPS is the underlying connection protocol, the A-MFA system can be used by any

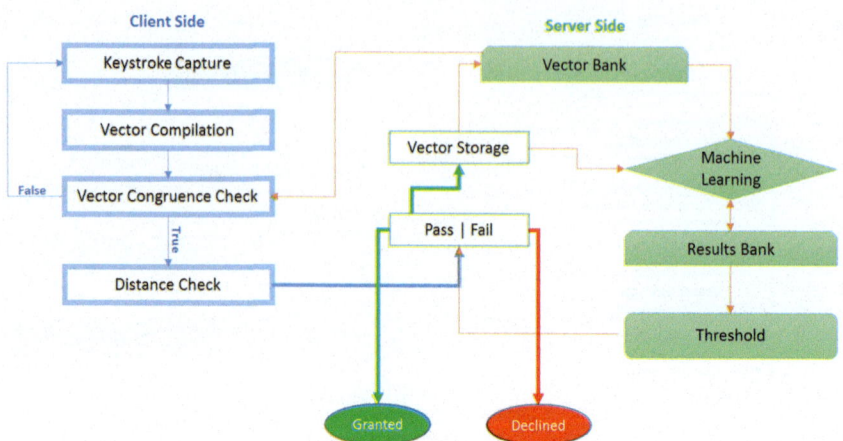

Fig. 7.25   Keystroke verification process flowchart

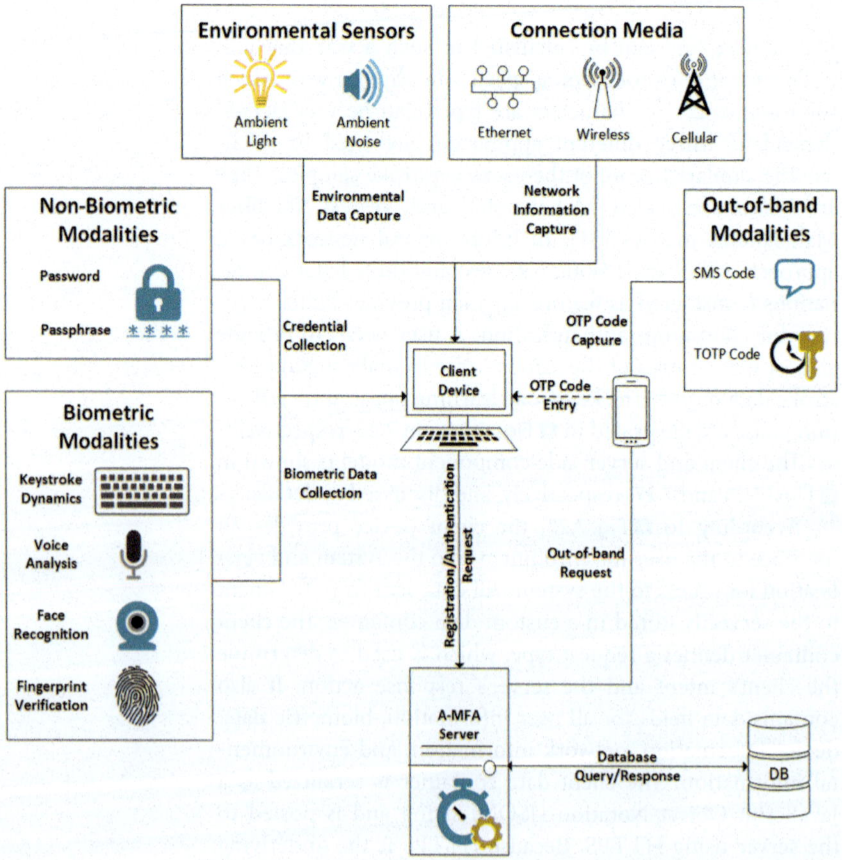

Fig. 7.26   The client component diagram for enrollment and authentication

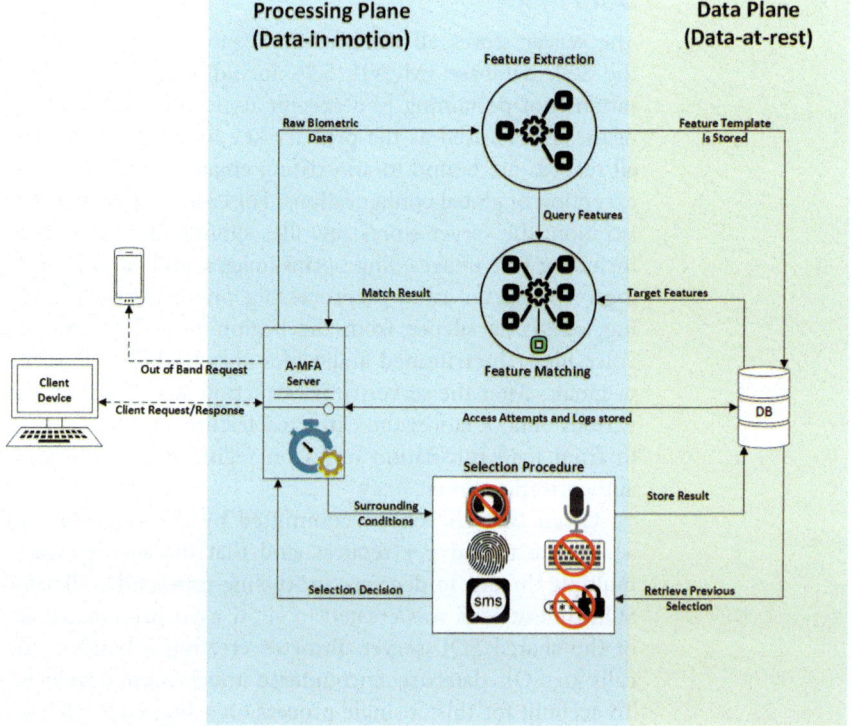

Fig. 7.27 The server-side diagram showing the logical separation of data processing and data storage tasks

device with network capabilities using either a native application or browser.

Processing Plane

Except for a small amount of preprocessing done by the client, the backend server handles most of the data processing. As with the client, the server uses a custom data container to package all responses that it will send back to the client. This container includes the requesting client's registration status, any action that the server might need to request from the client, values that define credentials that the user must supply, and in some cases information that is useful to the client. Information includes the Quick Response Code (QRCode) that contains a shared TOTP key, the keystroke phrase that needs to be used for keystroke recognition, and general purpose messages for relaying errors and other communications.

Data Plane

The server stores all information regarding every user in the SQL database (MySQL 5.7), including global configurations not pertaining to a specific user. The email address of the user is used as the primary key in the database, and all records are bound to an existing email address, with the exception of global configurations. For each user during registration, the server stores raw files submitted by the client including face images, fingerprint images, and voice recordings. The server runs preprocessing on the voice recording, removing silence from the beginning and end of the voice files. The trimmed audio files are saved as well as the originals. After the server runs extraction algorithms on the various data, it stores the extracted features in the database to front load calculation times on registration to decrease authentication times.

Given that all actions committed by the server are in a separate thread per request, and that the server creates multiple threads to decrease processing times, a handler for SQL transactions was created to allow asynchronous access of the shared SQL server. Prior to creating a handler, the calls to SQL database encountered transaction deadlocks. To account for this, a single process on a single thread handles committing all SQL transaction. Command objects are spawned from various threads and locations and are given to a First-In-First-Out (FIFO)-based queue for SQL commands. This causes the asynchronous calls to the shared SQL resource to be synchronized, solving deadlock issues.

Experimental Results of the A-MFA System

The A-MFA system is implemented in the Visual C#.NET application (Visual Studio 2015 Enterprise Edition). The client side application (all the implemented methods) is tested with Unit testing as part of the development process. To test the actual performance of the developed system in the diverse set of users and surrounding conditions, a synthetic dataset is created with 50 users' information. Detailed information regarding database creation is listed in the next subsection.

Creation of Synthetic Dataset

To test the accuracy of the A-MFA system, we need to have a good set of users' data to authenticate them multiple times (in different surrounding conditions). Several statistics (average and standard deviation values) are used to measure successful and unsuccessful authentication attempts. To accomplish these requirements, a synthetic dataset is generated to run hundreds of authentication requests simultaneously. The biometric data of each user are captured from different well-known datasets, which are listed below:

1. Face Dataset: In the A-MFA application, 15 images of the same person are required to run all face recognition experiments. Ten images are used for registration purpose and five images are used for authentication purpose. In order to use face images, images with same facial expression needed to be used. In this project, faces 94, faces 95, and faces 96 [32–34] datasets are used to generate face biometric data.

2. Fingerprint Dataset: In the MFA application, five fingerprint images of the same person are used to run all fingerprint recognition experiments. Of them, three images are used for registration purpose, and two images are used for authentication purpose. In A-MFA testing, CASIA Fingerprint Image Database Version 5.0 [40] was used which contains 20 k fingerprint images of 500 subjects. Other well-known datasets contain FVC 2000, FVC2002, and FVC 2004 [35–37]. But they contain very few subjects' fingerprint images that do not serve a testing purpose.

3. Voice Dataset: In the MFA application, three voice samples per user are required during registration, and two voice sample are required for authentication. CMU PDA dataset (▶ http://www.speech.cs.cmu.edu/databases/pda/) is used to create the required voice samples for the synthetic dataset.

4. Keystroke Dataset: In the implemented MFA, five keystrokes per user are required for registration (typing the same phrase). Three more keystrokes are required for authenticating the same set of users. For testing purpose, CMU dataset (▶ http://www.cs.cmu.edu/~keystroke/) is used which contains keystroke timing information of 51 subjects for a given text. Moreover, for each user, it has 400 times typing record which is a good base to test the performance of keystroke recognition problems.

Non-biometric data are generated programmatically to create user registration information to be stored in the test database. Passwords and passphrases are hashed using SHA-512 on the client side and hashed with B-Crypt in the server side. The communication between client and server are done through https protocol which is an end-to-end encrypted communication of data-in-motion.

System Evaluation Criteria

The following evaluation criteria are designed to evaluate the performance of the implemented MFA system.

(a) The number of successful and unsuccessful attempts while authenticating with the synthetic set of users. This metric will measure the accuracy of the system with three authentication factors. The test will be conducted multiple times with different sets of surrounding conditions and a different set of factors (2F and 3F) to calculate the average value of successful and unsuccessful attempts.

(b) To calculate the false acceptance rate of the A-MFA system, incorrect user data are used in 1500 authentication attempts to find the average rate of false acceptance for the generated dataset. This test is also repeated for different sets of surrounding conditions to get the overall accuracy of the system in all possible environment conditions.

(c) The individual accuracy of various biometric-based approaches is measured separately to see their actual performance over the generated dataset. This test helps us to tune the threshold parameters for all the adopted biometric methods to best fit with the implemented prototype.

(d) To find the total trustworthy values of the selected modalities in different selection schemes (adaptive selection, random selection, only biometric selection, only non-biometric selection, etc.)

The implemented prototype is tested against the criteria mentioned above and the results are reported in a later section.

Experimental Results

To meet the first evaluation criteria regarding successful and unsuccessful attempts of users, we try to authenticate a user multiple times (five times) and calculate the average values of successful and unsuccessful attempts. These attempts are calculated in different surrounding conditions to verify the efficiency of the implemented system. ◘ Table 7.9 displays the statistics in more details.

◘ Table 7.9 shows the statistics of successful attempts of the set of synthetic users (authenticating 10 times per user) using two-factor and three-factor of the Adaptive-MFA solution. With the good set of surrounding conditions, the efficiency of the solution is about 98% for two-factor and 95% for three-factor-based authentication for the valid sets of user data. This table helps in calculating the true positive and true negative rate of the implemented system.

To measure the false positive rate, the implemented system is tested with 1500 invalid requests (imposter data) and the statistics of the experiment is shown in ◘ Table 6.10. This experiment serves the purpose of second evaluation criteria.

According to ◘ Table 7.10, with light and noise in a good range, the success rate of imposter data is close to 0% for 2-Factor and 3-Factor cases. With only light or only noise

◘ **Table 7.9**   The successful attempts for valid user data in different surrounding conditions for two-factor and three-factor-based adaptive-MFA system

Surrounding conditions	Two-factor of A-MFA (%)	Three-factor of A-MFA (%)
Light and noise are in range	94	93
Light is in range	92	90
Noise is in range	90	88
None are in range	86	84

◘ **Table 7.10**   The successful attempts for invalid user data in different surrounding conditions for two-factor and three-factor-based adaptive-MFA system

Surrounding conditions	Two-factor of A-MFA (%)	Three-factor of A-MFA (%)
Light and noise are in range	0	0
Light is in range	0.5	0
Noise is in range	0.45	0
None are in range	3	0

condition, the accuracy of two-factor is very less and 0% for three-factor cases. When none of them are in range, 3% requests pass through for two-factor MFA cases. But using three factors, the success rate is 0% for all the possible cases of surroundings.

Finding the right threshold value for the biometric-based approaches is crucial toward the performance of these algorithms in authenticating the individuals. Hence, a thorough experiment is conducted to find the best set of a threshold value that minimizes both FAR and FRR values. Each of the biometric algorithms is tested with a previously mentioned synthetic dataset with varying threshold values to obtain EER (Equal Error Rate) values.

The FAR, FRR, and EER scores for face detection algorithms are shown in ◘ Fig. 7.28. It is important to be noted that Eigenface and Fisherface use the global model for the entire dataset. This model is trained on all currently enrolled face images. For these two models, a matching means fulfilling two criteria—a threshold score and a label (user id). It is possible that a face biometric will score below the given threshold value but has the wrong label, which results in a rejection. If the overall score is below the predefined threshold value and the labels match with the stored entity, these algorithms report a match for the provided image. Therefore, FAR (shown in red in ◘ Fig. 7.28) is very low for these two algorithms. For LBPH approach, it uses the local model and hence, a good balance between FAR and FRR values is required to find out the EER value for this algorithm.

The FAR, FRR, and EER values for fingerprint, keystroke, and voice algorithms are mentioned in ◘ Figs. 7.29, 7.30, and 7.31, respectively. In the case of fingerprint-based algorithms, the similarity scores are calculated between samples, where other three biometrics distance scores are calculated between samples.

The triggering of different reauthentication attempts in various environment conditions (device, media, and surroundings) and the selected authentication factors by the

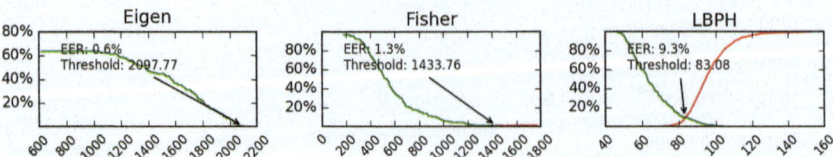

◘ **Fig. 7.28** Eigenface, Fisherface, and local binary pattern histogram-based face recognition performance

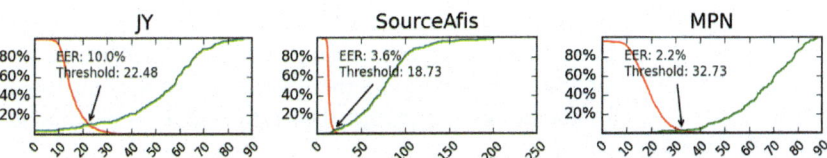

Fig. 7.29 Jiang-Yau, SourceAfis, and MPN fingerprint matching algorithms' performance

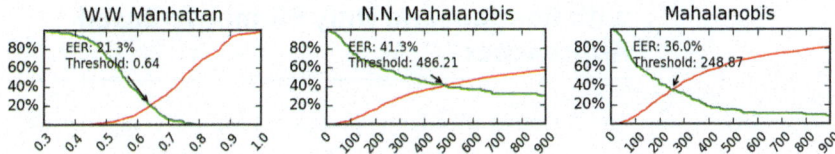

Fig. 7.30 Word-wise Manhattan, nearest neighbor Mahalanobis, and Mahalanobis approaches for keystroke distance metrics performance

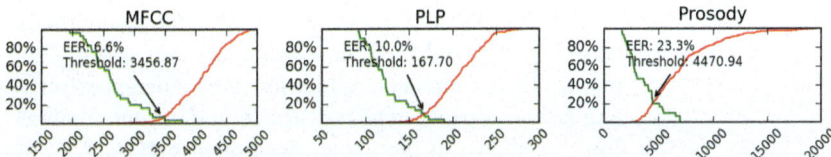

Fig. 7.31 MFCC, PLP, and Prosody features-based speaker recognition approaches performance

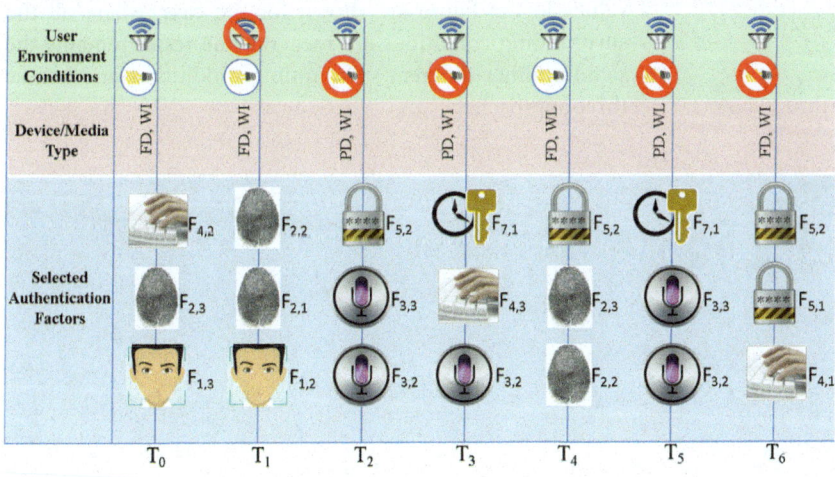

Fig. 7.32 Selected authentication factors for a user by adaptive-MFA in different environment conditions

A-MFA system is shown in ◘ Fig. 7.32. $F_{i,j}$ as shown in ◘ Fig. 7.32, signifies the i-th authentication modality with the j-th approach(feature) are considered as that triggering event sensing the device, media, and surroundings. From the figure, it is clear that various authentication factors are selected in different settings of device, media, and surrounding conditions and there is no pattern to compromise for the attackers to predict the next possible set of authentication factors.

Comparison of Adaptive Selection Approach with Random and Only Biometric-Based Approaches

The Adaptive selection approach for MFA is compared with the random selection and only biometric-based selection approaches in terms of total trustworthy values and accuracy rates. The histogram plot showing the comparison using cumulative trustworthy values is shown in Fig. 7.33.

The trustworthy values are calculated for 100 authentication requests and then those values are averaged to demonstrate their cumulative values for three different selection approaches. Various types of device and media combinations are shown in the figure. According to Fig. 7.33, the Adaptive-MFA approach clearly outperforms the other two approaches in different settings of the environment. These values in the figure are shown for the case, where all the surrounding conditions are met. But the result remains the same in other different surrounding conditions for all these three approaches.

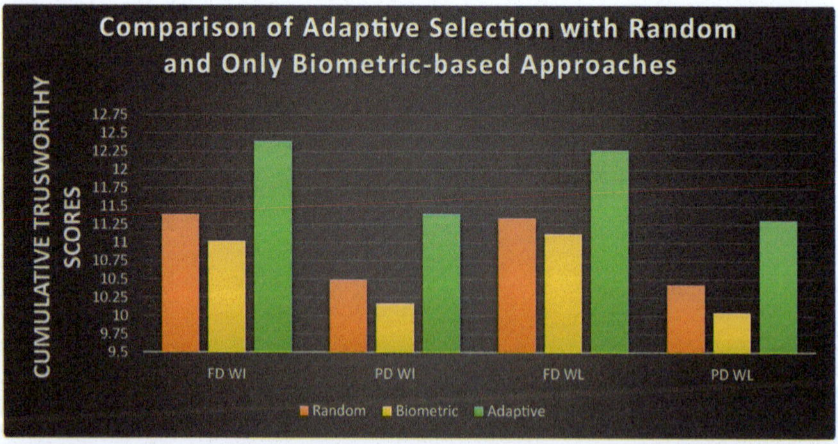

Fig. 7.33 Comparison among the adaptive selection, random selection, and only biometric-based selection approaches [38]

To compare the performance in authenticating valid users using random and only biometric-based selection approaches, 250 valid requests (authenticating each user 5 times) with different surrounding conditions are conducted. The results are shown in ◘ Tables 7.11 and 7.12.

According to ◘ Table 7.11, with the good surrounding conditions the accuracy using two factors is 91.5% and three factors is 90%, which is lower than the adaptive selection-based accuracies (◘ Table 7.8). However, when the surrounding conditions are not in the range, the random selection performs poorly in both two factors and three factors cases. For instance, when only light conditions are in range (second row), the accuracy rates of valid users drop to 62% for two-factor-based authentication and 56% for three-factor-based authentication. Because of randomly selecting the authentication approaches in various conditions, the selections choose some authentication modalities that have poor performance in a bad range of surrounding conditions.

◘ **Table 7.11** The accuracy rate for 250 valid requests using random selection approach

Surrounding conditions	Two-factor of A-MFA (%)	Three-factor of A-MFA (%)
Light and noise are in range	91.5	90
Only light is in range	62	56
Only noise is in range	66	48
None are in range	38	22

◘ **Table 7.12** The accuracy rate for 250 valid requests using only biometric-based selection approach

Surrounding conditions	Two-factor of A-MFA (%)	Three-factor of A-MFA (%)
Light and noise are in range	89	84
Only light is in range	56	38
Only noise is in range	48	36
None are in range	20	12

Similarly, ◻ Table 7.12 shows the accuracy of two-factor- and three-factor-based authentication for only biometric-based authentication modalities. In the good settings of surroundings, the accuracy rate is about 89% for two-factor-based and about 84% for three-factor-based approaches. For improper surrounding conditions, the only biometric-based selections suffer similarly as a random selection process. For example, when noise condition is not in range (second row), the accuracy of the only biometric-based approach is about 56% and 38% for two-factor and three-factor-based selections, respectively. ◻ Tables 7.11 and 7.12 also provide an estimate for FRR of the random and only biometric-based selection approaches.

In addition to that, 1000 invalid requests with various surrounding conditions are conducted to calculate the false acceptance rate of the system for the above-mentioned two selection approaches. These results are shown in ◻ Tables 7.13 and 7.14.

According to ◻ Table 7.13, with good surrounding conditions, about 6.5 and 1.5% false requests (two factors and

◻ **Table 7.13** The successful attempts for invalid user data in different surrounding conditions for two-factor and three-factor-based random selection in the A-MFA system

Surrounding conditions	Two-factor of A-MFA (%)	Three-factor of A-MFA (%)
Light and noise are in range	6.5	1.5
Only light is in range	5	0.8
Only noise is in range	6.5	1.3
None are in range	5	0.7

◻ **Table 7.14** The successful attempts for invalid user data in different surrounding conditions for two-factor and three-factor-based only biometric-based selection in the A-MFA system

Surrounding conditions	Two-factor of A-MFA (%)	Three-factor of A-MFA (%)
Light and noise are in range	3.4	0.7
Only light is in range	1.7	0.3
Only noise is in range	3.1	0.5
None are in range	2.5	0.25

three factors case, respectively) are passed through the authentication process. In different bad surrounding conditions, the rates are slightly lower compared to good surrounding conditions but those FAR values are higher than the adaptive selection scenario. Similar results are found for only biometric-based selection process as mentioned in ◘ Table 7.14.

Advantage over Other Existing MFA Approaches

The A-MFA approach can be compared with its current counterparts. The most commonly used MFA products are listed in ◘ Table 7.15. From the table, it is clear that the adaptive approach differs significantly from the other existing approaches. None of the other listed approaches uses an adaptive approach as part of their selection strategy. Many of these approaches choose static selection strategies that consider all the factors at the same time.

The A-MFA solution has several advantages over other existing MFA approaches. These advantages are summarized below:

(i)   By considering individual features of different authentication modalities, the search space of the problem becomes larger, which reduces the probability of selecting the same set of authentication modalities and consequently reduces the predictability of the system.

(ii)   The adaptive selection procedure is able to improve trust by selecting those particular features of the authentication modalities that have greater trustworthy values than the original modalities.

(iii) If the selected set of authentication modalities contain more than one feature for a single modality, then the user need not provide additional biometric samples to authenticate themselves. For example, if the selected set of features includes the lips and eyes (two different features of the face recognition modality (M_1)), then the same set of face image sample given by the user will be sufficient for the authentication purpose because it contains both of these features. Consequently, users have a more positive experience and usability is improved. In spite of appearing to be a single-factor authentication challenge, the model

Table 7.15 Qualitative comparison of different MFA products

	Microsoft Azure	Duo security	Authentify xFA	RSA secure ID	Adaptive MFA
Modality selection	Static	User's preference	User's preference	User's preference	Adaptive sensing environmental conditions
Surrounding factors	None	Location	None	Location	Location Light Noise
Biometric modalities	None	None	Finger Voice	Finger Iris	Face Voice Fingerprint Keystroke
Supported modalities	Non-biometric	Non-biometric	Biometric and non-biometric	Biometric and non-biometric	Biometric and non-biometric
Previous selection decision	Not considered	Not considered	Not Considered	Not Considered	Considered

presents a novel, advanced, and imperceptible multi-factor authentication system.

(iv)　The selection procedure is able to sense the surrounding ambiance and make selections accordingly. For example, if the luminance is less than 100 Lux, (i.e., outside the range of visibility), then modalities, dependent upon light (found in ◘ Table 7.3) cannot be selected. Similar conclusions can be made for other modalities that depend on the elements of S (given in ►Chap. 3).

(v)　Contrary to existing authentication strategies, the A-MFA method employs both biometric (physiological and behavioral) and non-biometric authentication modalities simultaneously.

(vi)　All existing multi-factor authentication approaches are mainly static and suffer from the repetitive selection of authentication modalities, whereas the selection system is dynamic and sensitive to the environment. Moreover, the selected sets of authentication factors chosen by the system are not repeated, making them unpredictable for the hackers. Consequently, the method makes online applications significantly more secure than the other existing multi-factor authentication strategies.

Future Trends

Multi-factor authentication is a growing trend in 2016, aimed at thwarting the growing number of password-based attacks, identity theft, and compromise of users' credentials. As the attacker sophistication and ability to subvert business security increases, an adaptive approach to security that selects the appropriate set of authentication factors by sensing the operating environment (connecting device, media, and existing surroundings) is an absolute necessity to protect users' identities on a large scale. Again, authenticating users with a diverse set of authentication factors in different authentication triggering events is a way to obfuscate the authentication process, thus preventing attackers from finding any exploitable pattern. This strategy provides a significant benefit and ensures the security of the MFA system.

This research work has been submitted for the United States Utility Patent (**Serial#: 14/968676**) [41] and the United States Provisional Patent Application (**Serial#: 62/262,626**). This novel work focuses on designing just-in-time authentication strategy using multiple authentication factors (modalities along with their feature components) in order to provide a trustworthy, resilient, and scalable solution for authentication. The trustworthy model computes the trust values for different authentication factors by considering several probabilistic constraints. In particular, it uses pairwise comparisons among various devices and media. The adaptive selection scheme makes intelligent decisions, choosing authentication factors at run-time by considering the performance, trustworthy values, and the history of the previous selection of the given factors. This approach also avoids repeated selections of the same set of authentication factors in successive reauthentications, thereby reducing the chance of establishing any recognizable patterns. Therefore, no prior information regarding the selected set of authentication factors is available for the attackers to exploit. Again, the selection mechanism considers individual features of a modality as different authentication factors and hence, the search space of the selection procedure becomes relatively large. This criterion ensures the selection of a non-repetitive set of authentication factors for different authentication triggering time. This adaptive selection approach would also run at different triggering events (time intervals which may vary with different devices and media combinations) for reauthentication of legitimate users.

This MFA approach opens the door for many possible applications where ensuring security is the topmost concern. Some of them are highlighted here:

1. Massive Open Online Courses (MOOC) where it is difficult to distinguish between the registered user and the actual user who is going to take the exams and homework problems. With the increase of MOOC among different universities, it is an urgent need to verify the students' identities in a secure manner. The robustness and scalability of A-MFA makes it a good solution for verifying students' identities because varying combinations of authentication factors can be generated based on the difficulty of types of the tasks (homework, quiz, midterm exam, project report, final exam).

2. Banking applications such as online fund transfer or online payment should be highly protected. A-MFA can be used to quickly verify legitimate users and allow them to complete the transaction successfully with more trusted settings. The amount of money can be a deciding factor such that stringent authentication factors may be required to be chosen to identify the users

successfully for transactions involving large monetary values, whereas less stringent authentication factors may be chosen for lower monetary amounts.

3. A-MFA can easily be integrated to secure access to all types of electronic medical records. These medical records are very confidential and sensitive to the patients. A-MFA can trigger different authentication factors by sensing the device, media, and surroundings of users, providing more robust and resilient authentication factors that provide higher accuracy in that particular environment.

4. A-MFA can be further extended for deployment at different levels of Internet Computing. Some of them are listed here:
   (a) Application level (financial applications, email/business/personal applications, social applications)
   (b) User level (root user, administrators, guest users)
   (c) Document level (pdf containing an application form, resume, a document containing proprietary information, image/video containing confidential and sensitive footage).

A-MFA Important Features

 **Continuous Identity Protection** A-MFA is always on, ensuring your identity before, during, and after you are connected to the system.

 **Just-in-time Authentication System** Authentication modalities are generated in real-time preventing malicious attackers from obtaining log-in data.

 **Machine Learning Approach** A-MFA utilize machine learning techniques to analyze user behavior and selects modalities.

 **Built with the future in mind** The A-MFA framework is robust and flexible enough to meet your current and future authentication needs.

 **Broad Range of Institutional Uses** A-MFA is suitable for many environments and scenarios such as online banking, healthcare networks, and online courses.

 **Cloud Identity Management** Deployment as a cloud service allows A-MFA to be easily integrated with existing applications and resources.

■ **Fig. 7.34** Important features of A-MFA system [38]

In this research work, ten different authentication modalities (along with their features) are considered and seven different authentication modalities are implemented in the pilot system. The system is thoroughly tested with a constructed set of synthetic data for all the considered authentication factors such that different sets of users are used to check the performance of the system. Several well-known, standard biometric datasets are used to create the synthetic user profiles.

Again, the developed MFA system will be adopted in a corporate system, and the trustworthy values of all the required authentication factors will be calculated to reflect the requirements of the organization. This adaptive-MFA model can also be adopted for the design of online Identity Management Eco System. It is anticipated that adaptive way of selecting the appropriate authentication factors to verify identities will provide a significant benefit toward the protection of users' sensitive information in the near future. Important features of A-MFA are shown in ◘ Fig. 7.34.

Chapter Summary

This chapter mainly focuses on a novel, just-in-time, and efficient authentication strategy using multiple authentication factors (modalities along with their several features) in order to provide a trustworthy, resilient, and scalable solution for authentication. The A-MFA discussed in this chapter initially finds the trustworthy values of every individual feature of different authentication modalities using a nonlinear optimization problem with probabilistic constraints that involve pair-wise comparisons among different devices and media. In the current implementation of A-MFA, seven authentication modalities (biometric and non-biometric modalities) have been considered. However, this approach can easily be extended to any number of authentication modalities. Then the trustworthy values of all combinations of features (authentication factors) have been calculated, which, in turn, significantly increases the total number of the possible set of authentication factors. Next, it selects the best set of authentication factors based on the connected device, operating medium, and existing surrounding influence in an adaptive selection. The adaptive selection scheme makes intelligent decisions, choosing authentication factors at run-time by considering the performance, trustworthy values, and the history of the previously selected authentication factors. This approach also avoids repeated selections of the same set of authentication factors in successive reauthentication attempts, thereby reducing the chance of establishing any recognizable pattern. Therefore, no *apriori* information regarding the selected set of authentication factors is available for the attackers to hack. Again, the selection mechanism considers individual features of a modality as different authentication factors and hence, the search space of the selection procedure becomes relatively large. This criterion ensures the selection of a non-repetitive set of authentication factors for different authentication triggering time. This adaptive selection approach would also run at different triggering times (time intervals which may vary with different devices and media combinations) for reauthentication of legitimate users.

Acknowledgements Some portion of this work is supported by the National Security Agency under Grant Number: H98230-15-1-0266. Points of view and opinions presented in this chapter are those of the author(s) and do not necessarily represent the position or policies of the National Security Agency or the United States. Authors would like to acknowledge the contribution of other members of the A-MFA research team who helped in running experiments, producing results, and providing feedback.

Questions

Question 1:
Why is multi-factor a better solution than single-factor-based authentication solution?

Question 2:
What is the authentication factor? What is the novel aspect of the authentication factor discussed in this chapter?

Question 3:
Define the trustworthy value of an authentication factor? How do error rates play a role in calculating the trustworthy factor?

Question 4:
Discuss in details the calculation of trustworthy values of combined authentication factors. Explain the calculation steps using a figure for a specific medium.

Question 5:
What are the two objectives that are addressed for the adaptive selection process? Describe them.

Question 6:
Why is the cardinality of the authentication factor incorporated in the formulation of the first objective for adaptive selection? What is the role of surrounding conditions in the adaptive selection procedure?

Question 7:
Which algorithms used in A-MFA applications are not affected by surrounding conditions and why?

Question 8:
Describe the Processing Plane and Data Plane of the server-side implementation of A-MFA?

Question 9:
Describe the evaluation criteria mentioned for A-MFA? How does adaptive selection perform better than its counterpart selection approaches?

Question 10:
Provide a qualitative analysis of Adaptive-MFA approach with other MFA products.

References

1. Rouse M (2015) Multifactor authentication (MFA). Available: ▶ http://searchsecurity.techtarget.com/definition/multifactorau-thentication-MFA, Mar 2015
2. Strom D (2014) Multifactor authentication examples and business case scenarios. Available: ▶ http://searchsecurity.techtarget.com/feature/The-fundamentals-of-MFA-The-business-case-for-multifactor-authentication, Oct 2014
3. Niinuma K, Jain AK (2010) Continuous user authentication using temporal information. In: SPIE defense, security, and sensing. International Society for Optics and Photonics, pp 76 670L–76 670L
4. Fathy ME, Patel VM, Yeh T, Zhang Y, Chellappa R, Davis LS (2014) Screen-based active user authentication. Pattern Recogn Lett 42:122–127
5. Primo A, Phoha VV, Kumar R, Serwadda A (2014) Context-aware active authentication using smartphone accelerometer measurements. In: Proceedings of the IEEE conference on computer vision and pattern recognition workshops, pp 98–105
6. Vielhauer C (2005) Biometric user authentication for IT security: from fundamentals to handwriting, vol 18. Springer Science & Business Media, Berlin
7. Nag AK, Dasgupta D, Deb K (2014) An adaptive approach for active multi-factor authentication. In: 9th annual symposium on information assurance (ASIA14), p 39
8. Nag AK, Dasgupta D (2014) An adaptive approach for continuous multi-factor authentication in an identity eco-system. In: ACM proceedings of the 9th annual cyber and information security research conference, pp 65–68
9. Nag AK, Roy A, Dasgupta D (2015) An adaptive approach towards the selection of multi-factor authentication. In: IEEE symposium series on computational intelligence, pp 463–472
10. Dasgupta D, Nag AK, Roy A (2016) Adaptive multi-factor authentication system. U.S. Patent Application No. 14/968, 676, Dec 2016
11. Duc NM, Minh BQ (2009) Your face is not your password face authentication hypassing lenovo–asus–toshiba. Black Hat Briefings
12. Hwang S, Lee H, Cho S (2006) Improving authentication accuracy of unfamiliar passwords with pauses and cues for keystroke dynamics-based authentication. In: Intelligence and security informatics. Springer, Berlin, pp 73–78
13. Hoog A (2016) SMS two-factor authentication security NIST statement on sms-based two-factor authentication: Whats it mean for CISOs? NowSecure, July 2016. Available: ▶ https://www.nowsecure.com/blog/2016/07/31/nist-statementon-sms-based-two-factor-authentication-what-s-it-mean-for-cisos/
14. Balaban D (2013) Understanding CAPTCHA-solving services in an economic context 3: evaluation of the human-based services, Mar 2013
15. Nag AK, Roy A, Dasgupta D (2015) An adaptive approach towards the selection of multi-factor authentication. IEEE symposium series on computational intelligence (SSCI), pp 463–472
16. Dasgupta D, Roy A, Nag Abhijit (2016) Toward the design of adaptive selection strategies for multi-factor authentication. Comput Secur 63:85–116

17. Griva I, Nash SG, Sofer A (2009) Linear and nonlinear optimization. SIAM

18. Arduino mega 2560 (2016) Available: ►https://www.arduino.cc/en/Main/arduinoBoardMega2560

19. Turk MA, Pentland AP (1991) Face recognition using eigenfaces. In: IEEE proceedings computer society conference on computer vision and pattern recognition (CVPR), pp 586–591

20. Jiang XY, Wong HS, Zhang D (2006) Face recognition based on 2D Fisherface approach. Pattern Recogn 39(4):707–710

21. Belhumeur PN, Hespanha JP, Kriegman DJ (1997) Eigenfaces versus Fisherfaces: recognition using class specific linear projection. IEEE Trans Pattern Anal Mach Intell 19(7):711–720

22. Ahonen T, Hadid A, Pietikainen M (2006) Face description with local binary patterns: application to face recognition. IEEE Trans Pattern Anal Mach Intell 28(12):2037–2041

23. Ratha NK, Chen S, Jain AK (1995) Adaptive flow orientation-based feature extraction in fingerprint images. Pattern Recogn 28(11):1657–1672

24. Vazan R (2009) Source AFIS—fingerprint recognition library for .net and experimentally for java

25. Jiang X, Yau WY (2000) Fingerprint minutiae matching based on the local and global structures. In: 15th international conference on pattern recognition. Proceedings, vol 2, pp 1038–1041

26. Medina-P´erez MA, García-Borroto M, Gutierrez-Rodr´ıguez AE, Altamirano-Robles L (2012) Improving fingerprint verification using minutiae triplets. Sensors 12(3):3418–3437

27. Davis S, Mermelstein P (1980) Comparison of parametric representations for monosyllabic word recognition in continuously spoken sentences. IEEE Trans Acoust Speech Signal Process 28(4):357–366

28. Hermansky H (1990) Perceptual linear predictive (PLP) analysis of speech. J Acoust Soc Am 87(4):1738–1752

29. Adami AG, Mihaescu R, Reynolds DA, Godfrey JJ (2003) Modeling prosodic dynamics for speaker recognition. In: Proceedings of IEEE international conference on acoustics, speech, and signal processing, vol 4, pp IV–788

30. Muda L, Begam M, Elamvazuthi I (2010) Voice recognition algorithms using Mel frequency Cepstral coefficient (MFCC) and dynamic time warping (DTW) techniques. arXiv preprint ►arXiv:1003.4083, 2010

31. Killourhy KS, Maxion RA (2009) Comparing anomaly-detection algorithms for keystroke dynamics. In: IEEE/IFIP international conference on dependable systems & networks, pp 125–134

32. Spacek L (1994) Faces 94: The University of Essex. Face recognition data set. Available: ►http://cswww.essex.ac.uk/mv/allfaces/faces94.html

33. Spacek L (1995) Faces 95: The University of Essex. Face recognition data set. Available: ►http://cswww.essex.ac.uk/mv/allfaces/faces95.html

34. Spacek L (1996) Faces 96: The university of Essex. Face recognition data set. Available: ►http://cswww.essex.ac.uk/mv/allfaces/faces96.html

35. Maio D, Maltoni D, Cappelli R, Wayman JL, Jain AK (2002) Fvc2000: Fingerprint verification competition. IEEE Trans Pattern Anal Mach Intell 24(3):402–412

36. Maio D, Maltoni D, Cappelli R, Wayman JL, Jain AK (2002) FVC 2002: second fingerprint verification competition. In: 16th international conference on pattern recognition, 2002. Proceedings, vol 3. IEEE, pp 811–814

37. Maio D, Maltoni D, Cappelli R, Wayman JL, Jain AK (2004) Fvc2004: third fingerprint verification competition. In: Biometric authentication. Springer, Berlin, pp 1–7

38. Dasgupta D, Nag A, Shrein JM, Swindle M (2017) Design and implementation of an adaptive multi-factor authentication (A-MFA) system. Technical report, The University of Memphis, Jan 2017

39. Simons RH, Bean AR (2008) Lighting engineering: applied calculations. Routledge, USA

40. Casia-fingerprint v5 (2010) Available: ▶http://biometrics.idealtest.org/

41. Dasgupta D, Nag AK, Roy A (2015) *Adaptive multi-factor authentication system*. U.S. Patent Application No. 14/968, 676

Service Part

© Springer International Publishing AG 2017
D. Dasgupta et al., *Advances in User Authentication*, Infosys Science Foundation Series,
DOI 10.1007/978-3-319-58808-7

Index

The manufacturer's authorised representative in the EU is Springer
Nature Customer Service Centre GmbH, Europaplatz 3, 69115 Heidelberg,
Germany. If you have any concerns regarding our products, please
contact ProductSafety@springernature.com

Printed and bound by CPI Group (UK) Ltd, Croydon, CR0 4YY

27/04/2026

02097569-0003